The City & Guilds textbook

Book 1

Electrical Installations

LEVEL 3 APPRENTICESHIP (5357)
LEVEL 2 TECHNICAL CERTIFICATE (8202)
LEVEL 2 DIPLOMA (2365)

▌ Peter Tanner

DYNAMIC LEARNING

HODDER EDUCATION
AN HACHETTE UK COMPANY

Orders: please contact Bookpoint Ltd, 130 Park Drive, Milton Park, Abingdon, Oxon OX14 4SE. Telephone: (44) 01235 827720. Fax: (44) 01235 400454. Email education@bookpoint.co.uk Lines are open from 9 a.m. to 5 p.m., Monday to Saturday, with a 24-hour message answering service. You can also order through our website: www.hoddereducation.co.uk

ISBN: 978 1 5104 3224 6

© The City and Guilds of London Institute, The Institution of Engineering and Technology and Hodder & Stoughton Limited 2018

Some text content in this book is published under licence from The City and Guilds of London Institute and The Institution of Engineering and Technology, London.

First published in 2018 by
Hodder Education,
An Hachette UK Company
Carmelite House
50 Victoria Embankment
London EC4Y 0DZ

www.hoddereducation.co.uk

Impression number   10 9 8 7 6 5 4 3

Year   2022 2021 2020 2019

Cover photo © AVD – Stock.adobe.com

City & Guilds and the City & Guilds logo are trade marks of The City and Guilds of London Institute. City & Guilds Logo © City & Guilds 2018

The IET logo is a registered trade mark of The Institution of Engineering and Technology.

Typeset in India.

Printed in India.

A catalogue record for this title is available from the British Library.

MIX
Paper from responsible sources
FSC™ C104740
www.fsc.org

Contents

Acknowledgements

This book draws on several earlier books that were published by City & Guilds, and we acknowledge and thank the writers of those books:

- Howard Carey
- Paul Cook
- James L Deans
- Paul Harris
- Andrew Hay-Ellis
- Trevor Pickard.

We would also like to thank everyone who has contributed to City & Guilds photoshoots. In particular, thanks to: Jules Selmes and Adam Giles; Andrew Buckley; Andy Jeffery, Ben King and students from Oaklands College, St Albans; Jaylec Electrical; Andrew Hay-Ellis, James L Deans and the staff at Trade Skills 4 U, Crawley; Jordan Hay-Ellis, Terry White, Katherine Hodges and Claire Owen.

Permission to reproduce extracts from BS 7671:2018 is granted by BSI Standards Limited (BSI) and the Institution of Engineering & Technology (IET). No other use of this material is permitted.

Picture credits

Every effort has been made to trace and acknowledge ownership of copyright. The publishers will be glad to make suitable arrangements with any copyright holders whom it has not been possible to contact.

Page vi Peter Tanner; Fig.1.3 © Robert/stock.adobe.com; Fig.1.4 © Alexlmx/stock.adobe.com; Fig.1.7 © SpeedKingz/Shutterstock.com; Fig.1.8 © Health and Safety Executive; Fig.1.9 City & Guilds; Fig.1.10 © Screwfix; Fig.1.11 © Shutterstock.com; Fig.1.15 L © stoonn/stock.adobe.com, R © Aleksander/stock.adobe.com; Fig.1.16–18 © Screwfix; Fig.1.19 L, M © Screwfix, R courtesy of Axminster Tool Centre Ltd; Fig.1.20 © Screwfix; Fig.1.21 L © Oleksandr Chub/Shutterstock.com, M © gezzeg/Shutterstock.com, R © Michele Cozzolino/Shutterstock.com; Fig.1.22–26 © Screwfix; Fig.1.28 © jamierogers1/Fotolia.com; Fig.1.31 © Cynthia Farmer/Shutterstock.com; Fig.1.33–34, 1.39 City & Guilds; Fig.1.42 TL, TM © test-meter.co.uk, TR, BL, BR City & Guilds; Fig.1.32 IET; Fig.2.20 © Olesia Bilkei/stock.adobe.com; Fig.2.21 © Oleksandr Moroz/stock.adobe.com; Fig.2.25 © Kadmy/stock.adobe.com; Fig.2.32 © Kewtech; Fig.2.56 © Gail Johnson/Shutterstock.com; Fig.2.57 Ralf Gosch/Shutterstock.com; Fig.2.66 City & Guilds; Fig.2.68 T © Robert_Chlopas/Shutterstock.com, M © BMPix /istockphoto.com, B © Coprid/Fotolia.com; Fig.2.69 Titan Products Ltd; Fig.2.86 © Lena Wurm/stock.adobe.com; Fig.3.1 © Health & Safety Executive (HSE)/ Comstock/stockbyte/ Getty Images (lightning image licensed separately via Getty Images); Fig.3.3 IET; Fig.3.6 City & Guilds; Fig.3.13 © Belinda Pretorius/Shutterstock.com; Fig.3.14 Deborah Benbrook/Shutterstock.com; Fig.3.15 © lertkaleepic/Shutterstock.com;

Fig.3.16 © foxaon1987/Shutterstock.com; Fig.3.18 © VikOI/Shutterstock.com; Fig.3.36 © Prysmian; Fig.3.37 © BSI; Fig.3.38 © Prysmian; Fig.3.39 © Marshall-Tufflex; Fig.3.41 City & Guilds; Fig.3.42 © Eldad Yitzhak/Shutterstock.com; Fig.3.43 © Schipkova Elena/Shutterstock.com; Fig.3.44 © sahua d/Shutterstock.com; Fig.3.45 © Marshall-Tufflex; Fig.3.47 City & Guilds; Fig.3.48 © Marshall-Tufflex; Fig.3.49–51 City & Guilds; Fig.3.54–56 © Marshall-Tufflex; Fig.3.57 L © Maxim/stock.adobe.com, R © Shutterstock.com; Fig.3.58 © remus20/stock.adobe.com; Fig.3.59, 3.62–64, 3.68, 3.70–71 City & Guilds; Fig.4.1 © Belinda Pretorius/Shutterstock.com; Fig.4.2 © Timothy Hodgkinson/Shutterstock.com; Fig.4.3 courtesy of Axminster Tool Centre Ltd; Fig.4.4 © Phovoir/Shutterstock.com; Fig.4.5 © Yanas/Shutterstock.com; Fig.4.6 courtesy of Axminster Tool Centre Ltd; Fig.4.7 L © Ilya Akinshin/Shutterstock.com, R © anaken2012/Shutterstock.com; Fig.4.8 © troy/Shutterstock.com; Fig.4.9 © Stocksnapper/Shutterstock.com; Fig.4.10 © Roblan/Shutterstock.com; Fig.4.11 City & Guilds; Fig.4.12 courtesy of Axminster Tool Centre Ltd; Fig.3.13 © Rynio Productions/Shutterstock.com; Fig.4.14 courtesy of Axminster Tool Centre Ltd; Fig.4.15 City & Guilds; Fig.4.16 courtesy of Axminster Tool Centre Ltd; Fig.4.17 City & Guilds; Fig.4.18 courtesy of Axminster Tool Centre Ltd; Fig.4.19 City & Guilds; Fig.4.20 © donatas1205/Shutterstock.com; Fig.4.21 City & Guilds; Fig.4.22 L © Shutterstock.com, M © dcwcreations/Shutterstock.com, R © Estwing Manufacturing Co.; Fig.4.23 L © Ruslan Semichev/Shutterstock.com, R © yevgeniy11/Shutterstock.com; Fig.4.24–25 City & Guilds; Fig.4.26 courtesy of Axminster Tool Centre Ltd; Fig.4.27 © Mahathir Mohd Yasin/Shutterstock.com; Fig.4.28 © Alexandr Makarov/Shutterstock.com; Fig.4.29 © -V-/Shutterstock.com; Fig.4.30 © Christopher Elwell/Shutterstock.com; Fig.4.31 © Prill/Shutterstock.com; Fig.4.32 © Dmitry Morgan/Shutterstock.com; Fig.4.33 © furtseff/Shutterstock.com; Fig.4.35 City & Guilds; Fig.4.82, 4.102 Andrew Hay-Ellis; Fig.4.103 © sauletas/Shutterstock.com; Fig.4.108 © Marshall-Tufflex; Fig.4.109 © Eldad Yitzhak/Shutterstock.com; Fig.4.114–120, 4.122–126, 4.131–137 City & Guilds; Fig.4.138 Andrew Hay-Ellis; Fig.4.139–218, 4.220–231, 4.234–241 City & Guilds; Fig.4.242–244 Andrew Hay-Ellis; Fig.4.245–264 City & Guilds; Fig.4.266–267 © Kewtech; Fig.4.269–270, 4.274–276, 4.279 City & Guilds; Fig.5.1 © Martin Sevcik/123RF; Fig.5.2 © Syda Productions/stock.adobe.com; Fig.5.4 © Kadmy/stock.adobe.com; Fig.5.6 © ratmaner/stock.adobe.com; Fig.5.7 © rosinka79/stock.adobe.com; Fig.5.8 © Rawpixel.com/stock.adobe.com; Fig.5.10 City & Guilds; Fig.5.13 © vinnstock/stock.adobe.com; Fig.5.14 © Bacho Foto/stock.adobe.com; pages 422–424 City & Guilds.

About the author

I started in the electrotechnical industry while still at school, chasing walls for my brother-in-law for a bit of pocket-money. This taught me quickly that if I took a career as an electrician I needed to progress as fast as I could.

Jobs in the industry were few and far between when I left school so after a spell in the armed forces, I gained a place as a sponsored trainee on the Construction Industry Training Board (CITB) training scheme. I attended block release at Guildford Technical College, where the CITB would find me work experience with 'local' employers. My first and only work experience placement was with a computer installation company located over 20 miles away so I had to cycle there every morning, but I was desperate to learn, and enjoyed my work.

Computer installations were very different in those days. Computers filled large rooms and needed massive armoured supply cables so the range of work I experienced was vast – from data cabling to all types of containment systems and low-voltage systems.

In the second year of my apprenticeship I found employment with a company where most of my work centred around the London area. The work was varied, from lift systems in well-known high-rise buildings to lightning protection on the sides of even higher ones!

On completion of my apprenticeship I worked for a short time as an intruder alarm installer, mainly in domestic dwellings – a role in which client relationships and handling information are very important.

Following this I began work with a company where I was involved in shop-fitting and restaurant and pub refurbishments. It wasn't long before I was managing jobs and gaining further qualifications through professional development. I was later seconded to the Property Services Agency, designing major installations within some of the best-known buildings in the UK.

A career-changing accident took me into teaching, where I truly found the rewards the industry has to offer. Seeing young trainees maturing into qualified electricians is a highly gratifying experience. I often see many of my old trainees when they attend further training and updated courses. Following their successes makes it all worthwhile.

I have worked with City & Guilds for over 20 years and represent them on a variety of industry committees, such as JPEL64, which is responsible for the production of BS 7671. I am passionate about using my extensive experience to maintain the high standards the industry expects.

Peter Tanner

March 2018

How to use this book

Throughout this book you will see the following features:

INDUSTRY TIPS and KEY FACTS are particularly useful pieces of advice that can assist you in your workplace or help you remember something important.

> ### INDUSTRY TIP
>
> Remember, when you are carrying out your practical tasks either in the workplace or your place of learning, everyone, including you, is responsible for safety.

> ### KEY FACT
>
> The internal angles of a right-angled triangle always add up to 180°.

KEY TERMS in bold purple in the text are explained in the margin to aid your understanding. (They are also explained in the Glossary at the back of the book.)

> ### KEY TERM
>
> **SI units:** The units of measurement adopted for international use by the Système International d'Unités.

HEALTH AND SAFETY boxes flag important points to keep yourself, colleagues and clients safe in the workplace. They also link to sections in the health and safety chapter for you to recap learning.

> ### HEALTH AND SAFETY
>
> An ammeter has a very low internal resistance and must never be connected across the supply.

ACTIVITIES help to test your understanding and learn from your colleagues' experiences.

> ### ACTIVITY
>
> A wall measures 6 cm on a 1:50 scale drawing. What is the true length of the wall?

IMPROVE YOUR MATHS items combine improving your understanding of electrical installations with practising or improving your maths skills.

IMPROVE YOUR ENGLISH items combine improving your understanding of electrical installations with practising or improving your English skills.

At the end of each chapter there are some TEST YOUR KNOWLEDGE questions and PRACTICAL TASKS. These are designed to identify any areas where you might need further training or revision. Answers to the questions are at the back of the book.

CHAPTER 1

HEALTH AND SAFETY AND INDUSTRY PRACTICES

INTRODUCTION

Every year, accidents at work result in the deaths of more than one hundred people in Great Britain, with several hundred thousand more being injured in the workplace. In 2016/17, Health and Safety Executive (HSE) statistics recorded 31.2 million working days being lost due to work-related illness and workplace injury.

Occupational health and safety affects all individuals and aspects of work in the complete range of working environments – hospitals, factories, schools, universities, commercial undertakings, manufacturing plants and offices. As well as the human cost in terms of pain, grief and suffering, accidents in the workplace also have a financial cost, such as lost income, insurance and production disturbance. The HSE put this figure at £14.9 billion for the year 2016/17. It is therefore important to identify, assess and control the activities that may cause harm in the workplace.

In addition to injuries to persons, construction site activities lead to many recorded incidents of land and air contamination impacting the environment we live in.

This chapter is designed to enable you to understand health, safety and environmental legislation and associated practices and procedures, when installing and maintaining electrotechnical systems and equipment. You will need this knowledge to underpin the application of health, safety and environmental legislation, practices and procedures.

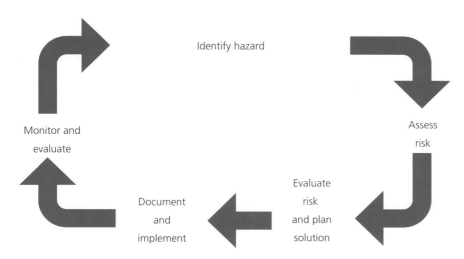

▲ Figure 1.1 The five steps for risk assessment

INDUSTRY TIP

You can see the most up-to-date statistics relating to accidents in the workplace by visiting the HSE website at: www.hse.gov.uk/statistics/index.htm

Learning objectives

This table shows how the topics in this chapter meet the outcomes of the different qualifications.

Topic	Electrotechnical Qualification (Installation) or (Maintenance) 5357 Unit 101-001	Level 2 Diploma in Electrical Installations (Buildings and Structures) 2365 Unit 201	Level 2 Technical Certificate in Electrical Installation 8202 Unit 201
1 The law: roles and responsibilities	1.1; 1.2		1.2
2 Health and safety legislation	1.1		1.1
3 Environmental legislation and dealing with waste	1.2; 2.5; 2.6; 2.7		1.1; 1.2; 5.1; 5.2; 5.3
4 Dealing with health and safety in the work environment	2.1; 2.2; 2.3; 2.4		3.2
5 Establishing a safe working environment including: – Hazards – Risk assessment and method statements – Safety signs and packaging signs – Hazardous material – Asbestos – Personal protective equipment – First aid facilities – Fire safety – Manual handling	2.2; 3.1; 3.2; 3.3; 3.4; 3.5; 3.6; 4.1; 4.2; 4.3; 4.4; 4.7; 4.8		2.2; 3.1; 3.3; 3.4; 3.5; 3.6; 3.7; 3.8
6 Using access equipment and working in confined spaces or excavations	3.7		2.1
7 Electrical safety on site including: – Safe use of power tools – Site supplies	4.3; 4.5		2.3; 4.2
8 Safe isolation	3.7; 3.8; 3.9		4.1; 4.3

1 THE LAW: ROLES AND RESPONSIBILITIES

KEY TERMS

Civil law: decides on a dispute between parties.

Criminal law: decides if someone is guilty of a criminal act.

Liability: a debt or other legal obligation in order to compensate for harm.

When something goes wrong, there are two sub-divisions of the law that apply to health and safety: civil law and criminal law.

- *Civil law* – deals with disputes between individuals, between organisations, or between individuals and organisations, in which compensation can be awarded to the victim. The civil court is concerned with liability and the extent of that liability rather than guilt or non-guilt.
  Example: If a sub-contractor has not been paid by a main contractor, the sub-contractor may make a claim against the main contractor through a civil court. The judge will examine the terms of the contract and settle the dispute based on those terms.
- *Criminal law* – the body of rules that regulates social behaviour and prohibits threatening, harming or other actions that may endanger the health, safety

and moral welfare of people. The rules are laid down by the government and are enacted by Acts of Parliament as **statutes**. The Health and Safety at Work etc. Act 1974 (HSW Act) is an example of criminal law. It is enforced by the Health and Safety Executive (HSE) or Local Authority environmental health officers.

In terms of health and safety, criminal law is based only on statute law, but civil law may be based on either common law or statute law.

- *Common law* – the body of law based on custom and decisions made by judges in courts. In health and safety, the legal definitions of terms such as 'negligence', 'duty of care', and 'so far as is reasonably practicable' are based on legal judgments and are part of common law. Common law usually uses past cases that set a precedent to help the judge decide an outcome.
- *Statute law* – the name given to law that has been laid down by Parliament as Acts of Parliament.

> **INDUSTRY TIP**
>
> In terms of common law, 'so far as is reasonably practicable' involves weighing a risk against the trouble, time and money needed to control it.

In summary, criminal law seeks to protect everyone in society and civil law seeks to recompense individuals, to make amends for loss or harm they have suffered (i.e. provide compensation).

The main legal requirements for health and safety at work

The HSW Act 1974 is the basis of all British health and safety law. It provides a comprehensive and integrated piece of legislation that sets out the general duties that employers have towards employees, contractors and members of the public, and that employees have towards themselves and each other. These duties are qualified in the HSW Act by the principle of 'so far as is reasonably practicable'.

What the law expects is what good management and common sense would lead employers to do anyway; that is, to look at what the risks are and take sensible measures to tackle those risks. The person(s) who is responsible for the risk and best placed to control that risk is usually designated as the **duty holder**.

The HSW Act lays down the general legal framework for health and safety in the workplace, with specific duties being contained in regulations, also called statutory instruments (SIs), which are also examples of laws approved by Parliament.

> **INDUSTRY TIP**
>
> The duty holder must be competent by formal training and experience and have sufficient knowledge to avoid danger. The appropriate level of competence will differ for different areas of work.

KEY TERM

Statute: a major written law passed by Parliament.

> **INDUSTRY TIP**
>
> You can access more information about the HSW Act 1974 at: www.hse.gov.uk/legislation/hswa.htm

KEY TERM

Duty holder: the person in control of the danger.

KEY TERM

Enabling Act: an Enabling Act allows the Secretary of State to make further laws (regulations) without the need to pass another Act of Parliament.

ACTIVITY

Think of any jobs you have had in the past, such as part-time work in the holidays. What do you think were your responsibilities for health and safety?

Individuals' responsibilities under health and safety legislation

The HSW Act, which is an **Enabling Act**, is based on the principle that those who create risks to employees or others in the course of carrying out work activities are responsible for controlling those risks.

The HSW Act places specific responsibilities on:

- employers
- the self-employed
- employees
- designers
- manufacturers and suppliers
- importers.

This section will deal with the responsibilities of employers, the self-employed and employees.

Responsibilities of employers and the self-employed

Under the main provisions of the HSW Act, employers and the self-employed have legal responsibilities in respect of the health and safety of their employees and other people (e.g. visitors and contractors) who may be affected by their undertaking and exposed to risks as a result. The employers' general duties are contained in Section 2 of the Act.

They are to ensure, 'so far as is reasonably practicable', the health, safety and welfare at work of all their employees, in particular:

- the provision of safe plant and systems of work
- the safe use, handling, storage and transport of articles and substances
- the provision of any required information, instruction, training or supervision
- a safe place of work including safe access and exit
- a safe working environment with adequate welfare facilities.

These duties apply to virtually everything in the workplace, including electrical systems and installations, plant and equipment. An employer does not have to take measures to avoid or reduce the risk if that is technically impossible or if the time, trouble or cost of the measures would be grossly disproportionate to the risk.

Responsibilities of employees

Employees are required to take reasonable care for the health and safety of themselves and others (including work colleagues, clients, members of the public and practically anyone who is affected). To achieve this aim, they have two main duties placed upon them:

- to take reasonable care for the health and safety of themselves and others who may be affected by their acts or omissions at work
- to co-operate with their employer and others to enable them to fulfil their legal obligations.

INDUSTRY TIP

Remember, when you are carrying out your practical tasks either in the workplace or your place of learning, everyone, including you, is responsible for safety.

In addition, there is a duty not to misuse or interfere with safety provisions. Most of the duties in the HSW Act and the general duties included in the Management of Health and Safety at Work Regulations 1999 (the Management Regulations) are expressed as goals or targets that are to be met 'so far as is reasonably practicable' or through exercising 'adequate control' or taking 'appropriate' (or 'reasonable') steps. This involves making judgements as to whether existing control measures are sufficient and, if not, deciding what else should be done to eliminate or reduce the risk. This risk assessment will be produced using Approved Codes of Practice (ACoP) and published standards, as well as HSE or industry guidance on good practice, where available.

INDUSTRY TIP

You can access Management of Health and Safety at Work Regulations 1999 (the Management Regulations) at: www.legislation.gov.uk/uksi/1999/3242/contents/made

2 HEALTH AND SAFETY LEGISLATION

When the HSW Act came into force, there were already some 30 statutes and 500 sets of regulations in place. The aim of the Health and Safety Commission (HSC) and the Health and Safety Executive (HSE) was to progressively replace the existing regulations with a system of regulation that expresses general duties, principles and goals, with any supporting detail set out in ACoPs and guidance.

Regulations

Statutory Instruments (SIs) are laws approved by Parliament. The regulations governing health and safety are usually made under the HSW Act, following proposals from the HSC/HSE. This applies to regulations based on European Commission (EC) Directives as well as those produced in Great Britain.

The HSW Act, and general duties in the Management Regulations, set goals and leave employers the freedom to decide how to control the risks they identify. Guidance and ACoPs give advice.

Some risks are considered so great or the proper control measures so costly that it would not be appropriate to leave employers to decide what to do about them. Regulations identify these risks and set out the specific action that must be taken. Often these requirements are absolute – they require something to be done, without qualification. The employer has no choice but to undertake whatever action is required to prevent injury, regardless of cost or effort.

Some regulations apply across all workplaces. Such regulations include the Manual Handling Operations Regulations 1992, which apply wherever things are moved by hand or bodily force, and the Health and Safety (Display Screen Equipment) Regulations 1992, which apply wherever visual display units (VDUs) are used. Other regulations apply to hazards unique to specific industries, such as mining or the nuclear industry.

STATUTORY INSTRUMENTS

1998 No. 2306

HEALTH AND SAFETY

▲ Figure 1.2 An example of the front page of a regulatory document

The following regulations apply across the full range of workplaces.

- **Control of Noise at Work Regulations 2005:** require employers to take action to protect employees from hearing damage.
- **Control of Substances Hazardous to Health (COSHH) Regulations 2002 (as amended):** require employers to assess the risks from hazardous substances and take appropriate precautions.
- **Electricity at Work Regulations 1989:** require people in control of electrical systems to ensure they are safe to use and maintained in a safe condition.
- **Health and Safety (Display Screen Equipment) Regulations 1992:** give specific requirements for the use of display equipment such as computer screens. This may affect the choice of lighting to reduce glare.
- **Health and Safety (First-Aid) Regulations 1981:** require employers to provide adequate and appropriate equipment, facilities and personnel to ensure their employees receive immediate attention if they are injured or taken ill at work. These regulations apply to all workplaces, including those with fewer than five employees, and to the self-employed.
- **Health and Safety Information for Employees Regulations 1989:** require employers to display a poster telling employees what they need to know about health and safety.
- **Management of Health and Safety at Work Regulations 1999 (as amended):** require employers to carry out risk assessments, make arrangements to implement necessary measures, appoint competent people and arrange for appropriate information and training.
- **Manual Handling Operations Regulations 1992:** cover the moving of objects by hand or bodily force.
- **Personal Protective Equipment at Work Regulations 1992 (as amended):** require employers to provide appropriate protective clothing and equipment for their employees.
- **Provision and Use of Work Equipment Regulations 1998:** require that equipment provided for use at work, including machinery, is suitable and safe.
- **Reporting of Injuries, Diseases and Dangerous Occurrences Regulations 2013:** require employers to notify the HSE of certain occupational injuries, diseases and dangerous events.
- **Workplace (Health, Safety and Welfare) Regulations 1992:** cover a wide range of basic health, safety and welfare issues such as ventilation, heating, lighting, workstations, seating and welfare facilities.

The following specific regulations cover particular areas, such as asbestos and lead.

- **Chemicals (Hazard Information and Packaging for Supply) Regulations 2002:** require suppliers to classify, label and package dangerous chemicals and provide safety data sheets for them.
- **Construction (Design and Management) Regulations 2015:** cover safe systems of work on construction sites.

- **Control of Asbestos Regulations 2012:** affect anyone who owns, occupies, manages or otherwise has responsibilities for the maintenance and repair of buildings that may contain asbestos.
- **Control of Lead at Work Regulations 2002:** impose duties on employers to carry out risk assessments, prevent or control exposure to lead and monitor the exposure of employees.
- **Control of Major Accident Hazards Regulations 1999 (as amended):** require those who manufacture, store or transport dangerous chemicals or explosives in certain quantities to notify the relevant authority.
- **Dangerous Substances and Explosive Atmospheres Regulations 2002:** require employers and the self-employed to carry out a risk assessment of work activities involving dangerous substances.
- **Work at Height Regulations 2005:** apply to all work at height where there is a risk of a fall liable to cause personal injury.

Approved Codes of Practice (ACoPs)

ACoPs offer practical examples of good practice. They were made under Section 16 of the HSW Act and have a special status. They give advice on how to comply with the law by, for example, providing a guide to what is reasonably practicable. For example, if regulations use words such as '**suitable**' and '**sufficient**', an ACoP can illustrate what is required in particular circumstances. If an employer is prosecuted for a breach of health and safety law, and it is proved that they have not followed the provisions of the relevant ACoP, a court can find them at fault unless they can show that they have complied with the law in some other way.

Guidance and non-statutory regulations

The HSE publishes guidance on a range of subjects. Guidance can be specific to the health and safety problems of an industry or to a particular process used in a number of industries. The main purposes of guidance are:

- to interpret and help people to understand what the law says
- to help people comply with the law
- to give technical advice.

Following guidance is not compulsory and employers are free to take other action, but if they do follow the guidance, they will normally be doing enough to comply with the law.

One very good example of guidance and non-statutory regulation is **BS 7671:2018** The IET Wiring Regulations, 18th Edition. If electrotechnical work is undertaken in accordance with BS 7671:2018, it is likely to meet the requirements of the Electricity at Work Regulations 1989, which deal with work with electrical equipment and systems.

BS 7671:2018 is the national standard in the UK for low-voltage electrical installations. The document is largely based on documents produced by the European Committee for Electrotechnical Standardization (CENELEC). The regulations deal with the design, selection, erection, inspection and testing of electrical installations operating at a voltage up to 1000 V AC.

KEY TERMS

Suitable: appropriate for a particular purpose or situation.

Sufficient: enough or adequate.

INDUSTRY TIP

BS 7671:2018 is the 18th edition of the Wiring Regulations and was published in July 2018. It can be obtained from the IET: https://electrical.theiet.org

BS 7671 is amended approximately every three years; the 18th edition will probably be amended two or three times before it is replaced with the 19th edition.

European law

In recent years, much of Great Britain's health and safety law has originated in Europe. Proposals from the European Commission (EC) may be agreed by member states, which are then responsible for making them part of their domestic law. Modern health and safety law in this country, including much of that from Europe, is based on the principle of risk assessment as required by the Management of Health and Safety at Work Regulations 1999.

Role of the HSE in enforcing health and safety legislation

Today, the HSE's aim is to prevent death, injury and ill health in United Kingdom workplaces and it has a number of ways of achieving this. Enforcing authorities may offer the duty holder information and advice, both face to face and in writing, or they may warn a duty holder that, in their opinion, the duty holder is failing to **comply** with the law.

In carrying out the HSE's enforcement role, inspectors appointed under the HSW Act can:

- enter premises at any reasonable time, accompanied by a police officer if necessary
- examine, investigate and require the premises to be left undisturbed
- take samples, photographs and, if necessary, dismantle and remove equipment or substances
- review relevant documents or information such as risk assessments, accident books, or similar
- seize, destroy or make harmless any substance or article
- issue enforcement notices and start prosecutions.

An inspector may serve one of three types of notice:

- a Prohibition Notice tells the duty holder to stop an activity immediately
- an Improvement Notice sets out action needed to remedy a situation and gives the duty holder a date by which they must complete the action
- a Crown Notice is issued under the same circumstances that would justify a Prohibition or Improvement Notice, but is only served on duty holders in Crown organisations such as government departments, the Forestry Commission or the Prison Service.

3 ENVIRONMENTAL LEGISLATION AND DEALING WITH WASTE

In the past, it was common for all **waste** produced on a construction site to be placed in a skip and for that waste to go to **landfill**. This practice has resulted in land **pollution** and groundwater pollution and even contributed to climate change due to the greenhouse gases that are emitted from landfill sites.

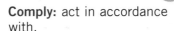

KEY TERM

Comply: act in accordance with.

KEY TERMS

Waste: something which has no further use to the person disposing of it.

Landfill: bury waste in large holes in the ground.

Pollution: contaminating the natural environment, causing change to that environment.

▲ Figure 1.3 A landfill site

European and UK laws have placed legal obligations on employers and operatives within all industry sectors to reduce waste, avoid pollution, reduce carbon emissions and **recycle** wherever possible.

What is waste?

Waste is quite difficult to define but, in general terms, it is any item that is thrown away because it is no longer useful or required by its owner. Electrical installation work generates many forms of waste, from packaging materials that come with new equipment and excess materials that cannot be saved for future use, such as part-used cable reels, to stripped-out materials and equipment, such as old light fittings and sockets and, of course, general building waste such as brick rubble and timber.

However we define waste, its disposal is governed by legislation. Previously, the majority of construction waste went to landfill sites without any thought to the potential impact of the buried materials on the **environment**.

European Union laws, that have been applied in the UK, have led to radical changes in waste handling and disposal. If you work within the construction industry, you need to have an understanding of those laws.

Legislation to protect the environment

The Department for Environment, Food and Rural Affairs (DEFRA) defines the environment as the land, water and air around us. Any pollution of land, water or the air will affect the quality of life for all organisms living within that environment.

The environment is under increasing pressure, not only because of our demand for resources, but also due to our need to dispose of waste.

Both of these can lead to pollution. There are several legislative documents that determine how we deal with waste and limit our impact on the environment:

- **Control of Pollution Act 1974 (COPA):** applies to activities such as waste disposal, water pollution, noise, atmospheric pollution and public health.

KEY TERMS

Recycle: to reuse the object or material.

Environment: the land, water and air around us.

INDUSTRY TIP

Visit DEFRA's website for further information at: www.defra.gov.uk

- **Environmental Protection Act 1990 (EPA):** makes provision for the improved control of pollution arising from certain industrial and other processes.
- **Environment Act 1995:** empowers the Environment Agency to control pollution and enhance the environment and conservation of natural resources.
- **Hazardous Waste Regulations 2005:** make provision for the controlled management of hazardous waste from the point of production to the final point of disposal or recovery.
- **Pollution Prevention and Control Act 1999:** requires a range of industrial installations to be regulated in which emissions to air, water and land, plus other environmental effects, are considered.
- **Waste Electrical and Electronic Equipment Regulations 2013 (WEEE Regulations):** require that producers (and with lesser obligations, distributors) of electrical and electronic equipment ('EEE') must be financially responsible for managing the waste.
- **Packaging (Essential Requirements) Regulations 2003:** relate to the management of packaging and packaging waste.

ACTIVITY

A four-storey block of flats built in the 1930s is to be totally refurbished. It still has the original imperial metal conduit (a metal pipe used to protect cables) but was rewired with PVC single-core cables in the late 1960s. List five different materials which will have to be disposed of in the proper manner.

Control of Pollution Act 1974 (COPA)

The aim of this Act is to deal with environmental issues including waste on land, water pollution, air pollution and noise pollution. If a person or organisation is found guilty under this Act, the maximum fine is £5000 plus £50 per day for each day the offence continues after conviction.

Local authorities require construction companies to apply for a permit under the Act prior to starting work. The construction company must carry out an analysis of the likely impact of noise and vibration on those in the surrounding area. The Act gives local authorities the power to impose restrictions on companies or individuals carrying out construction or demolition work, including imposing limits on noise levels and working times so as to avoid causing a nuisance to neighbours.

Environmental Protection Act 1990 (EPA)

The Environmental Protection Act (EPA) applies to England, Scotland and Wales. It deals with the disposal of controlled waste on land and sets out a framework for duty of care. The EPA specifically deals with:

- waste
- contaminated land
- **statutory nuisance.**

Controlled waste is domestic, commercial and industrial waste – in fact, all waste that is disposed of on the land. Under the EPA, it is an offence for anyone to deposit waste on any land unless a waste management licence authorising that deposit is held.

Land can be contaminated with naturally occurring substances, such as arsenic, by industrial processes, such as oil refining, or by fly tipping (illegally disposing of waste in undesignated sites).

Part 2A of the EPA works on the principle of the 'polluter pays'. The 'polluter' is defined as the person who caused the pollution or who 'knowingly permitted' the contamination. 'Knowingly permitted' not only applies to allowing the contamination to take place but also to having knowledge of the contamination and failing to deal with it. Where the polluter is unknown then the occupier or owner of the land is responsible.

Part 2A of the EPA applies where significant harm to the land has taken place or where the possibility of significant harm could take place or where rivers or groundwater are or could be affected.

The EPA also covers statutory nuisance and applies to any premises that may be detrimental to health or that cause a nuisance. This section is used by local authorities when dealing with antisocial behaviour, but it also applies to work procedures and covers such things as the emission of:

- dust
- steam
- smells
- **effluvia**
- noise.

When someone complains about any of the above, the local council must investigate. If the investigation reveals that a statutory nuisance does exist, a Notice of Abatement will be issued containing a list of steps that must be followed to reduce the nuisance. In the case of construction, this action could have a serious impact on the completion of the work.

Environment Act 1995

The Environment Act 1995 set new standards for environmental management and led to the creation of a number of agencies to oversee this management. The agencies created by this Act are:

- The Environment Agency
- The Scottish Environment Protection Agency
- The National Park Authorities.

The Act required that the Government prepare strategies on air pollution, national waste and the protection of hedgerows. The stated purpose of the Environment Agency is to 'enhance or protect the environment and promote sustainable development' and 'to create a better place for people and wildlife'. The Agency looks after everything from fishing rod licences to waste disposal, from flood defences to air pollution.

INDUSTRY TIP

Access the Environmental Protection Act (EPA) at: www.legislation.gov.uk/ukpga/1990/43/contents

KEY TERMS

Statutory nuisance: an unlawful interference with a person's right to use or enjoy land they have lawful access to.

Effluvia: emissions of gas, or odorous fumes given off by decaying waste.

INDUSTRY TIP

Access the Environment Act 1995 at: www.legislation.gov.uk/ukpga/1995/25/contents

The Environment Agency has been given the powers to:

- stop offending taking place
- restore and/or remediate, for which it will seek to recover the costs
- bring under regulatory control
- punish and/or deter, whether that be by criminal or civil sanctions.

The Environment Agency publishes all prosecutions and associated fines on their website and these range from a couple of thousand pounds for fishing without a licence to many hundreds of thousands of pounds for operating without a waste licence.

Hazardous Waste Regulations 2005

The Hazardous Waste Regulations set out a regime of control for the tracking and movement of hazardous waste, and deal with the production and disposal of that waste. Hazardous waste includes such items as:

- fluorescent tubes
- television sets
- fridges
- PC monitors
- batteries
- aerosols and paint
- contaminated soils.

When hazardous waste is moved from one location to another, a consignment note must be completed and passed to the licensed waste carrier. Hazardous waste must be kept separate from general waste. Electrical wholesalers generally run schemes whereby fluorescent tubes can be returned to them for safe disposal. It is a requirement of the Hazardous Waste Regulations that records are kept for a period of three years.

Pollution Prevention and Control Act 1999

According to this Act, industries that emit certain substances can only operate with a permit issued by the local authority or the Environment Agency. Included in the schedule of industries requiring a permit are those involved in metal and waste processing.

Waste Electrical and Electronic Equipment Regulations 2013 (WEEE Regulations)

The WEEE Regulations are the implementation of a European Directive to address the environmental impact of unwanted electrical and electronic equipment, namely to reduce the amount of **WEEE** that is sent to landfill sites and to encourage recycling, reuse and recovery before disposal in an environmentally friendly manner.

INDUSTRY TIP

Access the Hazardous Waste Regulations 2005 at: www.legislation.gov.uk/uksi/2005/894/contents/made

INDUSTRY TIP

Access the Pollution Prevention and Control Act 1999 at: www.legislation.gov.uk/ukpga/1999/24/contents

KEY TERM

WEEE: waste electrical and electronic equipment.

INDUSTRY TIP

Access the Waste Electrical and Electronic Equipment Regulations 2013 (WEEE Regulations) at: www.legislation.gov.uk/uksi/2013/3113/contents/made

INDUSTRY TIP

Remember, when you are undertaking practical activities and assessment, in college or on site, make sure you sort your waste materials for disposal.

You must comply with the WEEE Regulations if you manufacture, import, rebrand, distribute or dispose of electrical and electronic equipment. While it may seem obvious that manufacturers and distributors must comply, WEEE Regulations apply to *anyone* who disposes of such equipment.

As electricians frequently remove **redundant** electrical equipment and have **surplus** materials for disposal, compliance with the WEEE Regulations is an obligation that must be met.

Under the WEEE Regulations, electrical and electronic items are divided into 10 categories:

1 Large household appliances, for example refrigerators, fans and panel heaters
2 Small household appliances, such as vacuum cleaners and toasters
3 IT and telecommunications equipment
4 Consumer equipment, such as radios and televisions
5 Lighting equipment, for example fluorescent tubes and discharge lamps
6 Electrical and electronic tools, such as drills
7 Toys, leisure and sports equipment
8 Medical devices
9 Monitoring and control instruments, such as smoke detectors and thermostats
10 Automatic dispensers, for example vending machines.

Electricians most commonly deal with items in categories 5 and 9 but, at times, other categories may also apply.

▲ Figure 1.4 Examples of small electrical appliances

Bear in mind that some WEEE may also be classified as hazardous waste. Examples are: smoke detectors, which contain radioactive emitters; fluorescent tubes, which contain mercury and cadmium; as well as old discharge lighting control gear containing PCBs (polychlorinated biphenyls), which are hazardous to persons and the environment. If in doubt regarding any of these items, always seek advice.

KEY TERMS

Redundant: no longer needed or useful, even though the part may still function. For example, if the cables were removed from a conduit, but the conduit remained as its removal would create decorating problems, the conduit is classed as redundant.

Surplus: more than what is needed. For example, if 75 m of cable was used from a 100 m reel, the remaining 25 m is surplus.

ACTIVITY

Look around your home and identify five items that come under the WEEE Regulations.

INDUSTRY TIP

The WEEE Regulations will apply to you when disposing of any electrical equipment.

Packaging (Essential Requirements) Regulations 2003

These regulations require anyone, but generally manufacturers of products, who place packaging into the marketplace, to take certain steps to:

- minimise the amount of packaging used
- make sure packaging can be recovered (recycled)
- ensure that packaging has the minimum possible impact on the environment
- ensure that packaging does not contain high levels of hazardous materials or heavy metals.

Packaging has a very short life cycle. It is useful only from the time it leaves the manufacturer to when it arrives at the end user. While the Packaging (Essential Requirements) Regulations are not directly aimed at end users, they do reinforce the requirements for dealing with any waste product.

INDUSTRY TIP

Access the Packaging (Essential Requirements) Regulations 2003 at: www.legislation.gov.uk/uksi/2003/1941/contents/made

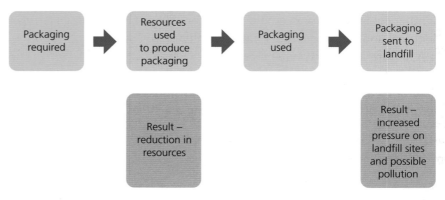

▲ Figure 1.5 A poor waste model

It is obvious that the poor waste model for packaging in Figure 1.5 is unsustainable. The sustainable waste model shown in Figure 1.6 is far better.

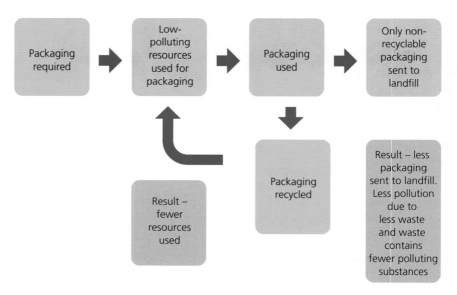

▲ Figure 1.6 A sustainable waste model

In the sustainable waste model, the packaging is made of materials that will cause minimum pollution in landfill sites. However, the materials are actually recyclable, thus cutting down waste and also reducing the demands on dwindling resources.

The key person in this cycle is the person who ensures that the waste product is recycled.

4 DEALING WITH HEALTH AND SAFETY IN THE WORK ENVIRONMENT

The role of safety culture

An organisation's **safety culture** can have as big an influence on safety outcomes as the safety management system itself, with safety culture being a subset of the overall organisational or company culture. Poor safety culture has contributed to many major incidents (e.g. the Piper Alpha oil platform disaster; the fire at King's Cross underground station; the sinking of the Herald of Free Enterprise passenger ferry; the passenger train crash at Clapham Junction; the Chernobyl disaster) and personal injuries. Success in this area normally comes from good leadership, good worker involvement and good communications.

By paying attention to human factors, forward-looking companies can identify and deal with potential hazards before they manifest themselves as accidents. This, coupled with legislation in the form of regulations and ACoPs that are easily understood and complied with, can have a positive effect on health and safety standards and help prevent or reduce accidents and incidents.

Procedures for handling injuries sustained on site

The type of accident that can occur in the workplace is dependent on the work activity being undertaken but can range from a cut finger to a **fatality**, or from a vehicle collision to the collapse of a structure. The person in control of the premises, such as a site supervisor, needs to be prepared to deal with all types of accidents to ensure that the injured person can be treated quickly and effectively and that all the legal obligations are met.

Having a well-established procedure that everyone on site is aware of and understands will enable the person in control of the premises to cope calmly and effectively when dealing with an accident. Good management following an accident will ensure that the injured person is attended to promptly, appropriate records are made, the accident is reported correctly and any lessons to be learned from the accident are understood and communicated to the workforce.

The procedures to be followed in the event of any accident or incident should be clear and specific to the project or site and should detail the following as a minimum:

KEY TERMS

Safety culture: 'the product of individual and group values, attitudes, perceptions, competencies, and patterns of behaviour that determine the commitment to, and the style and proficiency of, an organisation's health and safety management.' (Source: Health and Safety Executive)

Fatality: death.

ACTIVITY

How would you deal with a deeply cut hand sustained while cutting metal trunking?

- name of the appointed person(s) who will take control when someone is injured or falls ill
- name of the person(s) who will administer first aid
- location of the first aid boxes and name of the person(s) responsible for ensuring they are fully stocked
- course of action that must be followed by the appointed person who takes control in the event of an accident
- guidance on action to take after the accident
- how information should be recorded and by whom (F2508 RIDDOR Form, which can be found on the HSE website: www.hse.gov.uk/forms/incident/index.htm).

How to deal with electric shocks

If all of the correct requirements are met, precautions taken and training of staff undertaken, it is unlikely that an electrical accident will occur. However, procedures should be in place to deal with electric shock injury in the event of an accident. The recommended procedure for dealing with a person who has received a low-voltage shock is as follows:

- Raise the alarm (colleagues and a trained first-aider).
- Make sure the area is safe by switching off the electricity supply.
- Request colleagues to call an ambulance (999 or 112).
- If it is not possible to switch off the power supply, move the person away from the source of electricity by using a non-conductive item.
- Check if the person is responsive, whether their airway is clear and whether they are breathing.
- If the person is unconscious and breathing, move them into the recovery position.
- If they are unconscious and not breathing, start to give cardiopulmonary resuscitation (CPR):
  - CPR is undertaken by interlocking the hands and giving 30 chest compressions in the centre of the chest, between the two pectoral muscles, at a rate of about 100 pulses per minute.
  - Tilt the casualty's head back gently, by placing one hand on the forehead and the other under the chin, to open the airway and give two mouth-to-mouth breaths.
  - Repeat the cycle of 30 compressions to two breaths until either help arrives or the patient recovers.
- Any minor burns should be treated by placing a sterile dressing over the burn and securing with a bandage.

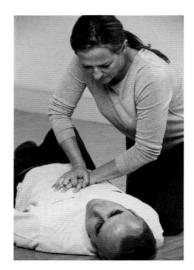

▲ Figure 1.7 CPR being performed

Procedures for recording accidents and near misses at work

An **accident** is defined by the HSE as 'any unplanned event that results in injury or ill health of people, or damage or loss to property, plant, materials or the environment, or a loss of a **business opportunity**'.

KEY TERMS

Accident: an unplanned event that results in injury or ill health, damage, or loss of business.

Business opportunity: in this context, the opportunity to make profit from the work or contract.

A **near miss** is an unplanned event that does not result in injury, illness or damage, but had the potential to do so. Normally, only a fortunate break in the chain of events prevents an injury, fatality or damage taking place. So, a near miss could be defined as any incident that could have resulted in an accident. The keeping of information on near misses is very important in helping to prevent accidents occurring. Research has shown that damage and near miss accidents occur much more frequently than injury accidents and therefore give an indication of hazards.

The Social Security Act 1975 specifically requires employers to keep information on accidents. This should be the Statutory Accident Book for all Employee Accidents (Form B1510, found on the HSE website: www.hse.gov.uk/forms/incident/index.htm) or equivalent. Each entry should be made on a separate page and the completed page securely stored to protect personal data (under the General Data Protection Regulation 2018). An entry may be made by the employee or by anyone acting on their behalf. This information should be kept for a period of not less than three years.

> ### INDUSTRY TIP
>
> You can access the Social Security Act 1975 at: www.legislation.gov.uk/ukpga/1975/14/contents/enacted

Reporting the incident

The reporting of certain types of **injury** and **incidents** to the enforcing authority (the HSE or the local authority) is a legal requirement under the Reporting of Injuries, Diseases and Dangerous Occurrences Regulations 2013 (RIDDOR). Failure to comply with these regulations is a criminal offence.

RIDDOR states that deaths, specified injuries (listed in the regulations) and injuries resulting in absence from work for over seven days, and dangerous occurrences (again, listed in the regulations) and occupational diseases must be reported. This seven-day period does not include the day of the accident, but does include weekends and rest days. The report must be made within 15 days of the accident.

It is the responsibility of employers or the person in control of premises to report these types of incidents. Reportable specified injuries include fractures, amputations, permanent loss of sight, crush injuries and serious burns. A dangerous occurrence is a 'near-miss' event (incident with the potential to cause harm). There are also special requirements for gas incidents. Accidents must be recorded, but not reported, where they result in a worker being incapacitated for more than three consecutive days.

The police and HSE have the right to investigate fatal accidents at work. Therefore, all fatal accidents must also be notified to the police. The police will often notify the HSE, but it is always a sensible precaution to ensure that the HSE has been notified.

▲ Figure 1.8 An accident book

Investigating accidents

There is nearly always something to be learned following an accident and ideally the causes of all accidents should be established regardless of whether injury or damage resulted. The level and nature of an investigation should reflect the significance of the event being investigated. The results of the accident investigation may lead to a review, possible amendment to the risk assessment and appropriate action to prevent similar accidents from occurring.

Keeping records

There are numerous records to keep following even a minor accident. Easily accessible records should be maintained for all accidents that have occurred. In addition to the legal requirements, accident information can help an organisation identify key risk areas within the business. The accident book must be kept for three years following the last entry. The HSE has an online reporting tool known as an F2508 which can be used to report various incidents, these can be found at www.hse.gov.uk/forms/incident/index.htm

The **F2508** for reportable incidents should be kept for a period of not less than three years from the date the accident occurred.

What to do in an accident or emergency

When an emergency situation occurs and the emergency services need to be called, be sure you know the following information:

- the address and location of the incident
- the nature of the incident, such as fire, injury etc.
- any difficulties the emergency services may encounter while trying to get to the incident, such as a high-rise building or in a field in a remote location
- any immediate dangers such as explosive materials, persons trapped etc.

Should an emergency occur that requires evacuation, it is essential you know what or where the designated escape route is. On construction sites, due to their ever-changing nature, designated escape routes may frequently move so always be sure you are familiar with them. What may have been a safe escape route one day could be a dangerous route the next.

In nearly all fire situations, the emergency services should be called, but some very small fires may be extinguished before they become too serious. Be sure you know how to tackle different fires and know where appropriate fire-fighting devices are located.

Emergencies

Emergency procedures are there to limit the damage to people and property caused by an incident. Although the most likely emergency to be dealt with is fire, there are many more emergency situations that need to be considered, including the following.

Electrical fire or explosion

Fires involving electricity are often caused by lack of care in the maintenance and use of electrical equipment and installations. The use of electrical equipment should be avoided in potentially flammable atmospheres as far as is possible. However, if the use of electrical equipment in these areas cannot be avoided, then equipment purchased in accordance with the Equipment and Protective Systems Intended for Use in Potentially Explosive Atmospheres Regulations 1996 must be used.

Escape of toxic fumes or gases

Some gases are poisonous and can be dangerous to life at very low concentrations. Some toxic gases have strong smells such as the distinctive 'rotten eggs' smell of hydrogen sulphide (H_2S). The measurements most often used for the concentration of toxic gases are parts per million (ppm) and parts per billion (ppb). More people die from toxic gas exposure than from explosions caused by the ignition of flammable gas. With toxic substances, the main concern is the effect on workers of exposure to even very low concentrations. These could be inhaled, ingested (swallowed) or absorbed through the skin. Since adverse effects can often result from cumulative, long-term exposure, it is important to measure not only the concentration of gas, but also the total time of exposure.

Gas explosion

A gas explosion is an explosion resulting from a gas leak in the presence of an ignition source. The main explosive gases are natural gas, methane, propane and butane because they are widely used for heating purposes in temporary and permanent situations. However, many other gases, such as hydrogen, are combustible and have caused explosions in the past.

The source of ignition can be anything from a naked flame to the electrical energy in a piece of equipment. Industrial gas explosions can be prevented with the use of intrinsic safety barriers to prevent ignition. The principle behind intrinsic safety is to ensure that the electrical and thermal energy from any electrical equipment in a hazardous area is kept low enough to prevent the ignition of flammable gas. Items such as electric motors would not be permitted in a hazardous area.

Employer and employee responsibilities

Employers have a duty of care to each of their employees. This duty rests solely with the employer and cannot be assigned to other persons – for example, consultants who offer advice on health and safety matters, or sub-contractors who are employed to undertake tasks within the company. All organisations should have a clear policy for the management of health and safety. The policy sets the direction for health and safety within the organisation, and its contents need to be clearly communicated to everyone within the organisation to ensure everyone understands what their responsibilities are in day-to-day operations. Everyone has responsibility for safety, but employers have the following duties under the HSW Act:

> **INDUSTRY TIP**
>
> Access the Equipment and Protective Systems Intended for Use in Potentially Explosive Atmospheres Regulations 1996 at: www.legislation.gov.uk/uksi/1996/192/made

- the health, safety and welfare at work of employees, and other workers whether they are contractors, casual, temporary, part time or trainees
- the health and safety of anyone who is allowed to use the organisation's equipment
- the health and safety of anyone who may be affected by the organisation's activities, i.e. the general public or adjacent organisations or neighbours.

Examples of matters under the control of an employer include:

- establishing policy to ensure that electrical equipment is purchased to an appropriate specification
- establishing policy to ensure that electrical equipment is properly maintained (including user inspection)
- implementing policy through the introduction of appropriate management systems
- on-going monitoring to confirm that the policy is properly implemented and remains fully relevant.

Employees have specific responsibilities under the HSW Act and these can be summarised as:

- to take reasonable care for the health and safety of themselves and others who may be affected by their acts or omissions at work
- to co-operate with their employer and others to enable them to fulfil their legal obligations
- not to interfere with or deliberately misuse anything provided in the interests of health and safety.

Examples of matters under the control of an employee include:

- adherence to company procedures and systems of work
- use of equipment in accordance with information and training provided
- not to use equipment that is faulty or damaged but to report it in accordance with company arrangements for dealing with defective equipment.

In summary, under the HSW Act employers have duties to ensure that appropriate management systems are established so that electrical work can be undertaken in a safe manner, while employees have a responsibility to comply fully with such management systems.

What to do if you have concerns about health and safety issues

ACTIVITY

Consider some situations where you could be at risk because the desire to get the work done is greater than the need to spend some time creating a safe situation. One example is disconnecting bonding before isolating the entire installation!

Employees are responsible for ensuring that the work they are required to undertake is carried out in a manner which is safe for themselves and other persons who may be affected by the work activity. They must undertake this work activity in accordance with the instruction or procedure provided by the employer.

If an employee has concerns about health and safety at work, or feels that there is a situation which they believe exceeds their level of responsibility, then these concerns should be raised with their supervisor or line manager. If the organisation has a recognised trades union and a safety representative has been appointed, it may be appropriate for the safety representative or the trades union official to be the first point of contact.

What to do if you have concerns about your work activity

The law requires employers and the self-employed to conduct their business in such a way as to ensure, 'so far as is reasonably practicable', that persons affected are not exposed to risks to their health or safety. This includes providing essential welfare facilities for employees.

If it is believed that an employer's (or someone else's) work activity is creating a risk or causing damage to health, then the concern should be raised with that employer or person responsible for the work activity. If no improvement is made and the safety or health risk continues, then it should be reported to the relevant enforcing authority, which should be asked to investigate.

There are two relevant enforcing authorities.

▼ Examples of workplaces covered by each enforcing authority

The HSE	The Local Authority Environmental Health Department
Mines	Shops
Factories	Hotels
Building sites	Restaurants, public houses and clubs
Nuclear installations	Leisure premises
Hospitals and nursing homes	Nurseries and playgroups
Gas, electricity and water systems	Offices (except government offices)
Schools and colleges	
Offshore installations	

Who to report concerns to

All companies need an effective internal system for reporting health, safety and welfare-related matters and this should be readily available to all employees, including the names and contact details of the relevant responsible persons.

The names, positions and duties of those people within the organisation who have responsibility for health and safety will normally be contained within the safety policy. This will include:

- managers (directors, site managers, supervisors)
- specialists (health and safety advisors, safety officers, first aiders, fire officers)
- employee representatives (trades union representatives, safety representatives)
- HSE Officers and Local Authority Environmental Health Officers.

HEALTH AND SAFETY

If you see someone carrying out a dangerous procedure, report it to your supervisor. It may not make you the most popular person but may save their, or another person's life, maybe even yours.

INDUSTRY TIP

Many places of work have notice boards detailing the contact information of the person responsible for health, safety and welfare issues. Other organisations have internal websites (intranets) detailing this information.

ACTIVITY

Look around your college or place of learning and identify any health and safety notices, such as first aid, detailing the person to contact.

5 ESTABLISHING A SAFE WORKING ENVIRONMENT

The control of **risks** is essential to the provision and maintenance of a safe and healthy workplace and ensures **compliance** with the relevant legal requirements.

Risk assessment starts with the need for **hazard** identification. Risk assessment is usually evaluated in terms of:

- the likelihood of something happening (i.e. whether a hazard is going to occur)
- the severity of outcome (i.e. how serious the resulting injury will be).

To control these hazardous situations, duty holders need to:

- find out what the hazards are
- decide how to prevent harm to health
- provide control measures to reduce harm to health
- make sure the control measures are used
- keep all control measures in good working order
- provide information, instruction and training for employees and others
- provide monitoring and health surveillance in appropriate cases
- plan for emergencies.

Common hazardous situations on site
Housekeeping

Housekeeping is one of the most important single items influencing safety within the workplace. Poor housekeeping not only causes an increase in the risk of fire, slips, trips and falls, but may also expose members of the public to risks created during building services engineers' work activities. The following are some examples of good housekeeping.

- Stairways, passages and gangways should be kept free from materials, electrical power leads and obstructions of every kind.
- Materials and equipment should be stored tidily so as not to cause obstruction and should be kept away from the edges of buildings, roofs, ladder access, stairways, floor openings and rising shafts.
- Tools should not be left where they may cause tripping or other hazards. Tools not in use should be placed in a tool belt or tool bag and should be collected and stored in an appropriate container at the end of each working day.
- Working areas should be kept clean and tidy. Scrap and rubbish must be removed regularly and placed in proper containers or disposal areas.
- Rooms and site accommodation should be kept clean. Soiled clothes, scraps of food, etc. should not be allowed to accumulate, especially around hot pipes or electric heaters.
- The spillage of oil or other substances must be contained and cleaned up immediately.
- All flammable liquids, liquefied petroleum gas (LPG) and gas cylinders must be stored properly.

KEY TERMS

Risk: the chance (large or small) of harm actually being done when things go wrong (e.g. risk of electric shock from faulty equipment).

Compliance: the act of carrying out a command or requirement.

Hazard: anything with the potential to cause harm (e.g. chemicals, working at height, a fault on electrical equipment).

INDUSTRY TIP

You must be able to identify hazards as part of a risk assessment. It should then be possible to eliminate, reduce, isolate or control the risk by the application of suitable control measures. The use of personal protective equipment (PPE) should be a last resort.

INDUSTRY TIP

When working on practical tasks in college or on site, always keep the work area clean and tidy. Dangerous working will mean you could fail.

▲ Figure 1.9 Stairs are a particular risk

▲ Figure 1.10 A tool belt is a good way to avoid leaving tools lying around

▲ Figure 1.11 Gas cylinders must be stored correctly

Slips, trips and falls

These are the most common hazards to people as they move around the workplace. They account for 29% of all major workplace injuries according to statistics provided by the HSE for 2016/17. Listed below are the main factors that can play a part in contributing to a slip-free and trip-free environment.

Flooring

- The workplace floor must be suitable for the type of work activity that will be taking place on it.
- Floors must be cleaned correctly to ensure they do not become slippery, or be of a non-slip type that keeps its slip-resistant properties.
- Flooring must be fitted correctly to ensure that there are no trip hazards and any non-slip coatings must be correctly applied.
- Floors must be maintained in good order to ensure that there are no trip hazards such as holes, uneven surfaces, curled-up carpet edges, or raised telephone or electrical sockets.

ACTIVITY

Check the HSE website at www.hse.gov.uk/statistics/causinj/index.htm to see the most common accidents. Then consider or discuss methods of preventing accidents in each of the categories shown.

KEY TERMS

Nosings: the front edges of steps. If they are not square, they can lead to people slipping down the step. Many steps are fitted with additional metal strips to avoid the step from wearing.

Contamination: the introduction of a substance that should not be there, such as water, oil or dust.

- Ramps, raised platforms and other changes of level should be avoided. If they cannot be avoided, they must be highlighted.

Stairs

Stairs should have:

- high-visibility, non-slip, square **nosings** on the step edges
- a suitable handrail
- steps of equal height
- steps of equal width.

Contamination

Most floors only become slippery once they become contaminated. If **contamination** can be prevented, the slip risk can be reduced or eliminated.

Contamination of a floor can be classed as anything that ends up on a floor, including rainwater, oil, grease, cardboard, product wrapping, dust, and so on. It can be a by-product of a poorly controlled work process or be due to bad weather conditions.

Cleaning

Cleaning and tidying are important in every workplace. It is not just a subject for the cleaning team. Everyone's aim should be to keep their workspace clear and deal with contamination such as spillages as soon as they occur.

The process of cleaning can itself create slip and trip hazards, especially for those entering the area being cleaned, including those undertaking the cleaning. Smooth floors left damp by a mop are likely to be extremely slippery. Trailing wires from a vacuum cleaner or polishing machine can present an additional trip hazard.

People often slip on floors that have been left wet after cleaning. Access to smooth wet floors should be restricted by using barriers, locking doors or cleaning in sections. Signs and warning cones only warn of a hazard – they do not prevent people from entering the area. If the water on the floor is not visible, signs and cones are usually ignored.

Human factors

How people act and behave in their work environments can affect the risk of slips and trips. Accidents can be reduced by:

- having a positive attitude toward health and safety – e.g. dealing with a spillage right away instead of waiting for someone else to deal with it
- wearing the correct footwear
- concentrating and avoiding distractions, such as being in a hurry, carrying large objects or using a mobile phone, all of which increase the risk of an accident.

Environmental factors

Environmental issues can affect the incidence of slips and trips. For example:

- Too much light on a shiny floor can cause glare and stop people from seeing hazards on floors and stairs.

- Rainwater on smooth surfaces inside or outside a building may create a slip hazard.
- Too little light will prevent people from seeing hazards on floors and stairs.
- Unfamiliar and loud noises can be distracting.

Footwear

Footwear can play an important part in preventing slips and trips. For example:

- The choice of footwear should take into account factors such as comfort and durability, as well as obvious safety features such as providing the protection required for the tasks undertaken.
- A good tread pattern on footwear is essential on wet surfaces.
- Sole tread patterns should not be allowed to become clogged with any waste or debris on the floor, as this makes them unsuitable for their purpose.
- Sole material type and hardness are key factors influencing safety.
- Footwear can perform differently in different situations – e.g. footwear that performs well in wet conditions might not be suitable where there are food spillages.

The requirements for a risk assessment

The HSW Act requires employers to provide health and safety systems and procedures in the workplace, and directors and managers of any company that employs more than five employees can be held personally responsible for failure to control health and safety. The general duties of Section 2 of the HSW Act imply that risk assessment is necessary. The aim of the Management of Health and Safety at Work Regulations 1999 (MHSWR) is to give more detail on how to achieve this and to encourage a more **systematic** and better organised approach to dealing with health and safety in all workplaces. The main objective of risk assessment is to determine the measures required by the organisation to comply with relevant health and safety legislation and so reduce the level of occupational injuries and ill health.

MHSWR Regulation 3 requires that:

'Every employer and self-employed person shall make a suitable and sufficient assessment of:

- the risks to the health and safety of his employees to which they are exposed while they are at work; and
- the risks to the health and safety of persons not in his employment arising out of or in connection with the conduct by him of his undertaking…'.

All organisations must have the freedom to prepare systems of work that match the risk potential of their particular work activity and which are practical in their application. Electrotechnical work operations are no different to other work disciplines in this respect and employers must examine the work activities and the workplace and introduce systems of work through a process of 'risk assessment' that takes account of the risks and the application of the health

> **HEALTH AND SAFETY**
> You may often hear the term RAMS, which stands for risk assessment and method statement.

> **KEY TERM**
>
> Systematic: working to a fixed plan or procedure.

> **INDUSTRY TIP**
>
> Section 3 of the Management of Health and Safety at Work Regulations 1999 (MHSWR) covering risk assessments can be accessed at: www.legislation.gov.uk/uksi/1999/3242/regulation/3/made

and safety legislation designed to manage those risks. An example of this would be carrying out safe isolation before working on electrical equipment.

The HSE promotes a risk assessment process entitled *Five Steps to Risk Assessment*. The five steps are as follows (and covered in further detail later in the section).

Step 1 Identify the hazards.
Step 2 Decide who may be harmed and how.
Step 3 Evaluate the risks and decide on precautions.
Step 4 Record the findings and implement them.
Step 5 Review the assessment and update if necessary.

It is important to remember that a risk assessment is a continual process, as the circumstances are constantly changing. Therefore, although the five steps are shown as a list, the process tends to be more of a cycle (Figure 1.12).

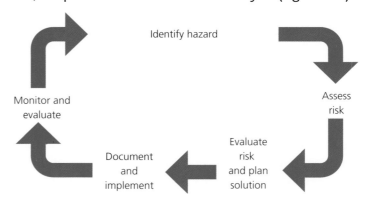

▲ Figure 1.12 The five steps for risk assessment

Completing a risk assessment

Step 1: Identify the hazard

Identifying the hazard is probably the easiest stage of the risk assessment. The person performing the risk assessment must understand fully what is involved in the task. So that the hazards are easier to identify, it is important not to overcomplicate analysis of the task.

The person doing the risk assessment should also consider the specific work environment, so it can be helpful to talk to others who have performed the task.

Remember a risk assessment must be specific to the exact task being performed in a certain place – generic risk assessments are only a starting point.

Step 2: Assess harm

When assessing the potential for harm, consider where the task is being performed – different locations may change the risk. For example, using a power tool in a public place involves a different set of risks compared to use of the power tool in a workshop or factory because the people in a place of work are more aware of risks than members of the public. Remember, the same task performed by different people may present different hazards, according to each person's knowledge, experience and work ethic.

Some hazards have different impacts depending on who is involved. For example, a person in a wheelchair may be at a particular risk from an activity, whereas a non-disabled person may be able to avoid the hazard. Having identified *who* may be at risk, the next stage is to think about *how* they could be harmed. Essentially, all the people at risk must be identified, as well as the potential harm that could be caused.

Because **young people** at work lack experience and maturity, special risk assessments need to be made, known as Young Person Risk Assessments.

Step 3: Evaluate risks

Once you have identified the hazard, established who may be affected and in what ways, you must plan how to remove the risks. At this stage, it is a good idea to rank the risks in terms of likelihood and severity of outcome. It may not be possible to remove some risks and so a level of control may be more appropriate, such as the use of barriers.

A simple way to rank the risks is to consider them in terms of impact versus likelihood of the risk happening. A complex numbering system can be used to score the risk: the higher the number the more important it is to reduce the risk. Alternatively, a simple 'traffic light' matrix can be used, as shown in Figure 1.13.

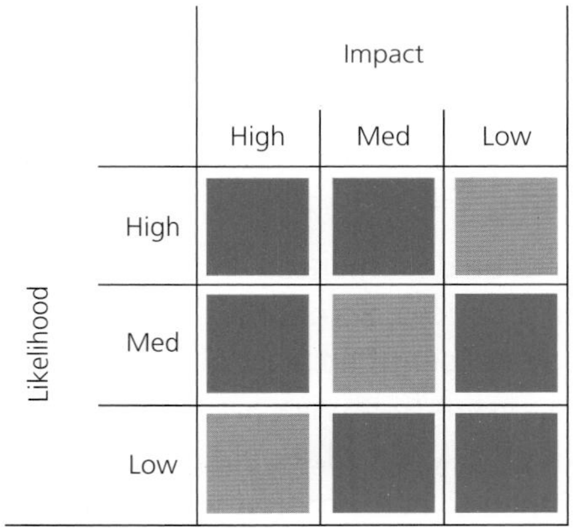

KEY TERM

Young people: people under the age of 18.

▲ Figure 1.13 A risk evaluation matrix

- *Red*: high risks that must be controlled. Work cannot proceed until these risks are removed or reduced.
- *Amber*: medium-level risks. The control measure is to ensure the safest working procedures and practices are being applied.
- *Green*: minor and unavoidable risks. There should be a plan for what to do if the hazard should occur.

It is possible for hazards to move from one level of risk to another as work is undertaken. Therefore, the risk assessment must be constantly monitored and evaluated.

Once the risks have been ranked into red, amber and green classifications, solutions need to be implemented. The approach to risk management includes a hierarchy of control, which should be approached in this sequence:

1 Elimination
2 Reduction
3 Enclosure
4 Remove persons
5 Reduce contact.

Risk management option 1, *Elimination*, is the most effective and should always be the first choice. Risk management option 5, *Reduce contact*, is the least effective, but is often chosen – it covers the use of personal protective equipment (PPE). Other control measures that can also be used include:

- protective clothing
- instruction
- training
- following a defined procedure for a task
- implementing a permit-to-work system
- information signs
- information sheets.

Step 4: Document and implement solution

Risk assessments are not very effective unless everyone knows about them. Therefore the documentation of hazards, risks and their control measures is essential.

Control measures need to be documented and people must be trained in any changes to working practices. Sometimes, while training is taking place, other potential risks may be identified. Part of the documentation can include **method statements** to ensure that work practices and safety measures are in place and followed correctly. A record should be kept of who receives training. In some cases, this is signed to confirm that workers have received training and understand the requirements now placed on them. This record can help employers to prove they are meeting the legal requirement to do everything reasonably practicable to keep people safe.

If the work is to be performed on a construction site, the Construction (Design and Management) Regulations 2015 (CDM) specify that the risk assessments and method statements must be retained in the site safety file. This file is referred to in most site safety induction training sessions and forms the basis of an employee site safety manual.

Step 5: Evaluate, review and update

Once a risk assessment has been performed, it should be checked regularly or before the task in question is performed to ensure the control measures being followed are effective. As well as being reviewed regularly, the risk assessment should be reassessed if the work or environment changes. Any changes that are required will become part of the future review process.

KEY TERM

Method statement: instructions on performing tasks safely in accordance with the risk assessment. (There is more detail on method statements later in this section.)

INDUSTRY TIP

Access the Construction (Design and Management) Regulations 2015 (CDM) at www.legislation.gov.uk/uksi/2007/320/contents/made

Control of high-level risks

If the risk level identified is high, the control measures may need to be more extensive. The aim of risk assessment is to reduce all residual risks to as low a level as reasonably practicable and use the resulting information to prepare control measures as necessary. Many hazards have specific regulations or guidance available to give advice on reducing the risk (e.g. lead, electricity, asbestos) and these should be referred to in the first instance, to obtain advice for the preparation of control measures.

Safe systems of work

A safe system of work is a work method that results from a systematic examination of the working process to identify the hazards and to specify work methods designed either to completely remove the hazards or to control and minimise the relevant risks. Section 2 of the Health and Safety at Work etc. Act 1974 (HSW Act) requires employers to provide safe **plant** and systems of work. Many of the regulations made under the HSW Act have more specific requirements for the provision of safe systems of work.

Safe systems of work should be developed by a **competent person** – that is, a person with sufficient training and experience or knowledge to assist with key aspects of safety management and compliance. Staff who are actively involved in the work process also have a valuable role to play in the development of the system. They can help to ensure that it is of practical benefit and that it will be applied diligently. All safe systems of work need to be monitored regularly to ensure that they are fully observed and effective, and updated as necessary. Safe systems of work are normally formal and documented, but may also be given as a verbal instruction. Examples of documented safe systems of work would be for asbestos removal, air-conditioning maintenance, working on live electrical equipment and portable appliance testing. Examples of verbal instructions could be as simple as 'ensure you clear the route before moving those ladders'.

Method statements

A method statement is a description of a safe system of work, presenting in a logical sequence, and generally in writing, a method of undertaking an activity without risk to health. The method statement should be clear, and illustrated where applicable by simple sketches. Statements are for the benefit of those carrying out the work and their immediate supervisors and should not be overcomplicated. The statement includes all the risks identified in the risk assessment and the measures to control those risks. The statement need be no longer than necessary to achieve the objective of safe working.

Employees should know what to do if the work process has changed and the method statement no longer achieves its aim. Method statements must be reviewed on a regular basis to accommodate any such changes in the work process.

KEY TERMS

Plant: tools and machinery.

Competent person: recognised term for someone with the necessary skills, knowledge and experience to manage health and safety in the workplace.

ACTIVITY

Clear instructions are important. Can you spot the difference between the two instructions below?

It needs to go over there, head for the door.

It needs to go over their head for the door.

ACTIVITY

Write a brief method statement for a simple task such as moving an awkwardly large but lightweight box from one end of a workshop to the other.

Method statement (safe system of work)				
Contract ref: RLF-0012-18 **Address:** Office 34, New Road, New Town. AB12 3CD	**Client Details:** Referb Contracts	**Start Date:** 12-05-18 **End Date:** 12-05-18	**Assessed by:** Lou Skinection **Modifications** Y [N]	**Associated risk assessment documents:** Doc. RA-12-2018 Doc MEWP-01-2016

Description of works:				
Replacement of wall luminaire 4.6 m above ground 1.1 m below accessible flat roof				

Labour requirements: 1 approved electrician with 2391 1 other for assistance	**Plant/equipment requirements:** Battery hammer drill Scaffold (in place) Harness Inspection and test equipment **Chemicals:** none	**Welfare facilities:** On site toilet block **First aid facilities:** Main site office (portacabin)	**Specialist training requirements:** Inspection and testing (2391) required

Work method:		Briefing record (tool box talk)	
<ul><li>Scaffold will be erected by others</li><li>Check scaffold</li><li>Isolate & secure circuit L1-04</li><li>Check isolation before full dismantle</li><li>Remove existing fitting</li><li>New fitting to be handled to position from flat roof</li><li>Existing fitting removed via flat roof</li><li>Install new luminaire</li><li>Check secure fixing and alignment</li><li>Test circuit</li><li>Test function & complete minor works</li></ul>		**Name**	**Signature**
		Lou Skinection	*L Skinection*
		Shaw Circuit	*S Circuit*

▲ Figure 1.14 A method statement should be clear and simple

Permit-to-work procedures

A **permit-to-work (PTW)** procedure is a specialised written safe system of work that ensures that potentially dangerous work is done safely. Examples include work in confined spaces, **hot work**, work with asbestos-based materials, work on pipelines with hazardous contents, or work on high-voltage electrical systems (above 1000 V) or complex low-voltage electrical systems.

A PTW procedure also serves as a means of communication between site or installation management, plant supervisors and operators, and those who carry out the hazardous work.

For example, if an electrician needed to work on the low-voltage side of a transformer controlled by a supply company, they would obtain a PTW that states that the supply company have isolated the transformer and the system is safe to work on.

A PTW should only be issued by a technically competent person, who is familiar with the system and equipment, and is authorised in writing by the employer to issue such documents.

Essential features of PTW systems are:

- clear identification of who may authorise particular jobs (and any limits to their authority) and who is responsible for specifying the necessary precautions

KEY TERMS

Permit-to-work (PTW): a written safe system of work produced to support the safe completion of potentially dangerous work and support communications between different persons.

Hot work: work that involves actual or potential sources of ignition, carried out in an area where there is a risk of fire or explosion (e.g. welding, flame cutting or grinding).

Audit: conduct a systematic review to make sure standards and management systems are being followed.

- training and instruction in the issue, use and closure of permits
- monitoring and **auditing** to ensure that the system works as intended
- clear identification of the types of work considered hazardous
- clear and standardised identification of tasks and risk assessments
- permitted task duration and any additional activity or control.

The effective operation of a PTW system requires involvement and co-operation from a number of persons. The procedure for issuing a PTW should be written and adhered to.

The hierarchy of control

The control of risk is essential to the provision of a safe working environment that complies with all of the relevant legal requirements. To maintain this safe working environment, any control measures (detailed below) that are produced require assessment and, if necessary, amendment or the introduction of new measures.

There is an acknowledged 'hierarchy of control', which is simply a list of measures designed to control risks that are considered in order of importance, effectiveness or priority. This control sequence normally begins with an extreme measure of control and ends with personal protective equipment as a last resort. If we consider a task such as working near a big hole in the ground, we consider:

1 *Eliminate* – does the hole need to be there, or can it be filled in?
2 *Substitute* – can the task be done somewhere else?
3 *Isolate* – e.g. by putting something over the hole to cover it
4 *Control* – e.g. by introducing guarding around the hole
5 *Manage* – e.g. by implementing a safe system of work near a hole
6 *Personal protective equipment* – e.g. working with a harness or rope.

Only once all of the measures numbered 1 to 5 above have been tried, and found ineffective in controlling the risks to a reasonably practical level, must point 6, personal protective equipment (PPE), be used.

Personal protective equipment (PPE)

The Personal Protective Equipment at Work Regulations 1992 identify the need for employers to provide suitable PPE for their employees to perform their work activities safely. The PPE Regulations cover all forms of protection, including clothing that is required to give protection against the weather. However, it does not necessarily include the provision of *all* protective equipment.

Hearing protection such as ear plugs and ear defenders are *not* covered by these regulations as they are covered by other specific regulations, as are respiratory devices. If you are working in environments where these items of PPE are required, other regulations and legislation also apply. Items such as cycle helmets and crash helmets that would normally be worn on public highways are also not covered by these regulations, even if the wearer is using a bicycle or motorbike as part of their job function.

ACTIVITY

Identify possible risks in using a ladder as access equipment when fixing a PIR luminaire 4 m high.

INDUSTRY TIP

Access the Personal Protective Equipment at Work Regulations 1992 at www.legislation.gov.uk/uksi/1992/2966/contents/made

▲ Figure 1.15 Personal protective equipment used in the industry includes gauntlets and spats

INDUSTRY TIP

Risk assessments can be used to identify the PPE required. The British Safety Industry Federation (www.bsif.co.uk) provides support as to the most suitable PPE for different functions.

According to the regulations, an employer has a legal duty to:

- provide the correct PPE
- ensure that it is used correctly
- ensure it is stored correctly
- train employees in the correct use.

The term 'employer' also applies to employment agencies, which must ensure that they comply with regulations.

When thinking about the suitability of PPE, ask these questions:

- Is the PPE appropriate, based on the risks identified?
- Does it prevent or adequately control the level of risk?
- Could it increase the risk or create a new risk?
- Can it be adjusted to fit different wearers?
- Is it suitable for the individual wearing it?
- Will wearing the PPE cause discomfort?
- Is it compatible with other PPE that has to be worn at the same time?

INDUSTRY TIP

You should always ensure you are wearing the correct PPE when working on practical tasks in college or on site. Many sites operate a card system for workers not wearing the appropriate PPE – the first time is a warning (yellow card); if there's a second offence, you are off site (red card).

Consider that, if the PPE is cumbersome, not practical or uncomfortable, the employee may not wear it. If they do wear it, they may not wear it correctly. For example:

- Sunglasses are popular but are not suitable eye protection for drilling or grinding.
- Safety boots may be tedious to lace up correctly but should not be left with loose laces.
- High visibility jackets do not make a suitable tool belt.

Types of PPE

INDUSTRY TIP

Employers are not allowed to charge employees for PPE. However, if an employee does not return items when they leave the employment, they can be charged.

PPE is about protecting people from harm. Certain tasks present different hazards to different parts of the body. The PPE Regulations consider different parts of the person, with each area requiring different protection. These areas are:

- eyes
- head

- breathing
- the body
- hands and arms
- feet and legs.

Eyes

Losing or damaging your sight can have a life-changing effect, not just on your ability to work, but on your standard and way of living. The types of hazard that can affect your eyes include:

- chemical splash
- metal swarf
- dust particles
- projectiles
- gas and vapour
- radiation (in welding).

Depending on the work to be undertaken, there are several types of PPE that provide suitable protection for the eyes (Figure 1.16):

- safety glasses
- goggles
- face shield
- visor.

▲ Figure 1.16 Different types of eye protection

Head

Concussion, caused by a simple knock to the head, can be fatal if not treated correctly. Therefore, head protection is vitally important. Head protection is intended to provide protection from:

- impact of falling objects
- impact of flying objects
- bumping your head on hard surfaces or sharp edges
- hair entanglement.

Depending on the work activity, the choice of PPE includes:

- general safety helmet
- working-at-height safety helmet
- safety helmet with in-built visor
- safety helmet with in-built ear defenders
- bump cap
- hair net.

▲ Figure 1.17 Types of head protection: hard hat (top), bump cap (bottom)

Breathing

The risk to breathing is not always immediately apparent – some illnesses that are caused by particle inhalation can take years to show, and a person's life may be limited by the time they receive a diagnosis. The types of hazard that must be considered include:

- dust
- vapour
- gas
- oxygen-deficient atmospheres.

The PPE for breathing hazards are:

- disposable filtering face pieces
- disposable respirators
- half- or full-face respirators
- air-fed helmets
- breathing apparatus.

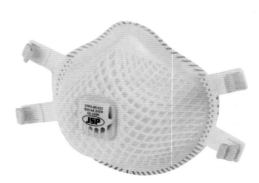

▲ Figure 1.18 Different types of facemasks

Protection of the body

General protection of the body is often overlooked but, in certain work activities, hazards that affect the body can be very important. These hazards include:

- temperature extremes
- adverse weather
- chemical splashes
- spray from pressure leaks or spray guns
- impact or penetration
- contaminated dust
- excessive wear
- electrical explosion
- entanglement.

There are several PPE options, including:

- overalls
- boiler suits, or bib and brace
- high-visibility clothing
- thermal underwear
- waterproof clothing
- leather aprons
- flash protection suits
- chain-mail aprons.

▲ Figure 1.19 Body protection: high-visibility jacket (left), bib and brace (centre) and apron (right)

Hands and arms

The hands and arms are easily injured, especially in electrical installation work, which tends to be mostly manual labour. The hazards to hands and arms include:

- abrasion
- temperature extremes
- cuts and punctures
- impact
- chemicals
- electric shock
- skin infection
- disease
- contamination.

Many injuries occur when people hold items that they are trying to cut or drill. In such cases, training in the correct cutting or drilling technique might be a better solution than PPE.

There are several PPE options for protecting hands and arms:

- gloves made of different material based on the protection needed, such as oil resistance, anti-cut, anti-slip
- gauntlets, which provide protection further up the arm.

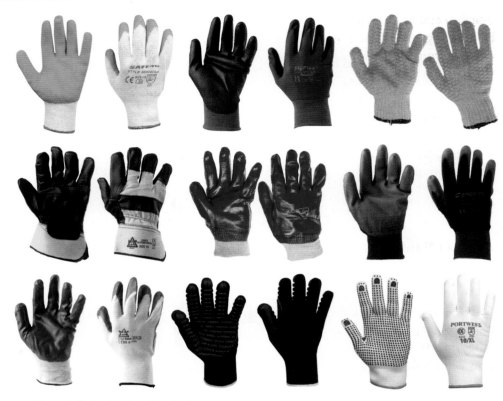

▲ Figure 1.20 A selection of work gloves

HEALTH AND SAFETY

The range of gloves available is vast. When choosing gloves, think of the task and what the gloves need to do such as grip, protect from chemicals, protect from cuts, and so on.

Feet and legs

Safety boots are the most common form of PPE among construction site workers. These protect against falling objects. However, feet and legs can be subject to other hazards including:

- wet conditions
- slipping
- cuts and punctures
- falling objects
- metal splashes from welding or cutting
- chemical splashes
- abrasion
- irritants such as dust or powders.

The use of safety boots does not address all these hazards. Other options include the use of:

- safety boots or shoes with protective toe caps and penetration resistant mid-sole
- boots or shoes with oil-resistant soles
- gaiters
- leggings
- overalls.

All construction sites now display safety boards at each entrance advising that people are not allowed on site unless they are wearing safety boots, high visibility clothing and a hard hat. Some companies are going further than this by ensuring workers do not wear shorts or go topless, due to the risk of skin damage from abrasion.

▲ Figure 1.21 Foot protection: shoes (left), rigger boots (centre) and safety boots (right)

Safety signs

The Health and Safety (Signs and Signals) Regulations require employers to display signage where risks to health and safety have not been removed by other means, such as a safe system of work. Many premises and construction sites will display a range of signs in addition to having safe systems in place.

Prohibition signs

These signs indicate an activity that must not be done. They are circular white signs with a red border and red cross bar.

| No access for unauthorised persons | No smoking in this building | In the event of fire do not use this lift | Do not drink |

▲ Figure 1.22 Various prohibition signs

Warning signs

These provide safety information and/or give warning of a hazard or danger. They are triangular yellow signs with a black border and symbol.

| Danger High voltage | Warning Trip hazard | Caution Men at work | Warning Falling objects |

▲ Figure 1.23 Warning signs

Mandatory signs

These signs give instructions that must be obeyed. They are circular blue signs with a white symbol.

▲ Figure 1.24 Mandatory signs

Advisory or safe condition signs

These give information about safety provision. They are square or rectangular signs with a white symbol.

▲ Figure 1.26 Advisory signs

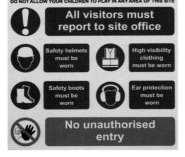

▲ Figure 1.25 An example of a site safety board

Symbols for hazardous substances

Hazardous substances are given a classification according to the **severity** and type of hazard they may present to people in the workplace. However, all over the world there are different laws on how to identify the hazardous properties of chemicals. The United Nations has therefore created the Globally Harmonised System of Classification and Labelling of Chemicals (GHS). The aim of the GHS is to have, worldwide, the same:

- criteria for classifying chemicals according to their health, environmental and physical hazards
- hazard communication requirements for labelling and safety data sheets.

The GHS is not a formal agreement, but is a non-legally binding international agreement. In Great Britain, the existing legislation is the Chemicals (Hazard Information and Packaging for Supply) Regulations 2009 (CHIPS), but this has gradually been replaced by the European Classification, Labelling and Packaging of Substances and Mixtures Regulations.

Examples of the GHS labelling system are shown in Figure 1.27.

KEY TERM

Severity: seriousness.

INDUSTRY TIP

You do not need to know all of the detail in the GHS as an electrician, but it is important that you are aware of it and how it might directly affect your process and work onsite.

Further information about the GHS can be found on the HSE's website: www.hse.gov.uk/chemical-classification/legal/background-directives-ghs.htm

▲ Figure 1.27 The hazard signs shown indicate substances that are (top row, left to right) corrosive, dangerous to the environment, gas under pressure, require caution (irritants), toxic; and (bottom row, left to right) long-term health hazards (causes of cancer), oxidising, explosive, flammable

Common hazardous substances

Hazardous substances at work may include the substances used directly in the work process – such as glue, paints, thinners, solvents and cleaning materials – and those produced by different work activities, such as welding fumes. Health hazards are always present during building services activities due to the nature of the activities and may include other hazards, such as vibration, dust (possibly including asbestos), cement and solvents.

The Control of Substances Hazardous to Health (COSHH) Regulations 2002 provide a framework that helps employers assess risk and monitor effective controls. A COSHH assessment is essential before work starts and should be updated as new substances are introduced. Hazardous substances include:

- any substance that gives off fumes that may cause headaches or respiratory irritation
- acids that cause skin burns or respiratory irritation (e.g. battery acid, metal-cleaning materials)
- solvents that cause skin and respiratory irritation (e.g. for PVC tubes and fittings)
- synthetic fibres that cause eye or skin irritation (e.g. thermal insulation, optical fibres)
- cement and wood dust that may cause eye irritation and respiratory irritation
- fumes and gases that cause respiratory irritation (e.g. soldering, brazing and welding fumes, overheating or burning PVC)
- asbestos.

The COSHH Regulations require employers to assess risk and ensure the prevention or adequate control of exposure to hazardous substances by measures other than the provision of personal protective equipment (PPE) 'so far as is reasonably practicable'.

As for any risk, the acknowledged hierarchy of control measures listed on page 31 should be used to assess the precautions to be taken when working with hazardous substances.

the acknowledged hierarchy of control measures listed on page 31

KEY TERMS

Toxic: Poisonous.

Hazardous substance: Something that can cause ill health to people.

INDUSTRY TIP

Access the Control of Substances Hazardous to Health (COSHH) Regulations 2002 at: www.legislation.gov.uk/uksi/2002/2677/introduction/made

HEALTH AND SAFETY

Make sure you can recognise a hazard and are able to know what you can do to reduce the risk before handling hazardous substances.

ACTIVITY

Think of some common electrical components that could contain substances hazardous to health, and produce a factsheet detailing precautions such as disposal methods or handling measures.

Asbestos encountered in the workplace

Asbestos is the single greatest cause of work-related deaths in the UK. It is a naturally occurring substance, which is obtained from the ground as a rock-like ore, normally through open-pit mining.

Asbestos was used extensively as a building material in the UK from the 1950s through to the mid-1980s. It was used for a variety of purposes and was ideal for fireproofing and insulation. Any building built before the year 2000 (houses, factories, offices, schools, hospitals, etc.) may contain asbestos.

Asbestos materials in good condition are safe, unless the asbestos fibres become airborne, which happens when materials are damaged due to demolition or remedial works on or in the vicinity of asbestos ceiling tiles, asbestos cement roofs and wall sheets, sprayed asbestos coatings on steel structures, and asbestos lagging. In older buildings the presence of asbestos in and around boilers and hot water pipes, and in structural fire protection, must always be anticipated when undertaking electrical work. It is difficult to identify asbestos by colour alone and laboratory tests are normally required for positive identification.

What to do if the presence of asbestos is suspected

If asbestos is discovered during a work activity, work should be stopped and the employer or duty holder informed immediately. The Control of Asbestos Regulations 2012 affect anyone who owns, occupies, manages or otherwise has responsibilities for the maintenance and repair of buildings that may contain asbestos. The regulations cover the need for a risk assessment, and the need for method statements for the removal and disposal of asbestos, air monitoring and the control measures required. These control measures include personal protective equipment and training.

Implications of being exposed to asbestos

Asbestos commonly comes in the form of chrysotile (white asbestos), amosite (brown asbestos) and crocidolite (blue asbestos). Chrysotile is the common form of asbestos and accounts for 90–95% of all asbestos in circulation. When **abraded** or drilled, asbestos produces a fine dust with fibres small enough to be taken into the lungs. Asbestos fibres are very sharp and can lead to mesothelioma (cancer of the lining of the lung), lung cancer, asbestosis (scarring of the lung), diffuse pleural thickening (thickening and hardening of the lung wall) and death.

The requirement for first aid facilities

In the event of injury or sudden illness, failure to provide correct first aid could result in death. Employers must ensure that an employee who is injured or taken ill at work receives immediate attention.

The Health and Safety (First-Aid) Regulations 1981 require employers to provide adequate and appropriate equipment, facilities and personnel to ensure their employees receive immediate attention if they are injured or taken ill at work. These Regulations apply to all workplaces, including those with fewer than five employees and to the self-employed.

What is 'adequate and appropriate' will depend on the circumstances in the workplace, and employers should carry out an assessment of first aid needs to determine what to provide. There are a number of factors to be considered:

- number of employees
- type of work being undertaken
- unusual hazards
- 'off-site' workers
- out of hours or shift patterns
- nearest emergency services.

This will determine whether trained first aiders are needed, what should be included in a first aid box and whether or not a first aid room is required. The HSE suggests at least one appointed first aider where a low-risk business, such as a shop or office, has fewer than 25 staff members.

The first aid box

There is no standard list of items that need to be contained in a first aid box and the contents depend on what the employer or the 'person in control of the site' assesses the needs to be. However, tablets or medicines should not be kept in first aid boxes. It is sensible to have an 'appointed person' to look after the first aid equipment and replenish when required.

Emergencies can occur at any time and there may well be an immediate need for items without having to search for them. Items in first aid boxes should not be taken for other uses. Items are needed for particular reasons and therefore misuse will decrease supplies.

Fire safety

Most fires are preventable and by adopting the right behaviours and procedures, prevention can easily be achieved.

How combustion takes place

A fire needs three elements to start:

- a source of ignition (heat)
- a source of fuel (something that burns)
- oxygen.

If any one of these elements is removed, a fire will not ignite or will cease to burn.

- Sources of ignition include heaters, lighting, naked flames, electrical equipment, smokers' materials (cigarettes, matches) and anything else that can get very hot or cause sparks.
- Sources of fuel include wood, paper, plastic, rubber or foam, loose packaging materials, waste rubbish, combustible liquids and furniture.
- Sources of oxygen include the air surrounding us.

A fire safety risk assessment (using the same approach as used in the health and safety risk assessment – see pages 26–28) should be used to determine

▲ Figure 1.28 A well-stocked first aid box

the risks. Based on the findings of the assessment, employers must ensure that adequate and appropriate fire safety measures are in place to minimise the risk of injury or loss of life in the event of a fire. These findings must be kept up to date.

Dangers of working with heat-producing equipment

The use of heat-producing equipment (hot work) is a common occurrence in building services work activities. Hot work is work that might generate sufficient heat, sparks or flame to cause a fire. It includes welding, flame cutting, soldering, brazing, grinding and other equipment that incorporates a flame, such as boilers or furnaces.

The flames, sparks and heat produced during the hot work are ignition sources that can cause fires and explosions in many different situations. For example:

- sparks produced during hot work can land on combustible materials and cause fires and explosions
- hot work performed on tanks and vessels with residual flammable substances and vapours can cause the tanks to explode.

To help prevent fire in the workplace, the risk assessment should identify what could cause a fire to start (the sources of ignition (heat or sparks), the substances that burn) and the people who may be at risk.

Once the risks have been identified, appropriate action can be taken to control them. Actions should be based on whether the risks can be avoided altogether or, if this is not possible, how they can be reduced. The checklist below will help with an appropriate action plan.

- Keep sources of ignition and flammable substances apart.
- Avoid accidental fires (e.g. make sure heaters cannot be knocked over).
- Ensure good housekeeping at all times (e.g. avoid build-up of rubbish that could burn).
- Consider how to detect fires and how to warn people quickly if a fire starts (e.g. install smoke alarms and fire alarms or bells).
- Have the correct fire-fighting equipment available.
- Keep fire exits and escape routes clearly marked and unobstructed at all times.
- Ensure workers receive appropriate training on procedures and fire drills.
- Review and update the risk assessment on a regular basis.

Procedures on discovery of fires on site

How people react in the event of fire depends on how well they have been prepared and trained for a fire emergency. It is therefore imperative that all employees (and visitors and contractors) are familiar with the company procedure to follow in the event of an emergency. A basic procedure is as follows.

1 On discovery of a fire, raise the alarm immediately.
2 If staff are trained and it is considered safe to do so, attempt to fight the fire using the equipment provided.

INDUSTRY TIP

The Regulatory Reform (Fire Safety) Order 2005 covers general fire safety in England and Wales and can be accessed at www.legislation.gov.uk/uksi/2005/1541/contents/made

ACTIVITY

Walk around your own house or place of work and make a list of potential sources of fire then apply the checklist above to them. As an example, electric fires can create a fire risk if clothes are draped over them.

3 If fire-fighting fails, evacuate (leave the area) immediately.
4 Ensure that no one is left in the fire area and close doors on exit in order to prevent the spread of fire.
5 Go straight to the designated assembly point. These points are specially chosen as they are in locations of safety and where the emergency services are not likely to be obstructed on arrival.

> **HEALTH AND SAFETY**
> Remember, NEVER ignore a fire alarm, even if it has gone off many times before that day. It may be the last one you ignore.

Different classifications of fires

All fires are grouped into classes, according to the types of materials that are burning.

Figure 1.29 is a guide to the different types of fire and the types of extinguisher that should be used.

Fire classification	Water	Foam	Powder	CO$_2$
Class A – Combustible materials such as paper, wood, cardboard and most plastics	✓	✓	✓	
Class B – Flammable or combustible liquids such as petrol, kerosene, paraffin, grease and oil		✓	✓	✓
Class C – Flammable gases, such as propane, butane and methane			✓	✓
Class D – Combustible metals, such as magnesium, titanium, potassium and sodium			Specialist dry powder	
Class F – Cooking oils and fats		Specialist wet chemical (yellow)		

▲ Figure 1.29 Fire classification and the correct extinguisher to be used

Fires involving equipment such as electrical circuits or electronic equipment are often referred to as Class E fires, although the category does not officially exist under the **BS EN 3** rating system. This is because electrical equipment is often the cause of the fire, rather than the actual type of fire. Most modern fire extinguishers specify on the label whether they should be used on electrical equipment. Normally carbon dioxide or dry powder are suitable agents for putting out a fire involving electricity.

Manual handling

Manual handling is one of the most common causes of injury at work and causes over a third of all workplace injuries. Manual handling injuries can occur almost anywhere in the workplace and heavy manual labour, awkward postures or previous or existing injury can increase the risk. Work-related manual handling injuries can have serious implications for both the employer and the person who has been injured.

> **KEY TERM**
> **Manual handling:** the movement of items by lifting, lowering, carrying, pushing or pulling by human effort alone.

Manual handling techniques

The introduction of the Manual Handling Operations Regulations 1992 saw a change from reliance on safe lifting techniques to an analysis, using risk assessment, of the need for manual handling. The regulations established a clear hierarchy of manual handling measures:

1 Avoid manual handling operations, so far as is reasonably practicable, by re-engineering the task to avoid moving the load or by mechanising the operation.
2 If manual handling cannot be avoided, a risk assessment should be made.
3 Reduce the risk of injury, so far as is reasonably practicable, either by the use of mechanical aids or by making improvements to the task (e.g. by using two persons), the load and the environment.

INDUSTRY TIP

Access the Manual Handling Operations Regulations 1992 at: www.legislation.gov.uk/uksi/1992/2793/contents/made

Stage 1 – Think before lifting

Stage 2 – Stand close to the load

Stage 3 – Bend your knees, not your back

Stage 4 – Grip the load at the base

Stage 5 – Straighten the knees

Stage 6 – Walk carefully

Stage 7 – Bend knees to lower the item

▲ Figure 1.30 Correct manual handling technique

How to manually handle loads using mechanical lifting aids

Even if mechanical handling methods are used to handle and transport equipment or materials, hazards may still be present in the four elements that make up the mechanical handling:

- *The handling equipment* – mechanical handling equipment must be suitable for the task, well maintained and inspected on a regular basis.
- *The load* – the load needs to be prepared in such a way as to minimise accidents, taking into account such things as security of the load, flammable materials and stability of the load.
- *The workplace* – if possible, the workplace should be designed to keep the workforce and the load apart.
- *The human element* – employees who use the equipment must be properly trained.

ACTIVITY

What should you check before moving an object from one place to another by manual handling?

INDUSTRY TIP

When you are manually handling a load, always make sure the route you will be taking is clear and all obstacles are removed. Never try to lift more than you can handle. Use safe handling techniques. Get someone to help you or use mechanical assistance to move the object.

⑥ USING ACCESS EQUIPMENT AND WORKING IN CONFINED SPACES OR EXCAVATONS

Working at height

Working at height remains one of the biggest causes of fatalities and major injuries within the construction industry with almost 25% of fatalities resulting from falls from ladders, stepladders and through fragile roofs (figures taken from HSE www.hse.gov.uk/statistics/pdf/fatalinjuries.pdf). Work at height means work in any place, including at or below ground level (e.g. in underground workings), where a person could fall a distance liable to cause injury.

The Work at Height Regulations 2005 require duty holders to ensure that:

- all work at height is properly planned and organised
- those involved in work at height are competent
- the risks from work at height are assessed and appropriate work equipment is selected and used
- the risks of working on or near fragile surfaces are properly managed
- the equipment used for work at height is properly inspected and maintained.

> **INDUSTRY TIP**
>
> Access the Work at Height Regulations 2005 at: www.legislation.gov.uk/uksi/2005/735/contents/made

Access equipment for different types of work at height

Many different types of access equipment are used in the building services industry, such as mobile elevated work platforms (MEWPs), ladders, mobile tower scaffolds, tube and fitting scaffolding, and personal suspension equipment (harnesses).

Mobile elevated work platforms

There is a wide range of MEWPs (vertical scissor lift, self-propelled boom, vehicle-mounted boom and trailer-mounted boom) and if any of these is to be used it is important to consider:

- the height from the ground
- whether the MEWP is appropriate for the job
- the ground conditions
- training of operators
- overhead hazards such as trees, steelwork or overhead cables
- the use of a restraint or fall arrest system
- closeness to passing traffic.

▲ Figure 1.31 A vertical scissor lift

Ladders

It is recommended that ladders are only used for low-risk, short-duration work (between 15 and 30 minutes depending upon the task). Common causes of falls from ladders are:

- *overreaching* – maintain three points of contact with the ladder
- *slipping from the ladder* – keep the rungs clean, wear non-slip footwear, maintain three points of contact with the ladder, make sure the rungs are horizontal
- *the ladder slips from its position* – position the ladder on a firm, level surface, secure the ladder top and bottom, check the ladder daily
- *the ladder breaks* – position the ladder properly using the 1:4 rule (four units up for every one unit out), do not exceed the maximum weight limit of the ladder, only carry light materials or tools.

Stepladders

Many of the common causes of falls from ladders can also be applied to stepladders. If stepladders are used, ensure that:

- they are suitable, in good condition and inspected before use
- they are sited on stable ground

▲ Figure 1.32 The correct angle for a ladder is 4:1

ACTIVITY

A wooden ladder is found to have a split in it. What action should you take?

- they are faced onto the work activity
- knees are never above the top tread of the stepladder
- the stepladder is open to the maximum
- wooden stepladders (or ladders) are not painted as this may hide defects.

Scaffolding

Some work activities, such as painting, roof work, window replacement or brickwork, are almost certainly more conveniently undertaken from a fixed external scaffold. Tube and fitting scaffolding must only be erected by competent people who have attended recognised training courses. Any alterations to the scaffolding must also be carried out by a competent person. Regular inspections of the scaffold must be made and recorded.

Safety checks and safe erection methods for access equipment

The Work at Height Regulations 2005 require that all scaffolding and equipment that is used for working at height, where a person could fall two metres or more, is inspected on a regular basis, using both formal and pre-use inspections to ensure that it is fit for use. A marking, coding or tagging system is a good method of indicating when the next formal inspection is due. However, regular pre-use checks must take place as well as formal inspections.

The following safety checks must be carried out.

▲ Figure 1.33 The correct use of a stepladder

Safety checks for MEWPs

MEWPs must only be operated by trained and competent persons, who must also be competent to carry out the following pre-use checks.

- The ground conditions must be suitable for the MEWP, with no risk of the MEWP becoming unstable or overturning.
- Check for overhead hazards such as trees, steelwork and overhead cables.
- Guard rails and toe boards must be in place.
- Outriggers should be fully extended and locked in position.
- The tyres must be properly inflated and the wheels immobilised.
- Check the controls to make sure they work as expected.
- Check the fluid and/or battery charge levels.
- Check that the descent alarm and horn are working.
- Check that the emergency or ground controls are working properly.

Safety checks for ladders and stepladders

Users must check that:

- ladders and stepladders are of the right classification (e.g. trade or industrial)
- the styles, rungs or steps are in good condition
- the feet are not missing, loose, damaged or worn
- rivets are in place and secure
- the locking bars are not bent or buckled.

▲ Figure 1.34 Tags to record that a ladder inspection has taken place

Safety checks for tube and fitting scaffolds

- Toe boards, guard rails and intermediate rails must be in place to prevent people and materials from falling.
- The scaffold must be on a stable surface and the uprights must be fitted with base plates and sole plates.
- Safe access and exit ladders must be in place and secured.

Safety checks for prefabricated mobile scaffold towers

Mobile scaffold towers are a convenient means of undertaking repetitive tasks in the building services industry. The erection and dismantling must only be undertaken by a competent person.

The following pre-use checks will ensure safe use.

- The maximum height-to-base ratios must not be exceeded.
- Diagonal bracing and stabilisers must not be damaged or bent.
- The brace claws must work properly.
- Internal access ladders must be in place.
- Wheels must be locked when work is in progress.
- The working platform must be boarded, with guard rails and toe boards fitted.
- The towers must be tied in windy conditions.
- The platform trap door must be operating correctly.
- Rivets must be checked visually to ensure they are in place and not damaged.

Working in excavations and confined spaces

Every year people are killed or seriously injured by collapses and falling materials while working in excavations. These excavations may be required in the course of ground source heating projects or the installation of drains and soakaways, septic tanks, electrical distribution networks and retaining structures.

How to prepare excavations for safe working

The hazards associated with excavations are:

- excavations collapsing and burying or injuring people working in them
- material falling from the sides into any excavation
- people or plant falling into excavations
- contact with underground services (e.g. electricity, high-pressure water and gas)
- undermining other structures
- exhaust fumes from petrol or diesel-engine equipment such as compressors or generators.

Planning is the key to the safety of any excavation. Before work commences a decision will be required on what temporary support will be needed and what precautions need to be taken. The equipment and precautions needed (trench sheets, props, baulks, etc.) should be available on site before work starts.

The sides of the excavation must be prevented from collapsing either by battering at an angle of between 5° and 45° or by shoring (supporting) the sides up with timber, sheeting or a proprietary support system. In **granular soils**, the angle of slope (**batter**) should be less than the natural angle of **repose** of the material being excavated. In wet ground, a considerably flatter slope will be required. Loose materials may fall from spoil heaps into the excavation, therefore edge protection should include toe boards or other means to protect against falling materials. Head protection should be worn in excavations.

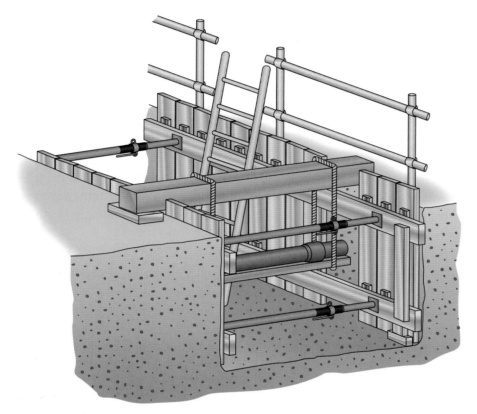

▲ Figure 1.35 Excavation showing support for services and safety guards

Making excavations safe

To prevent people from falling into an excavation, the edges should be protected in one of the following ways:

- with guard rails and toe boards inserted into the ground immediately next to the supported excavation side
- by a fabricated guard rail assembly that connects to the sides of the trench box
- by using the trench support system itself (e.g. using trench box extensions or trench sheets longer than the trench depth).

A competent person who fully understands the dangers and necessary precautions should inspect the excavation at the start of each shift and after any event that may have affected its strength or stability, or after a fall of rock or soil.

HEALTH AND SAFETY

Gas can accumulate in underground workings. Always beware of this and test the atmosphere in the underground workings first.

KEY TERMS

Granular soils: gravel, sand or silt (coarse-grained soil) with little or no clay content. Although some moist granular soils exhibit apparent cohesion (grains sticking together forming a solid), they have no cohesive strength. Granular soil cannot be moulded when moist and crumbles easily when dry.

Batter: the slope angle of the trench walls of an excavation, in relation to the horizontal surface, to prevent the walls collapsing.

Repose: the angle to the horizontal at which the material in the cut face is stable and does not fall away. Different materials have different angles of repose – for example, 90° for solid rock and 30° for sand.

A record of the inspections will be required and any faults that are found should be corrected immediately.

Working in a confined space

A confined space is a place which is substantially enclosed (though not always entirely) and where serious injury can occur from hazardous substances or conditions within the space or nearby (e.g. lack of oxygen). Examples include chambers, tanks, vessels, furnaces, boilers or cisterns, inspection chambers, pits, roof spaces and under suspended timber floors. Entry into a confined space requires the correct equipment, including PPE and a harness. Activities below ground require a descent control and rescue system.

The two main hazards associated with confined spaces are the presence of toxic or other dangerous substances and the absence of adequate oxygen.

The Confined Space Regulations 1997 detail the specific controls that are necessary when people enter confined spaces. These can be summarised as follows.

- Avoid entry to confined spaces (e.g. by doing the work from the outside).
- If entry to a confined space is unavoidable, follow a safe system of work which should include rigorous preparation, isolation, air testing and other precautions, and the use of a confined space entry permit.
- Put in place adequate emergency arrangements before the work starts.

Adequate ventilation, sufficient lighting, the provision of the correct PPE, emergency evacuation procedures and medical conditions must all be catered for. Lone working should not be allowed. Section 706 of **BS 7671:2018** The IET Wiring Regulations, 18th Edition gives the requirements for supplies to portable electrical equipment used in confined spaces that are restrictive and/or conducting.

7 ELECTRICAL SAFETY ON SITE

Modern living is shaped by electricity. It is a safe, clean and immensely powerful source of energy and is in use in every factory, office, workshop and home in the country. However, this energy source also has the potential to be very hazardous, with a possibility of death, if it is not handled with care. The Electricity at Work Regulations 1989 were made under the Health and Safety at Work etc. Act 1974 (HSW Act) and came into force on 1 April 1990.

Common electrical dangers on site

The 1989 Regulations are goal-setting in that they specify the objectives concerning the design, specification and construction of, and work activities on, electrical systems, in order to prevent injury caused by electricity. They do not specify the means for achieving these objectives. The 1989 Regulations are supported by a Memorandum of Guidance, HSR25 (2nd edition 2007).

Electrical injuries can be caused by a wide range of voltages, and are dependent upon individual circumstances, but the risk of injury is generally greater with higher voltages. Alternating current (AC) and direct current (DC) electrical supplies can cause a range of injuries including:

- electric shock
- electrical burns
- loss of muscle control
- fires arising from electrical causes
- arcing and explosion.

Electric shock

Electric shocks may arise either by direct contact with a live part or indirectly by contact with an exposed conductive part (e.g. a metal equipment case) that has become live as a result of a fault condition.

Faults can arise from a variety of sources:

- broken equipment case exposing internal bare live connections
- cracked equipment case causing 'tracking' from internal live parts to the external surface
- damaged supply cord insulation, exposing bare live conductors
- broken plug, exposing bare live connections.

The magnitude (size) and duration of the shock current are the two most significant factors determining the severity of an electric shock. The magnitude of the shock current will depend on the contact voltage and impedance (electrical resistance) of the shock path. A possible shock path always exists through ground contact (e.g. hand-to-feet). In this case:

shock path impedance = body impedance + any external impedance

A more dangerous situation is a hand-to-hand shock path, where the current flow is through the heart. This is when one hand is in contact with an exposed conductive part (e.g. an earthed metal equipment case), while the other simultaneously touches a live part. In this case the current will be limited only by the body **impedance**.

As the voltage increases, so the body impedance decreases, which increases the shock current. When the voltage decreases, the body impedance increases, which reduces the shock current. This has important implications concerning the voltage levels that are used in work situations and highlights the advantage of working with reduced low-voltage (110 V) systems or battery-operated hand tools.

At 230 V, the average person has a body impedance of approximately 1300 Ω. At mains voltage and frequency (230 V–50 Hz), currents as low as 50 milliamps (0.05 A) can prove fatal, particularly if flowing through the body for a few seconds.

KEY TERM

Impedance: measure of the opposition presented to an alternating current when a voltage is applied.

Electrical burn

Burns may arise due to:

- the passage of shock current through the body, particularly if at high voltage
- exposure to high-frequency radiation (e.g. from radio transmission antennas).

Loss of muscle control

People who experience an electric shock often get painful muscle spasms that can be strong enough to break bones or dislocate joints. This loss of muscle control often means the person cannot 'let go' or escape the electric shock. The person may fall if they are working at height or be thrown into nearby machinery and structures.

Fire

Electricity is believed to be a factor in the cause of many fires in domestic and commercial premises in Britain each year. One of the principal causes of such fires is wiring with defects such as insulation failure, the overloading of conductors, lack of electrical protection, or poor connections.

Arcing

Arcing frequently occurs due to a short circuit (conductive bridge between live parts) caused while working on live equipment (either intentionally or unintentionally). Arcing generates UV radiation, causing severe sunburn. Molten metal particles are also likely to be deposited on exposed skin surfaces.

Explosion

There are two main electrical causes of explosion:

- short circuit due to an equipment fault
- ignition of flammable vapours or liquids caused by sparks or high surface temperatures.

Controlling current flow

It is necessary to include devices in circuits to control current flow – that is, to switch the current on or off by making or breaking the circuit. This may be required:

- for functional purposes (to switch equipment on or off)
- for use in an emergency (switching off in the event of an accident)
- so that equipment can be switched off to prevent its use and allow maintenance work to be done safely on the mechanical parts
- to isolate a circuit, installation or piece of equipment to prevent the risk of shock where exposure to electrical parts and connections is likely for maintenance purposes.

The preparation of electrical equipment for maintenance purposes requires effective disconnection from all live supplies and the means for securing that disconnection (by locking off).

There is an important distinction between switching and isolation. **Switching** is cutting off the supply. **Isolation** is the secure disconnection and separation from all sources of electrical energy. A variety of control devices are available for switching, isolation or a combination of these functions, some incorporating protective devices. Before starting work on a piece of isolated equipment,

KEY TERMS

Switching: cutting off the supply of electricity, stopping equipment functioning.

Isolation: disconnecting and separating from all sources of electrical energy.

checks should be made to ensure that the circuit is dead, using an approved testing device.

In the case of portable equipment connected via a supply cord and plug, removal of the plug from the socket provides a ready means of isolation.

Electrical supply for tools and equipment

Portable electric tools can provide valuable assistance with much of the physical effort required in electrotechnical activities. These tools can use different sources of electrical supply (mains or battery) and different means of maintaining safety in relation to that electrical supply.

Basic equipment constructions, all aimed at preventing the risk of electric shock, are specified in **BS 2754** 1976: Construction of Electrical Equipment for Protection against Electric Shock, as summarised below.

Class I

The basic insulation may be an air gap and/or some form of insulating material. External conductive parts (e.g. the metal case) must be earthed by providing the supply through a three-core supply lead incorporating a protective conductor. The most important aspect of any periodic inspection or testing of Class I equipment is to check the integrity of this protective conductor. There is no recognised symbol for Class I equipment, though some appliances may show the symbol in Figure 1.36.

▲ Figure 1.36 This symbol may appear on Class I earthed items

Class II

Class II equipment has either no external conductive parts apart from fixing screws (insulation-encased equipment) or there is adequate insulation between any external conductive parts and the internal live parts to prevent the former becoming live as a result of an internal fault (metal-encased equipment). Periodic inspection or testing needs to focus on the integrity of the insulation. Class II equipment is identified by the symbol shown in Figure 1.37.

▲ Figure 1.37 This symbol may appear on Class II items

Class III

This method of protection is not designed to prevent shock but to reduce its severity and therefore make the shock more survivable. The supply (no greater than 50 V AC) must be provided from a separated extra-low-voltage (SELV) source such as a safety isolating transformer conforming to **BS EN 61558**, or a battery. Class III equipment is identified by the diamond symbol and the safety isolating symbol by two interlinked circles, as shown in Figure 1.38.

Another way of reducing the risk of electric shock is by using a reduced low-voltage system. This is *not* a Class III system but is a safer arrangement than using mains-operated (230 V) equipment because of the lower potential shock voltage. Supply is provided via a mains powered (230 V) step-down transformer with the centre point of the secondary winding connected to earth.

▲ Figure 1.38 This symbol may appear on Class III items

Low-voltage electrical supplies for tools and equipment

The human body's impedance increases with lower touch voltages, particularly below 50 V. Although 50 V can be dangerous in certain circumstances, if the system voltage can be reduced to around this level, then the magnitude (size) of any current flow through the body will be significantly reduced.

The most common low-voltage system of this type in use in the UK uses a 230/110 V double-wound step-down transformer with the secondary winding centre-tapped and connected to earth (CTE). While the supply voltage for equipment supplied from such a transformer is 110 V, the maximum voltage to earth is 55 V (63.5 V for a 110 V three-phase system). This is covered in **BS 7671: 2018** The IET Wiring Regulations, 18th Edition (Regulation 411.8 Reduced low-voltage systems). Overcurrent protection for the 110 V supply may be provided by fuses at the transformer's 110 V output terminals or by a thermal trip to detect excessive temperature rise in the secondary winding. The latter method is generally employed for portable units. Such reduced voltage supplies may be provided:

- through fixed installations (workshops, plant rooms, lift rooms or other areas where portable electrical equipment is in frequent use)
- through small portable transformers designed to supply individual portable tools. (See BS 4363: 1998 and BS 7375: 1996, which give a specification for the distribution of electricity on building and construction sites, based on the use of 230/110 V CTE transformers.)

One additional advantage of using a reduced low-voltage system is that this safeguard applies to all parts of the system on the load side of the transformer, including the flexible leads, as well as the tools, hand lamps, and so on.

Alternatively, cordless or battery-powered tools offer a convenient way of providing a powered hand tool without the inconvenience of using a mains supply and without the hazard of trailing power leads.

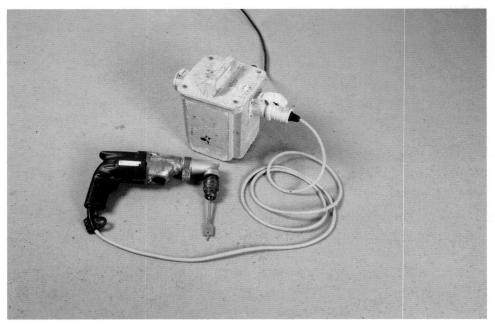

▲ Figure 1.39 Portable CTE transformer and drill

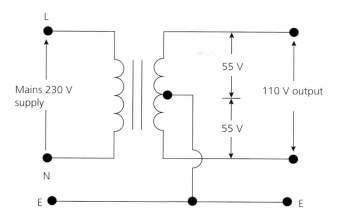

▲ Figure 1.40 Schematic diagram for a 110/55 V transformer

How to conduct a visual inspection of portable electrical equipment

To maintain the safety and integrity of tools and equipment, regular in-service inspection and testing should be undertaken to confirm that the equipment remains in a safe condition. Portable appliance testing (PAT) is the term used to describe the examination of electrical appliances and equipment to ensure they are safe to use. There are three categories of in-service inspection or testing for portable tools and equipment:

- user checks (pre-use inspections) – users play an important role by checking equipment before use for signs of damage or obvious defects liable to affect safety in use
- periodic formal inspection or checks
- periodic combined inspection and testing.

The user checks are a visual check only. They should deal with the following inspection requirements separately for the equipment, supply lead and plug.

Visual inspection of equipment:

- Equipment should be manufactured to relevant standards (BS or BS EN).
- The casing should have no visible damage and be free from dents and cracks.
- The switches should operate correctly.
- The supply lead should be secure and correctly connected.
- Equipment should be suitable for the task.

Visual inspection of supply leads:

- Supply leads should be manufactured to relevant standards (BS or BS EN).
- They should be suitable for the environment.
- They should be free from cuts or fraying.
- There should be no visible exposed conductor insulation (damaged sheath) or exposed live conductors.
- There should be no signs of damage to the cord sheath.
- There should be no joints evident.

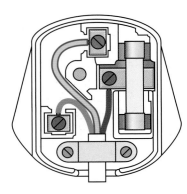

▲ Figure 1.41 A correctly wired 13 A plug

Visual inspection of supply lead plugs:

- Supply lead plugs should be manufactured to relevant standards (BS or BS EN).
- The insulation should have no damage.
- The cord or cable connections should be correct and secure.

The user checks must be backed up by periodic inspection and, where appropriate, testing. At that time, a thorough examination is undertaken by a nominated person, competent for that purpose. Schedules giving details of inspection and maintenance periods, together with records of the inspection, should form part of the procedure.

What to do when portable electrical equipment fails visual inspection

Most electrical safety defects can be found by visual examination, but some types of defect, such as a broken earth conductor, can only be found by testing. However, it is essential to understand that visual examination is an essential part of the process because some types of electrical safety defect cannot be detected by testing alone.

A relatively brief user check, based upon simple training and the use of a brief checklist, is a very useful part of any electrical maintenance regime. If the user checks detailed above are carried out, 95% of all faults will be identified and the appropriate action taken. No record is needed if there are no defects found. However, if equipment is found to be unsafe, it must be removed from service immediately. It must be labelled to show that it must not be used and the fault must be reported to a responsible person.

⑧ SAFE ISOLATION

Isolation can be very complex due to the differing industrial, commercial and domestic working environments, some of which require experience and knowledge of the system processes. This section deals with a basic practical procedure for isolation and for the securing of isolation. It also looks at the reasons for safe isolation and the potential risks involved during the isolation process.

Safe operating procedures for the isolation of plant and machinery during both electrical and mechanical maintenance must be prepared and followed. Wherever possible, electrical isolators should be fitted with a means by which the isolating mechanism can be locked in the open/off position. If this is not possible, an agreed procedure must be followed for the removal and storage of fuse links. Regulation 12 of the Electricity at Work Regulations 1989 requires that, where necessary to prevent danger, suitable means (including, where appropriate, methods of identifying circuits) must be available for:

- cutting off the supply of electrical energy to any electrical equipment
- the isolation of any electrical equipment.

The aim of Regulation 12 is to ensure that work can be undertaken on an electrical system without danger, in compliance with Regulation 13 (work when equipment has been made dead).

Terms used in Regulation 12 are defined below.

- *Cutting off the supply* – Depending on the equipment and the circumstances, this may be no more than normal functional switching (on/off) or emergency switching by means of a stop button or a trip switch.
- *Isolation* – The disconnection and separation of the electrical equipment from every source of electrical energy in such a way that this disconnection and separation is secure.
- *From every source of electrical energy* – Many accidents occur due to a failure to isolate all sources of supply to or within equipment (e.g. control and auxiliary supplies, uninterruptable power supply (UPS) systems or parallel circuit arrangements giving rise to back feeds).
- *Secure* – Keeping safe; achieved by locking off with a safety lock (i.e. a lock with a unique key). The posting of a warning notice also serves to alert others to the isolation.

How to undertake a basic practical procedure for isolation

Being able to perform safe isolation is a key skill every electrician needs. Be sure to do this under supervision by a skilled person such as your tutor.

Gather together equipment

You will need the following equipment for this task:

- a voltage indicator which has been manufactured and maintained in accordance with Health and Safety Executive (HSE) Guidance Note GS38
- a proving unit compatible with the voltage indicator
- a lock and/or multi-lock system (there are many types of lock available)
- warning notices which identify the work being carried out
- relevant personal protective equipment (PPE) that adheres to all site PPE rules.

The equipment shown in Figures 1.43 and 1.44 can be used to isolate various main switches and isolators. To isolate individual circuit breakers with suitable locks and locking aids, you should consult the manufacturer's guidance.

When working on or near electrical equipment and circuits, it is important to ensure that:

- the correct point of isolation is identified
- an appropriate means of isolation is used
- the supply cannot inadvertently be reinstated while the work is in progress
- caution notices are applied at the point(s) of isolation
- conductors are proved to be dead at the point of work before they are touched
- safety barriers are erected as appropriate when working in an area that is open to other people.

INDUSTRY TIP

Access Regulations 12 and 13 of the Electricity at Work Regulations 1989 at www.legislation.gov.uk/uksi/1989/635/made

ACTIVITY

Check your approved voltage indicator for any damage and for compliance with GS38.

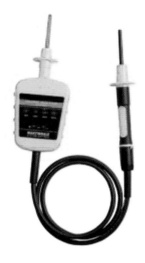

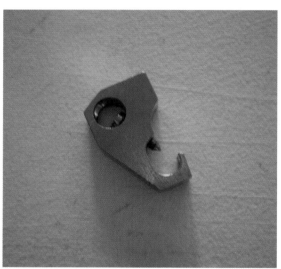

▲ Figure 1.42 A range of pieces of isolating equipment

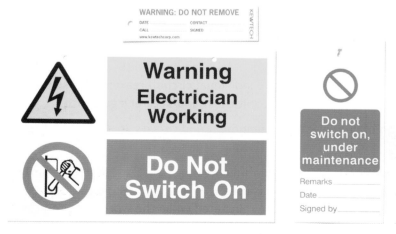

▲ Figure 1.43 Warning notices

Carry out the practical isolation

The method of isolation is outlined below.

1 *Identify* – identify equipment or circuit to be worked on and point(s) of isolation.
2 *Isolate* – switch off, isolate and lock off (secure) equipment or circuit in an appropriate manner. Retain the key and post caution signs with details of work being carried out.
3 *Check* – check the condition of the voltage indicator leads and probes. Confirm that the voltage indicator is functioning correctly by using a proving unit.
4 *Test* – using the voltage indicator, test the outgoing terminals of the isolation switch. Take precautions against adjacent live parts where necessary.
   During single-phase isolation, there are three tests to be carried out (L = line, N = neutral, E = earth):
   L – N   L – E   N – E
   During three-phase isolation, there are 10 possible tests (if the neutral is present):
   L1 – N   L2 – N   L3 – N
   L1 – E   L2 – E   L3 – E
   L1 – L2   L1 – L3   L2 – L3
   N – E

5 *Prove* – using the voltage indicator and proving unit, prove that the voltage indicator is still functioning correctly.
6 *Confirm* – confirm that the isolation is secure and the correct equipment has been isolated. This can be achieved by operating functional switching for the isolated circuit(s). The relevant inspection and testing can now be carried out.

ACTIVITY
Discuss why the N or E probe is connected before the Line when carrying out this test.

Reinstate the supply

When the 'dead' electrical work is completed, you must ensure that all electrical barriers and enclosures are in place and that it is safe to switch on the isolated circuit.

1 Remove the locking device and danger/warning signs.
2 Reinstate the supply.
3 Carry out system checks to ensure that the equipment is working correctly.

Implications of safe isolation

When you isolate an electricity supply, there will be disruption. So, careful planning should precede isolation of circuits. For example, when isolating a section of a nursing home where elderly residents live, you will need to consult the nursing home staff, to consider all the possible consequences of isolation and to prepare a procedure.

The following questions are useful:

1 How will the isolation affect the staff and other personnel? For example, think about loss of power to lifts, heating and other essential systems.

2 How could the isolation affect the residents and clients? For example, some residents may rely on oxygen, medical drips and ripple beds to aid circulation. These critical systems usually have battery back-up facilities for short durations.

3 How could the isolation affect the members of the public? For example, fire alarms, nurse call systems, emergency lighting and other systems may stop working.

4 How can an isolation affect systems? For example, IT programs and data systems could be affected; timing devices could be disrupted. In this scenario, you must make the employers, employees, clients, residents and members of the public aware of the planned isolation.

Alternative electrical back-up supplies may be required in the form of generators or uninterruptable power supply systems.

Risks involved in isolation procedures

Before any isolation is carried out, you must assess the risks involved. This section deals with the practical implications and the risks involved during the isolation procedure, if risk assessment and method statements are not followed.

Who is at risk and why?

If isolation is *not* carried out safely, what are the possible risks when performing inspection and testing tasks?

Risks to you

Risks to you might include:

- *shock* – touching a line conductor (e.g. if isolation is not secure)
- *burns* – resulting from touching a line conductor and earth, or arcing
- *arcing* – due to a short circuit between live conductors, or an earth fault between a line conductor and earth
- *explosion* – arcing in certain environmental conditions (e.g. in the presence of airborne dust particles or gases) may cause an explosion.

Things that can cause risk to you include:

- inadequate information to enable safe or effective inspection and testing (i.e. lack of diagrams, legends or charts)
- poor knowledge of the system you are working on (and so not meeting the competence requirements of Regulation 16 of EAWR)
- insufficient risk assessment
- inadequate test instruments (not manufactured or maintained to the standards of GS38).

Risks to other tradespersons, customers and clients

Risks to other tradespersons, customers and clients might include:

- switching off electrical circuits – for example, switching off a heating system might cause hypothermia (resulting from being too cold); if lifts stop, people may be trapped

- applying potentially dangerous test voltages and currents
- access to open distribution boards and consumer units
- loss of production, service or equipment – for example, essential supplies, or lights for access.

Risks to members of the public

Risks to members of the public might include prolonged loss of essential power supply, causing problems, for example, with safety and evacuation systems, such as fire alarms and emergency exit and corridor lighting.

Risks to buildings and systems within buildings

Risks to buildings and systems within buildings might involve applying excessive voltages to sensitive electronic equipment, for example:

- computers and associated IT equipment
- residual current devices (RCDs) and residual current operated circuit breakers with integral overcurrent protection (RCBOs)
- heating controls
- surge protection devices.

There might also be risk of loss of data and communications systems.

> **HEALTH AND SAFETY**
> Although safety services usually have back-up supplies such as batteries, these may only last for a few hours. Other safety or standby systems may have generator back-up, but this will also require isolation, leaving the building without any safety systems.

ACTIVITY

Write down a list of the risks associated with isolation and the effects isolation can have on people, livestock, systems and buildings. You could think about:

- Who is at risk if inspection and testing is not carried out correctly?
- What might happen if you need to switch off a socket outlet circuit – for example, in a hospital?
- What must you do if you encounter a computer server that requires a permanent supply and you need to switch off the main supply to enable safe testing procedures?

Test your knowledge

1 Which of the following describes the HSW Act?

  A Non-statutory.

  B Code of Practice.

  C Enabling Act.

  D Guidance Note.

2 Which regulations control waste electrical products?

  A EAWR.

  B WEEE.

  C HSWA.

  D PUWER.

3 Who is responsible for providing PPE on a construction site?

  A Employees.

  B Health and Safety Executive.

  C Local Buildings Authority.

  D Employers.

4 For how long after the last entry date must an accident book be kept?

  A 1 year.

  B 3 years.

  C 5 years.

  D 10 years.

5 What does RAMS stand for?

  A Risk Addressing Method Scheme.

  B Rules Assessing Meticulous Safety.

  C Risk Assessment Method Statements.

  D Rules Addressing Meticulous Statements.

6 What is considered as the last resort when considering risk reduction?

  A Personal protective equipment.

  B Elimination of the hazard.

  C Enclosing the problem.

  D Reducing contact.

7 Which of the following would be covered by COSHH Regulations?

  A Power tools.

  B Adhesives.

  C Manual handling.

  D Working at heights.

8 What are the **three** key elements needed for a fire to start?

  A Ignition, fuel, oxygen.

  B Electricity, ignition, fuel.

C Fuel, oxygen, water.

D Water, ignition, fuel.

9 What is the common supply voltage supplied by transformers on site, used for power tools?

A 50 V AC

B 110 V AC

C 230 V AC

D 400 V AC

10 Where would safe isolation be proven, using a voltage indicator, when isolating a single circuit within an installation?

A On the supply side of the installation main switch.

B On the load side of the installation main switch.

C On the supply side of the circuit protective device.

D On the load side of the circuit protective device.

11 Describe, using examples, who the duty holder is within an empty house being rewired by several electricians and apprentices.

12 List **three** statutory regulations intended to protect the environment.

13 Describe the **five** steps used to complete a risk assessment.

14 List the **five** stages, in sequence, for the hierarchy of control when approaching risk management.

15 Explain what should be visually inspected when carrying out a safety check to the supply leads of a power drill used on site.

Practical task

In your training centre or place of work, assume that work has to be carried out involving the blocking of a staircase by a platform or scaffold.

Carry out a risk assessment in relation to the loss of the staircase including measures that need to be taken to ensure safe operation of the building in normal and emergency situations.

CHAPTER 2

SCIENTIFIC AND MATHEMATICAL PRINCIPLES OF ELECTRICAL INSTALLATION

INTRODUCTION

Working with electricity safely and effectively involves a good understanding of maths. Using formulae and understanding mathematical principles are important throughout, from designing circuits to inspection and testing.

Learning objectives

This table shows how the topics in this chapter meet the outcomes of the different qualifications.

Topic	Electrotechnical Qualification (installation) or (maintenance) 5357-003	Level 2 Diploma in Electrical Installations (Buildings and Structures) 2365 Unit 202 (602) Principles of Electrical Science	Level 2 Technical Certificate in Electrical Installation 8202 Unit 202 Electrical Scientific Principles and Technologies
1 Mathematical principles	1.1		1.1; 1.2; 1.3
2 Units of measurement	2.1; 2.2; 2.3		Across all outcomes
3 Basic mechanics – Levers – Pulleys – Force – Energy – Power – Efficiency	3.1; 3.2; 3.3; 3.4		N/A
4 Basic electrical theory – Electrons – Conductors and insulators – Resistivity and Ohm's law – Temperature effects – Series and parallel circuits – Power – Current effects	4.1; 4.2; 4.3; 4.4; 4.5; 4.6; 4.7; 4.8		2.1; 2.2; 2.3; 2.4
5 Magnetism and electricity – Bar magnets – Current and magnetic fields – Solenoids – Force on a conductor – Producing a sinusoidal waveform – Induced e.m.f. – Sources of e.m.f. – Waveforms – Generating electricity – Basic supplies – Transformers	5.1; 5.2; 5.3; 5.4		3.1; 3.2; 3.3; 3.4
6 Electronic components	6.1; 6.2		4.1; 4.2

1 MATHEMATICAL PRINCIPLES

Indices

Indices are used to replace repetitive multiplications. For example, $10 \times 10 \times 10 = 1000$, so the calculation could be written easily by saying 10^3, which means ten multiplied by itself twice, or three lots of ten multiplied together.

Where indices are negative, the value becomes a fraction. For example:

$$5^{-1} = \frac{1}{5}$$

$$\text{or } 5^{-2} = \frac{1}{25}$$

$$\text{or } 5^{-3} = \frac{1}{125}$$

Most calculators will have a ($\boxed{x^2}$) button to square a number and scientific calculators also have a button ($\boxed{x^y}$), which allows a number to be raised to any power or index. For example, to calculate 5^5, use buttons $\boxed{5}\,\boxed{x^y}\,\boxed{5} = 3125$. This is much easier than keying $5 \times 5 \times 5 \times 5 \times 5$.

Generally, in electrical science and principles, large values are used, such as thousands of watts or millions of ohms. Other aspects of electrical work deal with tiny amounts, such as millionths of an ampere or thousandths of an ohm. This can become a problem in calculations, as errors may occur if the correct number of zeros is not entered into the calculator. Instead of inserting the actual number with lots of zeros, we use 'to the power of ten'.

The 'power of' numbers are given names that are explained in the table opposite. There is less chance making an error using this method.

▼ Numbers expressed as indices (to the power of 10)

Actual number	Number shown to the power of 10	Prefix used
1 000 000 000 000	10^{12}	tera (T)
1 000 000 000	10^9	giga (G)
1 000 000	10^6	mega (M)
1000	10^3	kilo (k)
100	10^2	hecto (h)
10	10^1	deka (da)
0.1	10^{-1}	deci (d)
0.01	10^{-2}	centi (c)
0.001	10^{-3}	milli (m)
0.000001	10^{-6}	micro (μ)
0.000000001	10^{-9}	nano (n)
0.000000000001	10^{-12}	pico (p)

KEY TERM

Indices: the plural of index.

KEY FACT

Any number which is raised to the power of zero (0) has a value of 1, so $10^0 = 1$ and $25^0 = 1$ and $3.142^0 = 1$ and so on.

INDUSTRY TIP

When you are writing down 'values of' in milli or mega units, always make sure that mega is a capital M and milli is a lower case m. It has to be clear what unit you are using.

KEY TERMS

Algebra: The branch of mathematics that uses letters and symbols to represent numbers, to express rules and formulae in general terms.

Transposition: Rearranging a formula to make the unknown you need to find, the subject of the formula.

To perform this on a calculator, use the button marked **EXP** (it may also be marked **×10**), and then insert the index (the 'to the power of' number).

To perform a complex calculation such as 325 giga × 5 micro ÷ 12 mega, you would need to insert a lot of zeros before and after the decimal point, like this:

$$\frac{325\,000\,000\,000 \times 0.000005}{12\,000\,000}$$

To make it easier to perform on a calculator, use the indices and the **EXP** button, so the formula becomes:

$$\frac{325 \times 10^9 \times 5 \times 10^{-6}}{12 \times 10^6}$$

So, on a calculator **325** **EXP** **9** **×** **5** **EXP** **−** **6** **÷** **12** **EXP** **6** **=** will give the result 0.14.

If you try calculating $25 \times 10^6 \times 2 \times 10^3$, depending on your calculator, the result obtained may be 5 with a 10 in the right of the screen. If you press the button marked **ENG** or **SHIFT ENG**, you will see the 'to the power of' number change to 50 to the power of 9, which is equal to 50 G, or, if you keep pressing the **ENG** button, you may eventually get 50 000 000 000. All the numbers displayed have the same values, just represented in different ways.

Transposing basic formulae

In electrical science we use **algebra** all the time when dealing with formulae. For example, Ohm's law states that a voltage can be determined by multiplying current by resistance. Instead of writing it out in full, we use letters and symbols to represent unknown values, which are known as variables. Therefore $V = I \times R$ is a basic use of algebra. Algebra can be used to show relationships between different quantities.

If, for example, we want to find the total cost, b, of four bars of chocolate, and the price of one bar is a, the formula would be:

$$b = 4a$$

Many formulae are used in electrical science. It is much easier to remember one particular formula in one particular way. For example:

$$R = \frac{\rho l}{A}$$

where:
R = resistance, in ohms (Ω)
ρ = resistivity value of a particular material (Ωm)
l = length of a cable conductor (m)
A = cross-sectional area of the conductor (m^3).
(This formula is explained in more detail on page 98.)

The above formula could be rearranged, to determine the formula for A (see page 68). This is called **transposition**.

To learn the rules of transposition, you need to consider three types of formula:

- those that use addition and subtraction
- those that use multiplication and division
- those involving both (mixed formulae).

The following methods follow simplified mathematical rules and will give you the ability to transpose any formulae you come across at Levels 2 and 3.

Transposing formulae involving addition and subtraction

The rules of transposition for addition and subtraction

1 The unknown you want to find must be on its own on one side of the equals sign.
2 The unknown should not have a minus sign in front of it.
3 Any unknown that moves over the equals sign has the sign in front of it changed from addition to subtraction or from subtraction to addition.
4 Any unknown that does not have a sign (positive or negative) in front of it is assumed to be positive.

Consider how to transpose this formula to find c.

$$a + b + c - d = e$$

The unknown c has a plus sign (+) in front of it so it stays where it is and the other unknowns around it need to move. The unknowns a and e have no sign in front of them so assume they are not being subtracted from anything.

So, to transpose it:

$a + b + c - d = e$	move d over the equals sign
$a + b + c = e + d$	$-d$ changes to $+d$ when moved
$a + c = e + d - b$	move b, changing the + to −
$c = e + d - b - a$	move a (changing the sign) to leave c alone
$c = e + d - b - a$	This is the finished, transposed formula.

ACTIVITY

There are many formulae that will have to be transposed as you progress through the course. Get as much practice as you can. Start with all the formulae you can think of with three items, then four and so on.

At all times, the equal sign acts as the centre of a set of balance scales and the formula remains balanced (Figure 2.1). If numbers are substituted for the letters, this might give:

$10 + 20 + 15 - 5 = 40$	move numbers away from 15 over the equals sign
$10 + 20 + 15 = 40 + 5$	see that the formula is balanced as it equals 45 on each side
$10 + 15 = 40 + 5 - 20$	keep the balance and move the 20
$15 = 40 + 5 - 20 - 10$	move 10 to leave 15 alone
$15 = 40 + 5 - 20 - 10$	This is the finished, transposed formula.

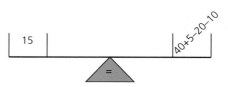

▲ Figure 2.1 Formulae involving addition and subtraction remain balanced over the equals sign.

Transposing formulae involving multiplication and division

Formulae involving division often include parts that are made up of letters and numbers written above and below a line, as fractions. The number above the line is called the numerator, and the number below is the denominator. A denominator is the number of parts a whole is divided into, a numerator is the number of these parts that we are dealing with. So, thinking of ¾ of a cake, the cake is divided into four equal parts, and we have three of them.

$$\frac{\text{numerator}}{\text{denominator}} = \frac{\text{number of parts we have}}{\text{number of equal parts in the whole}}$$

The rules of transposition for multiplication and division

1 The unknown you want to find must be on its own on one side of the equals sign.
2 The unknown to be found must be on its own, not part of a fraction.
3 Any unknown moved over the equal sign changes from top to bottom or bottom to top.
4 Any unknown or number can be written as a fraction by writing it over a denominator of 1, for example, $4 = {}^4/_1$.

INDUSTRY TIP

Where a formula shows two unknowns together with no symbol in between, they should be multiplied, so ρl means $\rho \times l$.

ACTIVITY

Transpose each of the following formulae to find the unknown indicated.

$E = V - I_a R_a$ Find V.

$E = \Phi N$ Find N.

$P = I^2 R$ Find I.

$E = \dfrac{2P\Phi NZ}{A}$ Find N.

$P = V \times I \times \cos\theta$ Find $\cos\theta$.

The following formula can be transposed, to find a formula for A:

$$R = \frac{\rho l}{A}$$

The A is at the bottom and it needs to be at the top. This is done by moving it over the equals sign. The letter R can be written as a fraction, as ${}^R/_1$. So moving A over the equals sign gives:

$$\frac{RA}{1} = \frac{\rho l}{1}$$

Now all the unknowns are at the top. Remember that anything divided or multiplied by 1 remains the same, but writing the 1s in helps us to remember that there is a top and bottom. Now divide both sides by R.

$$\frac{A}{1} = \frac{\rho l}{R}$$

R has now been moved over the equals sign to leave A alone, so:

$$A = \frac{\rho l}{R}$$

This can once again be demonstrated by using numbers to see how balance is maintained:

$$50 = \frac{20 \times 5}{2}$$

Insert a 1 to show 50 as a fraction:

$$\frac{50}{1} = \frac{20 \times 5}{2}$$

Move the 2 over the equals sign from bottom to top:

$$\frac{50 \times 2}{1} = \frac{20 \times 5}{1}$$

To keep the balance, move the 50 over the equals sign from top to bottom:

$$\frac{2}{1} = \frac{20 \times 5}{50}$$

Then to finish off, remove the 1 from the left-hand side to get:

$$2 = \frac{20 \times 5}{50}$$

Job done!

With practice, this routine becomes second nature. You just need to remember the rules.

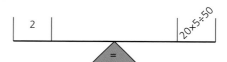

▲ Figure 2.2 Formulae involving multiplication and division remain balanced over the equals sign

Transposing mixed formulae

This requires combining all the rules. Sometimes numbers or unknowns need to be combined, so they are grouped in brackets. This effectively makes each group act as if it is a single number or unknown. For example, to determine the unknown *d* from the following formula:

$$\frac{(a + b) \times c \times d}{e} = f$$

As *d* is at the top, it needs to be left where it is and the other unknowns are moved. As the (*a* + *b*) is in brackets, the whole thing can be moved together and treated as a single unknown. Remember, *f* is also over 1.

$$\frac{(a + b) \times c \times d}{e} = f$$

ACTIVITY

Make sure you practise using as many formulae as you can. Make up your own, and test them out by substituting numbers.

Move $(a + b)$ over the equals sign, so:

$$\frac{c \times d}{e} = \frac{f}{(a+b)}$$

And:

$$\frac{d}{e} = \frac{f}{(a+b) \times c}$$

So finally:

$$\frac{d}{1} = \frac{f \times e}{(a+b) \times c}$$

Or:

$$d = \frac{f \times e}{(a+b) \times c}$$

Once again, to prove this with numbers:

$$\frac{(3+2) \times 4 \times 10}{25} = 8$$

The (3 + 2) is treated as a single number (i.e. 5), so:

$$\frac{4 \times 10}{25} = \frac{8}{(3+2)}$$

Then:

$$\frac{10}{25} = \frac{8}{(3+2) \times 4}$$

And finally:

$$10 = \frac{8 \times 25}{(3+2) \times 4}$$

There is one further rule to remember:

Whatever you do to one side of the formula, you must keep the balance on the other.

To transpose or rearrange the formula below to make b the subject:

$$\sqrt{a^2 + b^2} = c$$

KEY FACT

This method of transposition may be slightly different to the one that you learnt at school, although both methods will give you the same result. You may have learned to carry out the same operation to both sides of the equation. Use the method that suits you best.

the square root must be eliminated because it locks in b. The opposite of taking a square root is to square, so remove square root and square c, so:

$$a^2 + b^2 = c^2$$

Applying the rules gives:

$$+ b^2 = c^2 - a^2$$

As the unknown needed is b, not b^2, take the square root on both sides:

$$b = \sqrt{c^2 - a^2}$$

Area, volume and density

When planning or designing electrical installations, it is important to know how to calculate areas and volumes such as those of rooms. In addition, the density of materials, such as air or water, is important when we need to consider the heating of spaces or water.

Area

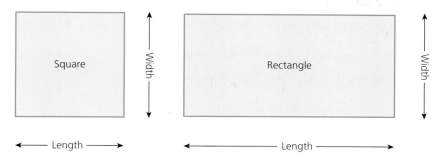

▲ Figure 2.3 Properties of a square and a rectangle

The area of a square or rectangle is calculated by multiplying the length by the width.

$$\text{area (m}^2) = \text{length (m)} \times \text{width (m)}$$

So if a rectangle had a length of 3 m and a width of 1.5 m, its area can be calculated by:

$$\text{area (m}^2) = 3 \times 1.5 = 4.5 \text{ m}^2$$

In order to calculate the area of a circle, we need to apply:

$$\text{area (m}^2) = \pi r^2 \text{ or } \pi \times \text{radius}^2 \text{ (m)}$$

where π is **pi**, which is equal to 3.14...

So, if a circle had a radius of 3 m, its area would be:

$$\text{area (m}^2) = 3.14 \times 3^2 = 28.26 \text{ m}^2$$

KEY TERM

Pi: a number which is the ratio of a circle's circumference to its diameter and is therefore a constant value used to determine the properties of a circle. Pi is also used to determine some electrical properties because electricity generators revolve in a circle. Although pi is often expressed as 3.14, the number of digits after the decimal point is infinite (never ending) so using the pi key on a calculator provides a more accurate result than simply keying in 3.14.

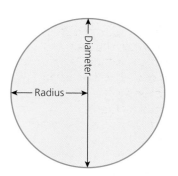

▲ Figure 2.4 Properties of a circle

Volume

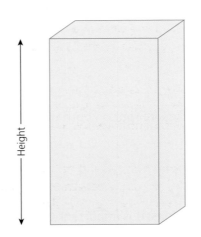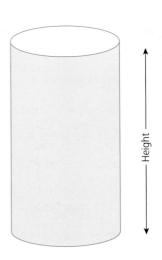

▲ Figure 2.5 A cuboid and a cylinder

The volume of a space is calculated by taking the area of its side and multiplying it by the height – but instead of doing two calculations, the values can all be combined into one calculation.

For a cuboid (rectangular side) or cube (square side):

$$\text{volume (m}^3) = \text{length (m)} \times \text{width (m)} \times \text{height (m)}$$

For a cylinder (circular side)

$$\text{volume (m}^3) = \pi \times \text{radius}^2 \text{ (m)} \times \text{height (m)}$$

So, the volume of a room measuring 12 m × 6 m with a height of 2.4 m can be calculated by:

$$\text{volume (m}^3) = 12 \times 6 \times 2.4 = 172.8 \text{ m}^3$$

Density

The density of a material is the mass divided by the volume, and is measured in kg/m³. Commonly, we use the known density of a material to calculate its mass in a given space.

For example, water has a density of approximately 1000 kg per m³ at 25 °C. (Density changes with temperature.)

Using this, we can calculate the mass of water in a tank by:

$$\text{mass (kg)} = \text{density (kg/m}^3) \times \text{volume (m}^3)$$

So, if we need to determine the mass of water in a tank measuring 3 m × 2 m × 0.5 m, we calculate (remembering the density of water is 1000 kg/m³):

$$\text{mass (kg)} = 1000 \times 3 \times 2 \times 0.5 = 3000 \text{ kg}$$

KEY FACT

While the above calculations use the SI unit of length, the metre (m), other units may be used such as centimetres (cm) or millimetres (mm). However, you should never use a mixture of units in the same calculation. (See page 79 for more about SI units.)

Triangles and Pythagoras' theorem

Triangles are used to quantify electrical values. Later you will explore power triangles and phasor diagrams as a way of determining circuit values. To help with these, you need to understand basic principles of trigonometry (see page 74) and **Pythagoras' theorem**.

The triangle in Figure 2.6 is a right-angled triangle. Given the lengths of any two sides of a right-angled triangle, you can use Pythagoras' theorem to find the length of the third side. The **hypotenuse** of a right-angled triangle is always opposite the right angle.

Pythagoras was a scholar in Ancient Greece, who discovered that the square of the length of the hypotenuse is equal to the square of the length of side a added to the square of the length of side b. Or, to express this as a formula:

$$a^2 + b^2 = h^2$$

or:

$$\sqrt{a^2 + b^2} = h$$

EXAMPLE

The length of the hypotenuse for the triangle in Figure 2.7 can therefore be calculated by applying Pythagoras' theorem.

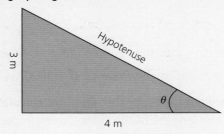

▲ Figure 2.7 Applying Pythagoras' theorem

As:

$$\sqrt{a^2 + b^2} = h$$

Then:

$$\sqrt{3^2 + 4^2} = 5\,\text{m} \text{ (the length of the hypotenuse)}$$

Trigonometry

Trigonometry is used extensively in engineering and construction technology as well as many other sciences. Without trigonometry we would not be able to establish the heights of hills, mountains and buildings or the distance to stars and other planets.

INDUSTRY TIP

Remember Pythagoras' theorem only applies to right-angled triangles.

KEY TERMS

Pythagoras' theorem: algebraically, this states that for a right-angled triangle with sides of lengths a, b and h, where h is the length of the hypotenuse: $a^2 + b^2 = h^2$.

Hypotenuse: The longest side of a right-angled triangle, which is opposite the right angle.

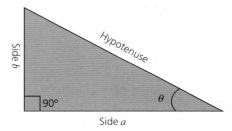

▲ Figure 2.6 A right-angled triangle

INDUSTRY TIP

You will find calculations based on triangles and Pythagoras' theorem all through electrical science work. These will include power triangles, lighting calculations and power factor correction among others.

KEY TERM

Trigonometry: the mathematical study of the relationships between the lengths and angles of triangles.

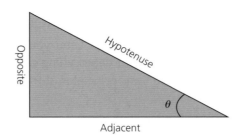

▲ Figure 2.8 The three sides of a right-angled triangle are: adjacent to the angle theta (θ), opposite the angle theta (θ) and the hypotenuse, which is opposite the right angle.

Consider this explanation to help you understand the relationships in trigonometry. Looking at a right-angled triangle, if any of the sides increase or decrease in length, the angle θ changes accordingly. (Unknown angles are often given the symbol theta, θ, from the Greek alphabet.)

There are three formulae to remember when it comes to trigonometry.

$$\sin\theta = \frac{\text{opposite}}{\text{hypotenuse}}$$

$$\cos\theta = \frac{\text{adjacent}}{\text{hypotenuse}}$$

$$\tan\theta = \frac{\text{opposite}}{\text{adjacent}}$$

Some people remember them by the mnemonic SOH CAH TOA.

The sine, cosine and tangent ratios for the angle are calculated by using the relationships between the lengths of the sides. For example, the triangle shown in Figure 2.9 has the following dimensions:

- adjacent = 3 m
- opposite = 4 m
- hypotenuse = 5 m.

Using these values you can work out the sine, cosine and tangent values of the angle.

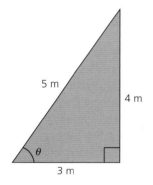

▲ Figure 2.9

KEY FACT

The internal angles of a right-angled triangle always add up to 180°.

$$\sin\theta = \frac{\text{opposite}}{\text{hypotenuse}} = \frac{4}{5} = 0.8$$

$$\cos\theta = \frac{\text{adjacent}}{\text{hypotenuse}} = \frac{3}{5} = 0.6$$

$$\tan\theta = \frac{\text{opposite}}{\text{adjacent}} = \frac{4}{3} = 1.333$$

With these ratios, the angle θ can be found using a calculator:

- $\sin^{-1} 0.8 = 53.1°$
- $\cos^{-1} 0.6 = 53.1°$
- $\tan^{-1} 1.333 = 53.1°$

Each of the ratios determined by using the different lengths relates to the same angle. We can use these different ratios to determine any missing value from the triangle.

You need to select the right formula to do this. Choose the one that uses the information you already have and that gives you what you need to find. For example, if you knew the length of the hypotenuse and the angle, but needed to determine the length of the opposite, you would choose the formula that uses all three. This is the sine formula.

Values of sine, cosine and tangent are ratios that have been calculated for every possible angle. Many years ago, before calculators, these factors were found from books of mathematical tables. These days, all of the different possible values are programmed into your scientific calculator.

When using the **SIN**, **COS** and **TAN** functions on a calculator, pressing each button directly will provide the sine, cosine or tangent value (ratio) for the given angle. For example, **SIN 45** will give a value of 0.7071, which means an angle of 45° has a sine value of 0.7071. To find a value of angle from a calculated or given sine, cosine or tangent, you will need to use the **SHIFT** or **2nd Function** button followed by the $\sin^{-1}$, $\cos^{-1}$ and $\tan^{-1}$ symbols. With this book, if you see the function $\cos^{-1}$, $\tan^{-1}$ or $\sin^{-1}$, the **SHIFT** or **2nd Function** button is required for that calcuation.

Sine

To determine the length of the hypotenuse of the triangle in Figure 2.10, as you know the value of the angle and the length of the opposite side, use the trigonometric function sine as it includes the three known or needed ingredients.

Use the formula:

$$\sin \theta = \frac{o}{h}$$

Transpose the formula:

$$h = \frac{o}{\sin \theta}$$

so:

$$h = \frac{7}{\sin 35°} = 12.2 \, \text{m}$$

ACTIVITY

Sine, cosine and tangent are the three main trigonometric ratios. Identify three others.

INDUSTRY TIP

Calculators do not all work in the same way. Therefore, any calculator key sequence suggested is only an example of what would be required on a standard scientific (non-programmable) calculator. If the key sequence does not give the expected result on your calculator, either ask your tutor for advice or refer to the manual for your calculator. Ensure your calculator is set to DEG (for degrees), not RAD or GRAD or your results will be very different. Radians and gradians are other methods of measuring angles. Be familiar with your calculator.

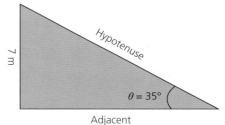

▲ Figure 2.10

EXAMPLE

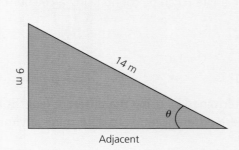

▲ Figure 2.11

Now, use the sine function to determine the value of the required angle in Figure 2.11. Then:

$$\sin \theta = \frac{o}{h}$$

so:

$$\sin \theta = \frac{9}{14} = 0.642$$

This is not the actual value of the angle, but the sine of the angle. In order to find the angle from this, use the $\sin^{-1}$ function by pressing the **SHIFT** button or second function button, depending on your calculator.

On the calculator, press **SHIFT** **SIN** **0.642** **=**

So the angle is found as:

$$\sin^{-1} 0.642 = 40°$$

The sequence **9** **÷** **14** **=** **sin⁻¹** **=** will give the angle as 40°. This answer is more accurate as the calculator remembers all of the values after the decimal point in the value of the sine of the angle.

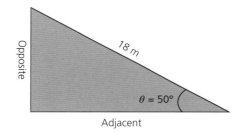

▲ Figure 2.12

Cosine

To determine the length of the side adjacent to θ in the triangle in Figure 2.12, use cosine as you know the size of the angle and the length of the hypotenuse, and you need to find the length of the adjacent side.

Use the formula:

$$\cos \theta = \frac{a}{h}$$

Transpose the formula:

$$a = \cos \theta \times h$$

so:

$$a = \cos 50° \times 18 = 11.57 \text{ m}$$

EXAMPLE

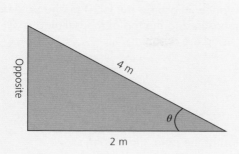

▲ Figure 2.13

To determine the angle θ in Figure 2.13, use the formula:

$$\cos \theta = \frac{a}{h}$$

So:

$$\cos \theta = \frac{2}{4} = 0.5$$

Then $\cos^{-1} 0.5 = 60°$.

Tangent

Where calculations in trigonometry do not involve the hypotenuse of a triangle, use the tangent function.

EXAMPLE

Imagine you need to determine the height of a building. With the use of a very simple theodolite made from a protractor, straw and a pole of known length, the height can be determined by measuring a distance from the building and measuring the angle, by looking through the straw, to the top of the building.

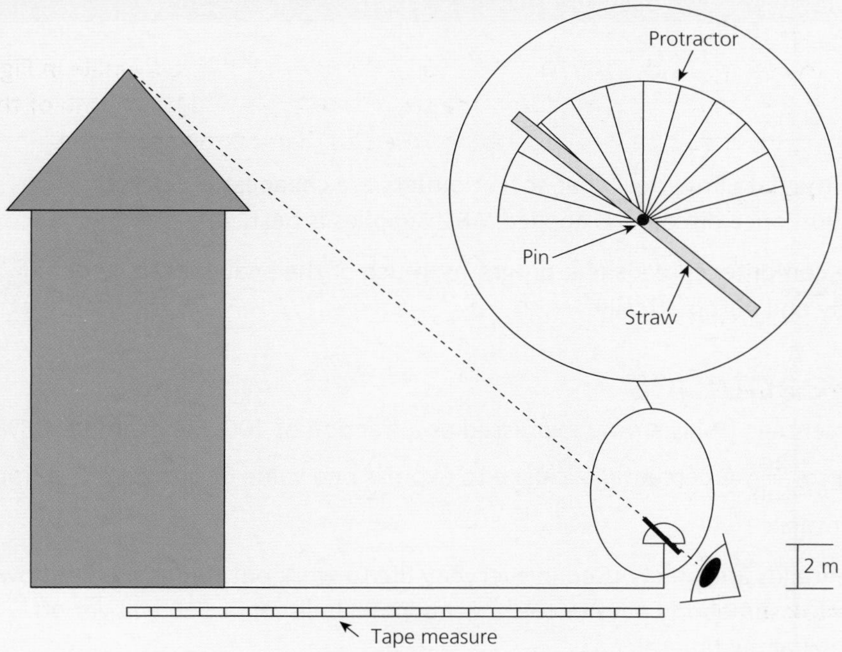

Protractor

Pin

Straw

2 m

Tape measure

▲ Figure 2.14 Using trigonometry to measure the height of a building

INDUSTRY TIP

The use of tangent should be encouraged when the opposite and adjacent sides are known. Don't waste your time unnecessarily calculating the hypotenuse.

Assuming that the distance measured out is 20 m, the pole being used is 2 m long and the angle measured to the top of the building is 70°, the height of the building is found by using the formula:

$$\tan\theta = \frac{o}{a}$$

Transpose the formula:

$$\tan\theta \times a = o$$

Therefore, from the top of the pole, the height is:

$$\tan 70° \times 20 = 54.9\ m$$

The overall height (rounded up) is 57 m, once the pole length of 2 m is included.

Statistics

Statistics is the branch of mathematics that includes collecting, organising and interpreting data. In the electrotechnical sector, data may be used in situations such as:

- cable selection tables
- maximum permitted values of earth fault loop impedance
- quotations and tendering
- billing
- quantity surveying
- monitoring job progression.

Statistics are best summarised in table form and can be presented as graphs. Below is an example of how data may be presented and analysed.

Wholesaler	Cost of socket per unit (£)	Discount offered (%)	Cost after discount (£)
ABC Supplies	2.52	20	2.02
Sockets R Us	2.45	8	2.25
Lecky World	3.65	30	2.56

From the data, it is clear that socket outlets are cheapest at Sockets R Us but, once discount is applied, ABC Supplies is best.

Using computer spreadsheet programs, much of the analysis can be done quickly and automatically.

Percentages

A percentage (%) is a ratio expressed as a fraction of 100. For example, 35% is equal to $\frac{35}{100}$. A percentage is used to express one value or quantity as a fraction of another.

Percentages are widely used in everyday life to work out things such as how much tax somebody pays or to show how much discount you will get off material at a wholesaler.

EXAMPLE

If you have to pay 20% of your earnings each month in tax, and you earn £1400.00 per month you can work out the amount of tax to pay by calculating:

$$\frac{1400 \times 20}{100} = 280 \text{ or } £280.00$$

A shorter way to do this is to show the percentage as a decimal multiplier, so 20% becomes $\frac{20}{100} = 0.2$

so £1400.00 × 0.2 = £280.00

If you have to pay an extra 20% VAT on goods, and the goods cost £28.00 excluding VAT, you need to add 20% of £28.00 to the cost, so

$\frac{28 \times 20}{100} = £5.60$. So you need to add £5.60 to the ex-VAT price of £28.00, so the cost including VAT is £33.60.

Once again, a shorter way to do this is to use a decimal multiplier. This time we need to add 20% so we use a multiplier of 1.2. This essentially adds 20% to the initial cost as it multiplies the value by 1, keeping the £28, and adds a further 0.2 representing the 20% increase.

The use of percentages as multipliers is very common in the electrotechnical industry as they are often used to determine temperature changes and effects when designing electrical installations.

INDUSTRY TIP

Multipliers are often referred to as factors in publications such as BS 7671:2018 and are commonly used in circuit design calculations. (See https://electrical.theiet.org)

ACTIVITY

A computer in a sale has a 30% discount. The full price is £429.99. What is the sale price?

② UNITS OF MEASUREMENT USED IN ELECTRICAL INSTALLATION WORK

SI units

As the world of science, and electrical science in particular, developed, it became necessary to agree on some form of standardisation so that scientists could understand one another's work and share their ideas.

The system of **SI units** (short for *Système International d'Unités*) is internationally recognised and based on the metric system. The alternative, based on the imperial system of measurement and still used in the USA (US imperial), is far more complex and less elegant, as the non-electrical comparisons below between SI and imperial demonstrate.

Imperial

1 ton = 20 cwt (hundred-weight) = 2240 lb (pounds weight)
      = 35 840 oz (ounces)

Metric

1 tonne = 1000 kg (kilograms) = 1 000 000 g (grams)

There are seven base SI units, which then generate many derived units.

KEY TERM

SI units: The units of measurement adopted for international use by the Système International d'Unités.

ACTIVITY

There are seven base SI units in total. The seventh unit is mainly used in chemistry to measure chemical substances and is not relevant to electricity at this level. What is it called?

▼ Base SI units

Quantity	Quantity symbol	Unit name	Unit symbol
Current	I	ampere	A
Length	l	metre	m
Luminous intensity	L	candela	cd
Mass	m	kilogram	kg
Temperature	T	kelvin	K
Time	t	second	s

Electrical SI units

In electrical applications, units of measurement are standardised. You must always ensure that any calculation or formula uses the base SI unit or the derived units shown below. If a calculation uses a different unit, the result will be incorrect.

Standardisation of units of measurement is essential to be sure that everybody knows the correct measurements to use. There have been many instances over the years where very complex and large international projects have gone disastrously wrong because Europe uses metric units and America uses imperial.

INDUSTRY TIP

Energy is normally measured in joules (J), but for many purposes this unit is too small so the kilowatt hour (kWh) is used. This is the unit used by the electricity meter at your home to measure how much energy you have used. For example, a 1 kW heater running for 1 hour will use 1 kWh or 1 Unit of electricity.

INDUSTRY TIP

Sometimes, you will see a U used as the symbol for voltage. This is because U is used in BS 7671:2018 The IET Wiring Regulations, 18th Edition and other European standards to represent the voltage between two parts of an electrical system.

The following table lists the most commonly used electrical units of measurement.

▼ Most commonly used electrical units of measurement

Quantity	Quantity symbol	Unit name	Unit symbol
Area	A	square metre	m^2
Capacitance	C	farad	F
Charge	Q	coulomb	C
Energy (work)	W	joule	J
Force	F	newton	N
Frequency	f	hertz	Hz
Impedance	Z	ohm	Ω
Inductance	L	henry	H
Magnetic flux	Φ	weber	Wb
Magnetic flux density	B	tesla	T
Potential difference	V	volt	V
Power	P	watt	W
Reactance	X	ohm	Ω
Resistance	R	ohm	Ω
Resistivity	ρ	ohm-metre	Ωm

ACTIVITY

Look in Part 2 of BS 7671: 2018 in the section called 'Symbols Used in the Regulations'. Find out what is represented by:

1 U

2 U_0

3 U_{oc}.

Measuring electrical quantities in circuits

In order to measure the correct values, ammeters and voltmeters need to be connected differently.

As ammeters measure the current flowing through a circuit, they are connected in series with the load.

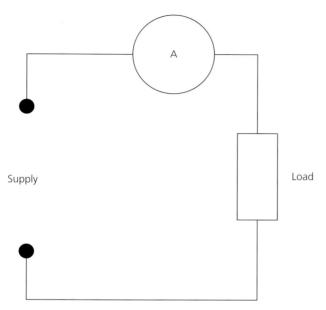

▲ Figure 2.15 An ammeter (A) *must* be connected in series

As voltmeters read the potential difference across the load, they are connected in parallel.

ACTIVITY

Why will a voltmeter not affect the circuit parameters (conditions) if connected across a resistor of low value, but will affect the readings if the resistor is of a high value?

INDUSTRY TIP

Instruments may be either **analogue** or **digital**. Analogue instruments normally have a needle or pointer which follows the input value (analogous). A digital meter has a numeric display.

INDUSTRY TIP

As voltmeters have very high resistance they can be connected in parallel across the supply.

KEY TERMS

Analogue: instruments that use magnetism to measure electrical quantities.

Digital: instruments that use electronic components to measure electricity.

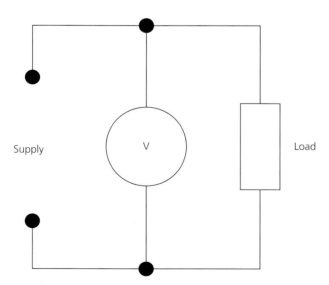

▲ Figure 2.16 Volmeter (V) connection in parallel

Wattmeters measure the power in a circuit, which is determined by the current and voltage. As a result, they are connected in parallel and series, allowing measurement of both properties.

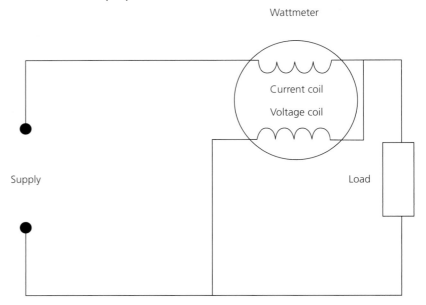

▲ Figure 2.17 Wattmeter connection

Unlike the meters above, ohmmeters must only be used when a circuit is *not* energised. Ohmmeters use a small internal current and voltage to determine resistance and must be connected in parallel to the given resistance, as shown below.

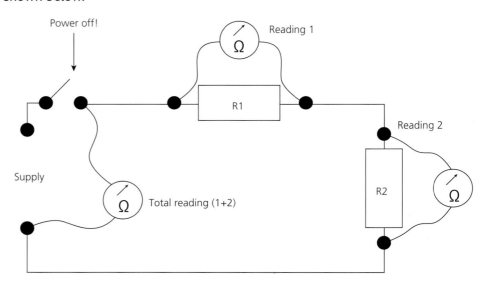

▲ Figure 2.18 Three different ohmmeter readings with series resistors

3 BASIC MECHANICS

Understanding basic mechanics will help you to link the mechanical properties and electrical quantities that enable electricity to be turned into a useful form of energy, linking physical work and electrical power. It may also help you to understand everyday objects, from seesaws to cranes and from claw hammers to wire cutters.

Measuring mechanical loads

It is important to understand the ways in which mechanical loads are measured. You need to understand the differences between mass and weight as well as energy and power. Energy and power can be mechanical as well as electrical.

The fundamental relationship between the mass and the weight is defined by Newton's Second Law and can be expressed as:

$$F = ma$$

where:
F = force (N)
m = mass (kg)
a = acceleration (m/s^2).

Weight

Weight is the gravitational force acting on a body mass. Newton's Second Law can be related to weight as a force due to gravity and can be expressed as:

$$W = mg$$

where:
W = weight (N)
m = mass (kg)
g = acceleration of gravity (on Earth this is 9.81 m/s^2).

Mass

Mass is the comparison of an amount of material measured against a known value. The SI unit of mass is the kilogram. At the International Bureau of Weights and Measures, located in Sevres, France, a 1 kg mass of platinum is retained as the criterion for determining mass.

As mass is determined by balancing one body against another, gravitational pull does not make any difference to the value. Mass is the same in outer space as it is on Earth.

> **KEY FACT**
>
> There is no weight in space, just mass.

Acceleration

Acceleration is the rate of change of velocity with time.

$$a = \frac{\mathrm{d}v}{\mathrm{d}t}$$

where:

a = acceleration (m/s^2)
v = velocity (m/s)
t = time (s).

As acceleration is a measurement of the rate of change of speed, in metres per second (m/s) for every second, the unit is metres per second per second or metres per second squared (m/s^2).

Calculating mechanical advantage of using levers

Archimedes claimed that, if he was able to stand in the appropriate place, he could use a lever to move the Earth. Although it is not actually possible to lift the Earth, the principle is a good illustration of the power of levers.

How levers work

The total turning moment (**torque**) applied to the ends of a **lever** must be equal but opposite. Therefore:

> force of load to be lifted × distance of load from fulcrum =
> force of effort × distance of effort from fulcrum

Torque is the product of the force (N) applied to a lever and the distance (m), from the **fulcrum**, of the point where the force is applied. It is measured in newton-metres (Nm).

Levers are divided into three classes, as described below.

Class 1 levers

▲ Figure 2.20 Children playing on a seesaw – a class 1 lever

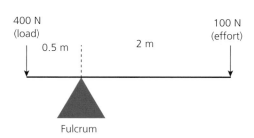

▲ Figure 2.19 Example of a class 1 lever

In a class 1 lever, the force and the load are on different sides of the fulcrum. The effectiveness of the force is affected by the distance of the point of application of the force from the fulcrum. You measure the torque by multiplying the force by its distance from the fulcrum. So, the greater the distance, the greater the effect of the force. In the example above, force applied downwards can be four times less than the load as the lever on the force (effort) side of the fulcrum is four times longer than the lever on the load side. Examples of class 1 levers are seesaws, crowbars and claw hammers when used to lift nails.

The advantage gained by a lever is referred to as **mechanical advantage (MA)**, which is calculated as:

$$MA = \frac{load}{effort} = \frac{400}{100} = 4$$

Class 2 levers

In a class 2 lever, the load is between the point of effort and the fulcrum. The calculations remain the same, but the load is more limited because it is between the two points, and the effort and the desired movement are both in the same direction. An example of a class 2 lever is a wheelbarrow.

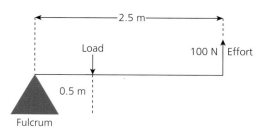

▲ Figure 2.22 Example of a class 2 lever

In the example in Figure 2.22 above, the force of 100 N is applied at a distance of 2.5 m from the fulcrum and the load is applied 0.5 m from the fulcrum.

The load that can be lifted is calculated as follows:

force of load × distance of load from fulcrum
= force of effort × distance of effort from fulcrum

So:

$$0.5 \text{ m} \times load = 2.5 \text{ m} \times 100 \text{ N}$$

Transposing:

$$load = \frac{2.5 \times 100}{0.5} = 500 \text{ N}$$

To find the mechanical advantage:

$$MA = \frac{load}{effort} = \frac{500}{100} = 5$$

Therefore the mechanical advantage is 5.

KEY TERM

Mechanical advantage: a measure of the force gained by using a tool.

▲ Figure 2.21 A wheelbarrow is a class 2 lever

KEY FACT

Levers and **pulleys** are aids for lifting where the load is too much for an unaided human to lift.

ACTIVITY

What class of lever is a pair of pliers?

Class 3 levers

Class 3 levers are slightly different to class 2 levers in that the load is at the end opposite the fulcrum. Examples of class 3 levers include fishing rods, tweezers and the human arm.

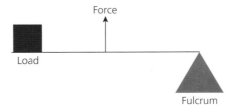

▲ Figure 2.23 Example of a class 3 lever

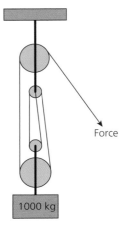

▲ Figure 2.24 Example of a pulley system

Pulleys

A **pulley** is an effective way of gaining a mechanical advantage when lifting an object. By running a rope through a four-pulley system, a mechanical advantage of four is gained.

EXAMPLE

Determine the force required to raise the mass of 1000 kg.

As load has been described as a mass, first determine the downward force of the load. So:

$$\text{force} = \text{mass} \times \text{gravity}$$

$$F = 1000 \times 9.81 = 9810 \text{ N}$$

(Remember: Gravity is acceleration which, on Earth, averages 9.81 m/s²)

As the pulley system has four ropes compared to the one pulling rope, the mechanical advantage is 4:1. So, rearranging the formula for mechanical advantage:

$$\frac{9810}{4} = 2452 \text{ N}$$

So a downward force (or effort) of 2452 newtons is required to raise the load. Although the pulley gives a mechanical advantage, the pulling rope will need to be pulled four times further than the load is raised. This means, to raise the load by 1 m, the pulling rope needs to be pulled 4 m.

Gears

Gears are used to transmit torque from one rotating part to another. Gears are made of two or more cogwheels often having different diameters and therefore different numbers of teeth. Gears are used to change speed or effort needed to produce torque. For example, if a cog containing 20 teeth was connected to a cog containing 100 teeth and the smaller cog was driven by an electric motor rotating 40 times a second, we can determine the rotations per second of the large cog by expressing the gear ratio. So:

▲ Figure 2.25 Cogs and gears of different sizes

$$\frac{100}{20} = 5 \text{ giving a 5:1 ratio}$$

So if the small gear rotates five times more than the larger one, the number of times the larger wheel will rotate is:

$$\frac{40}{5} = 8 \text{ times per second}$$

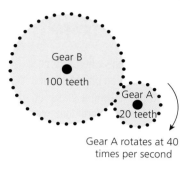

Gear A rotates at 40 times per second

▲ Figure 2.26 Gears

Calculating quantities of mechanical loads
Force

Force can be defined as the strength or energy of physical action or movement. In practice, it can be any influence that causes an object with a mass to undergo a change in velocity, including any movement from rest, or a change in direction or geometrical construction (deformation).

Simply, force is also the 'weight' an object has due to the force of gravity pulling it onto the ground beneath it.

The SI unit for force, represented by the symbol F, is the newton (N). Since mass is measured in kilograms (kg) and acceleration is measured in metres per second per second (m/s^2), then:

force in newtons = mass × acceleration, so $F = ma$ (N)

The force exerted on a mass by gravity can be determined for a mass of 1 kg, as acceleration caused by **standard gravity** on Earth is 9.81 m/s^2, so:

$$1 \text{ kg} \times 9.81 \text{ m/s}^2 = 9.81 \text{ newtons}$$

This can also be described as the weight of an object.

EXAMPLES
Calculate how much force is needed to overcome gravity and lift a 500 kg object:

$F = 500 \times 9.81 = 4905$ N

Ignoring the downward force of gravity, if the above mass of 500 kg is accelerated horizontally at 1 m/s^2, the force working horizontally on the mass is:

$F = 500 \times 1 = 500$ N

Work and energy

If force is applied to a body and that results in movement, then work has been done. This applies to forces that lift or push objects or twist them.

KEY TERM

Standard gravity: the average gravity on Earth. (The actual value can vary depending on where you are.)

KEY FACT

On Earth, acceleration due to gravity is 9.81 m/s^2.

When a force moves an object in the same direction as the force exerted, the work done is equal to the distance moved multiplied by the force exerted:

$$\text{work} = \text{force} \times \text{distance}$$

Or, to include the values used to determine force:

$$\text{work} = \text{mass} \times \text{gravity} \times \text{distance}$$

Mechanical work is measured in joules (J). (Newton-metres (Nm) can be used for mechanical work, but are also used as a measurement for torque.) Other units of work or energy, which are not SI units but commonly used for specific applications, include:

- kilowatt hour (kWh), used to measure electrical energy by electricity supply companies
- calorie, often used as a measure of food energy
- BTU (British Thermal Unit), which is often used for heat source applications such as burning gas.

EXAMPLE

If a mass of 100 kg is lifted 10 m, calculate the work done.

Force = weight = mass × gravity = 100 × 9.81 = 981 newtons = 981 N

Work done = force × distance = 981 × 10 = 9810 joules = 9810 J

If the same mass is doubled, the work done is:

Force = weight = mass × gravity = 200 × 9.81 = 1962 N or 1.962 kN

Work done = force × distance = 1962 × 10 = 19620 joules

Energy can exist in many forms but is categorised in two main groups:

- kinetic energy, which is energy of motion, such as a rotating machine
- potential energy, which is 'stored' energy due to gravity or in a spring.

For example, the potential energy of gravity keeps a mass on the ground. If the mass is to be raised, then a machine uses kinetic energy. If the input of kinetic energy ceases, the potential energy tries to bring the mass back down to the ground.

Power

Power is defined as the rate of doing work – that is, work done divided by the time taken to carry out that work. The unit of power is joules per second (J/s), which is equivalent to a watt.

$$\text{average power} = \frac{\text{work done}}{\text{time taken}} \text{ (measured in joules per second or watts)}$$

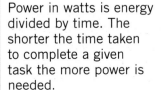

EXAMPLE

The output (mechanical) power required for a motor to raise a mass of 1000 kg to a height of 5 m above the ground in 1 minute is calculated as:

$$\text{power} = \frac{\text{mass} \times \text{gravity} \times \text{distance}}{\text{time}}$$

So:

$$\text{power} = \frac{1000 \times 9.81 \times 5}{60} = 817.5 \text{ watts}$$

If the same motor raised the same load in 10 seconds, the output power required by the motor would be:

$$\text{power} = \frac{\text{mass} \times \text{gravity} \times \text{distance}}{\text{time}}$$

So:

$$\text{power} = \frac{1000 \times 9.81 \times 5}{10} = 4905 \text{ watts}$$

Although the same amount of energy is used, no matter how quickly the task is carried out, the power required to do the work more quickly is increased due to the shorter time.

Calculating the efficiency of machines

The law of conservation of energy states that energy cannot be created or destroyed. However, when energy is changed from one format to another, the energy does not fully transfer from one type to another. The remaining energy is converted into other forms, such as noise and heat, which are common causes of loss during energy transfer. Such losses, or wasted energy, are common in any mechanical process.

The **efficiency** of a mechanical system can be defined as the ratio of output power compared to the input power and is rated as a percentage. This can be expressed as:

$$\% \text{ efficiency} = \frac{\text{output power}}{\text{input power}} \times 100$$

It is more common for efficiency to be expressed in terms of power rather than energy although energy could substitute power in this formula.

EXAMPLE

If a 200 kW output machine has an input power of 220 kW, the machine efficiency is:

$$\% \text{ efficiency} = \frac{200}{220} \times 100 = 90.9\%$$

KEY TERM

Efficiency: the amount of useful work by a machine compared to the power supplied, due to losses and expressed as a percentage.

INDUSTRY TIP

The losses in a machine are copper losses in the windings, bearing loss (friction), iron losses in the rotor, and frame and windage losses as the rotor rubs against the surrounding air. There will also be brush losses in machines with brushes.

ACTIVITY

A 4 kW electric motor drives a hoist. If the motor is 75% efficient and the hoist is 55% efficient, what is the output power of the hoist?

The passage of an electric current represents a flow of power or energy. When current flows in a circuit, power loss occurs in the conductors due to the conductor resistance. This manifests itself as heat dissipation and voltage drop.

Power in basic electrical circuits

Like mechanical power, electrical power is also measured in watts and is also a measure of energy used over a period of time. Before you explore electrical power, you must be able to understand basic electrical principles. Therefore electrical power will be covered in greater detail on pages 102–104.

4 BASIC ELECTRICAL THEORY: THE RELATIONSHIP BETWEEN RESISTANCE, RESISTIVITY, VOLTAGE, CURRENT AND POWER

The basic principles of electricity have been studied for centuries and what is now common electrical theory was once groundbreaking new discoveries. You need to understand the basic principles, including atomic composition, in order to work safely with electricity, magnetism and electrochemical reactions, and to progress in the industry.

Atomic theory

In order to understand where electricity comes from and what it is, it is necessary to understand a small amount of **atomic theory**.

Atoms are very small particles that are sometimes arranged as molecules. An atom is not solid but is made up of smaller particles, separated by space. The centre of an atom is the nucleus, which is made up of various particles including protons and neutrons. Protons are positively charged and neutrons have no charge. Orbiting the atom are the **electrons**, which are negatively charged.

The atoms that make up different materials have different numbers of electrons. In the steady state an atom has equal numbers of protons and electrons, and this leaves the atom electrically neutral.

Atoms in solids and liquids are more tightly bound together than those in gases. The diagram (left) shows the simplified structure of copper, which is often used to conduct electricity. The representation is two-dimensional when in fact the actual atom is three-dimensional. Where there are two or more electrons orbiting a nucleus, their orbiting paths are known as shells. The electron paths (shells) form an elliptical orbit.

Reaction of atoms

Different atoms have different numbers of electrons. Copper has 29 electrons and 29 protons. The outer shell is weakly held in orbit and can break free, causing random movement of the outer electron among other copper atoms.

KEY TERMS

Atomic theory: the study of atoms and electrons (also referred to as electron theory).

Electron: the orbiting part of an atom, which is negatively charged.

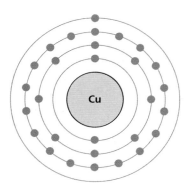

▲ Figure 2.27 A simplified copper atom structure with electrons

KEY FACT

Atoms are bound by an electrical force, whereas molecules are bound by a chemical force.

The loss of an electron causes an atom to become positively charged. It is known as a positive ion. Positive ions attract electrons, causing electron movement. Negative ions have more electrons orbiting them than protons in the nucleus.

The movement of electrons throughout a material is random but, by the laws of electric charge, like charges repel and unlike charges attract.

Figure 2.28 shows the random movement of electrons in a conducting material. The inset shows the electrons orbiting the positively charged nucleus. Electrons in an outer shell are released, as the force of attraction from the nucleus is weak, and the electron moves toward the next nucleus.

ACTIVITY

Copper has 29 electrons and 29 protons. Using textbooks or the internet, find how many there are for:

- carbon
- aluminium
- silicon
- gold.

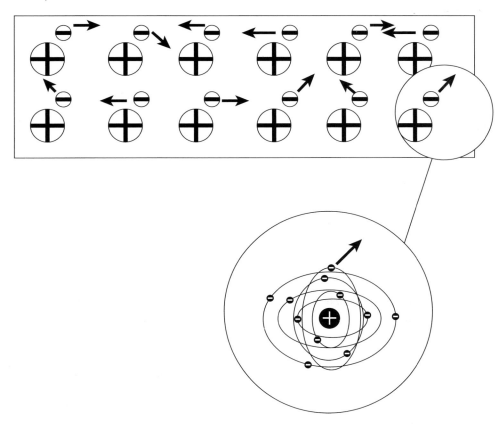

▲ Figure 2.28 Random movement of electrons in a conducting material

Flow of electrons

Random free electrons can be configured if the conducting material is connected to a battery. The free electrons are attracted to the positive plate and repelled by the negative plate. This causes the electrons to drift, in a conducting material, from the negative terminal of the battery to the positive terminal. As positive ions are unable to drift in solids, every time an electron leaves the negative terminal, one enters the positive terminal. This flow of electrons is electric current.

KEY FACT

Electrons orbit the nucleus of an atom. Electricity is the flow of these electrons as they move from atom to atom. Electron is the Greek word for amber. Amber is a material that is easily electrified by static. Rubbing wool on amber can charge the amber, which can then release an electric charge when held against another material or person.

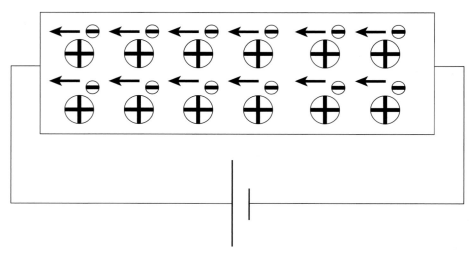

▲ Figure 2.29 Electrons are attracted in one direction when a source of energy is connected to the material

KEY TERM

Closed circuit: a complete circuit connected to a source of energy. If the circuit contains a switch and the switch is switched off, it becomes an open circuit.

INDUSTRY TIP

Conventional current direction is actually opposite to actual current direction. Current flows from negative to positive but, in our industry, we still refer to conventional current flow.

The flow of current in one direction is called direct current (DC), which is the main current form referred to in this chapter.

INDUSTRY TIP

These equations may look strange, but do not let them put you off. The letters stand for numbers.

In order for electric current to flow, there must be a **closed circuit**. Once the circuit is opened, the drift of electrons is immediately stopped and the current flow ceases.

Current direction

Electric current flows from a negative terminal to a positive terminal. However, before atoms and electrons were understood, scientists believed that electricity was a fluid and flowed from positive to negative. This is known as conventional current direction and, although it is now understood that electrons flow from negative to positive, we still refer to conventional current direction as being positive to negative.

How many electrons make one ampere?

The flow of electrons in one direction is known as charge, which is measured in coulombs (C).

As electron flow is electrical current and current is measured in amperes (A), how is one converted to the other?

French physicist Charles Coulomb (1736–1806) determined that 1 coulomb of charge is equal to 6.24×10^{18} electrons. That is equal to 6 240 000 000 000 000 000 electrons! So if that many electrons flowed through a material such as copper and past an electron counter, that would be equal to 1 coulomb. He also determined that if the drift of electrons was at a rate of 1 coulomb per second, the resulting current would be 1 ampere.

Therefore a current of 1 ampere flowing indicates a charge of 1 coulomb per second, giving:

$$Q = It \text{ or } I = \frac{Q}{t}$$

where:

Q = charge transferred, in coulombs (C)

I = current, in amperes (A)

t = time, in seconds (s).

EXAMPLE

If a total charge to be transferred is 750 C in one minute, calculate the current flow.

Using the formula $Q = It$, calculate the current:

$$I = \frac{Q}{t}$$

Since 1 minute = 60 seconds (s): $\frac{750}{60} = 12.5\ A$

The current flow is therefore 12.5 A for 60 seconds to give the total charge of 750 C.

If a current of 25 A was to flow for 2 minutes in the above circuit arrangement, the total charge would be calculated like this.

Since 2 minutes = 120 s and $Q = It$:

$$Q = It = 25 \times 120 = 3000\ C$$

Therefore the total charge would be 3000 coulombs.

KEY FACT

Where a formula shows two symbols together with no mathematical symbol, it means they must be multiplied. So $Q = It$ simply means $Q = I \times t$.

ACTIVITY

Remember that the SI unit of time is the second, not the minute, hour or day. Use the internet to find the definition of the second.

Insulators and conductors

As has been described previously, the movement of free electrons constitutes the flow of electric current and since the atomic structure varies from material to material, some will allow electron movement better than others when an external voltage is applied. Where the freedom of electrons to move is high, the material will act as a good conductor of electricity.

Examples of **conductors** are:

- aluminium and copper (used in cables and overhead line conductors)
- brass (used in plug pins and terminals)
- carbon (motor brushes)
- mercury (discharge lamps and special contacts)
- sodium (discharge lamps)
- tungsten (lamp filaments).

However, where the material's atomic structure is such that there is minimal electron movement, there will be negligible current flow and the material will act as an insulator.

Examples of **insulators** are:

- thermoplastics and polyethylene (cable insulation)
- glass and porcelain (overhead line conductor support insulators)
- rubber (mats, gloves and shrouding for live working).

KEY TERMS

Conductors: materials that allow the movement of electrons and therefore current.

Insulators: materials that resist the flow of electrons and therefore current.

It must be stressed, however, that the level of insulation afforded by an insulator can be severely reduced by:

- damage (cracks, splits, etc.)
- deterioration (cracks, splits, etc. due to ageing)
- contamination (water, salt spray, chemicals, etc.).

Cable components and types

Electric cables used for electrical installations in industrial, commercial and domestic situations come in a wide range of sizes, materials and types. Electric cables used for the long-distance transmission of electrical energy (400 kV and 275 kV) are normally buried in the ground or suspended on towers or pylons and the cables used for the more local distribution of electrical energy around the country (132 kV, 66 kV, 33 kV and 11 kV) are buried in the ground or suspended on towers or pylons or on wooden poles.

Cables generally consist of three major components:

- the current-carrying material (conductors)
- electrical insulation (normally colour or number coded for identification)
- a protective outer covering called a sheath (this is not present on some single-core cables).

The makeup of individual cables varies according to the application for which they are to be used. The construction and material are determined by three main factors:

- working voltage, which determines the thickness of the insulation
- current-carrying capacity, which determines the cross-sectional area of the conductor(s)
- environmental conditions – such as requirement for mechanical protection, temperature, water and chemical protection – which ultimately determine the form and composition of the outer protective sheath.

The current-carrying conductors of an electricity cable are normally made of copper or sometimes aluminium, either of stranded or solid construction.

Resistance and resistivity

As we have seen, the flow of electric current in a material is related to its atomic structure and the freedom of electrons to move. Where the freedom of electrons to move is high, the material will act as a good conductor of electricity when an external voltage if applied. However, where a material's atomic structure is such that there is minimal electron movement, there will be **negligible** current flow and the material will act as an insulator.

Resistance in an electrical circuit is encountered in two forms: continuity resistance (in conductors) and insulation resistance (in insulators).

Continuity resistance of a conductor

This is the end-to-end resistance of an electrical conductor or the resistance between two points in an electrical circuit. In the case of an electrical conductor, the continuity resistance needs to be as low as possible to enable the maximum amount of current to flow through the conductor. Continuity resistance is generally measured on an ohmmeter. Measured values can be very small and therefore may be quoted in milliohms (mΩ).

The resistance of a conductor is based on a number of variable factors:

- type of conductor
- physical dimensions
- temperature.

In order to understand the resistance of a conductor, it is necessary to be aware of these factors.

During the design and any inspection and testing of an electrical installation it is always important to keep these factors in mind.

Designers will factor in temperature increases for conductors because the temperature of a conductor under load will increase, resulting in increased conductor resistance.

Inspectors, while testing circuit resistances under no-load conditions, need to understand that the actual resistance will be greater when the circuit is carrying a full load current. They will need to factor in the temperature increases.

Effects of physical dimensions: length

If a cube of conducting material, with a resistance between two opposite faces of R ohms, is joined in a line of equal cubes, then the overall resistance is that of all the cubes added together. This arrangement is known as resistances in series or a series circuit, for example:

$R_{t_{series}} = R_1 + R_2 + R_3 + R_4 + R_5 = 5 \times R = 5R$, as shown in the diagram.

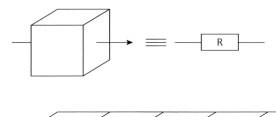

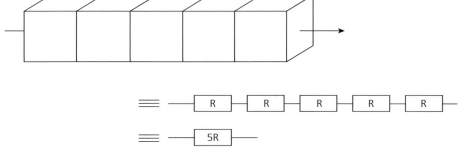

▲ Figure 2.30 The effect of increasing length on resistance

ACTIVITY

Practise doing series and parallel combination calculations, finding the unknown in a group of resistors, as well as the total resistance.

EXAMPLE

If there are seven pieces of conducting material (resistors) connected in a series arrangement, the formula would be:

$$R_{t_{series}} = R_1 + R_2 + R_3 + R_4 + R_5 + R_6 + R_7$$

If the value of each resistor is the same, the total is $7 \times R = 7R$.

If $R = 2\ \Omega$, $7R = 14\ \Omega$.

If the resistors have different values, they are still added together.

If $R_1 = 2\ \Omega$, $R_2 = 3\ \Omega$, $R_3 = 4\ \Omega$, $R_4 = 6\ \Omega$, $R_5 = 9\ \Omega$, $R_6 = 3\ \Omega$, $R_7 = 1\ \Omega$, then:

$$R_{t_{series}} = 2 + 3 + 4 + 6 + 9 + 3 + 1 = 28\ \Omega$$

As conductors in cables are generally the same diameter throughout their length, each metre of cable is like one of the cubes in the diagram above. The longer the conductor, the higher the resistance.

EXAMPLE

If a conductor in a cable has a resistance of 0.01 Ω per metre (Ω/m), calculate the resistance of 120 m.

As the total resistance $(R_t) = \Omega/m \times length$, then $0.01 \times 120 = 1.2\ \Omega$

Effects of physical dimensions: cross-sectional area

If a pipe had a cross-sectional area of 150 mm² and water was poured into it, the flow of water would be restricted by the cross-sectional area of the pipe as only so much water could pass through it at one time. If the pipe diameter was increased so that the cross-sectional area was 300 mm², twice the amount of water would be able to flow. Current flowing through a cable is much the same. The smaller the cross-sectional area of a conductor, the greater the resistance. If the cross-sectional area is increased, the resistance decreases proportionally. This effect is the same as having resistances in parallel, which is covered on page 101.

By increasing the cross-sectional area by a factor of 4, the resistance is reduced accordingly.

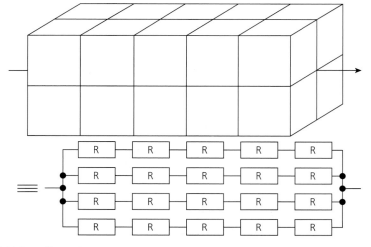

▲ Figure 2.31 The effect of increasing the cross-sectional area

EXAMPLE

If the resistance of 100 m of a particular cable is 0.3 Ω, calculate the resistance for 1 km of the same cable, assuming the temperature remains constant.

1 km = 1000 m

Therefore, if the resistance of 100 m = 0.3 Ω:

the resistance of 1000 m = 10 × the resistance of 100 m = 10 × 0.3 = 3 Ω

If the cross-sectional area of the cable is doubled in size, the resistance would halve, so:

$$\frac{3 \, \Omega}{2} = 1.5 \, \Omega$$

KEY FACT

If something doubles in size it is two times (× 2) as large. If it halves in size it is divided by two (÷ 2).

Cable manufacturers' data provides resistance values for cable conductors according to the conductor material and cross-sectional area, usually expressed in mΩ/m at a specified temperature. If the end-to-end cable length is known, the end-to-end resistance can then be calculated.

Insulation resistance

This is the resistance measured across the electrical insulation surrounding a conductor, from the outside of the cable through to the conductor or through the insulation separating conductors. In this case the resistance needs to be very high to prevent current leakage from the circuit or a short circuit between adjacent circuit conductors. Insulation resistance is measured on an insulation resistance tester, giving readings in megohms (MΩ). Older instruments provided with a scale and pointer display usually have the scale marked with a range of 0–∞ MΩ. Since the measured value will invariably be very high, the pointer is likely to swing full scale to the ∞ (infinity) mark. This does not mean that the insulation resistance is infinity! The measurement is beyond the scale range of that particular instrument and, to be correct, the reading should be recorded as greater than the highest marked scale value or maximum of scale range if known – for example, > 99 MΩ (greater than 99 MΩ).

INDUSTRY TIP

'Megger' is a trade name and should not be used instead of 'insulation resistance tester'.

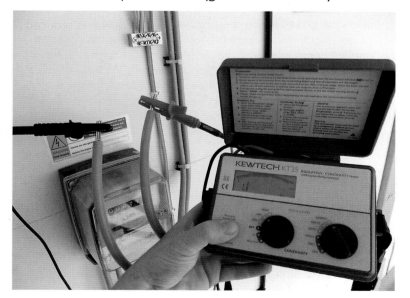

▲ Figure 2.32 Testing between live conductors

ACTIVITY

It would be very difficult to measure the resistance across the opposite faces of a one-millimetre cube of copper. It would be difficult to hold and varying contact pressure would vary the resistance. What would be the mass of a one-metre cube of copper?

INDUSTRY TIP

Silver is a better conductor than copper but not as strong and obviously much more expensive.

INDUSTRY TIP

Values of resistivity for materials used as conductors are always based on resistance at 20 °C.

ACTIVITY

Put these conductors in ascending order of resistivity: copper, steel, aluminium, brass, silver, gold.

A high reading (typically > 1 MΩ) indicates a good standard of electrical insulation around live parts. This will result in negligible current leakage and so the circuit is described as 'healthy'. A low reading (e.g. less than 0.5 MΩ) indicates deterioration in the effectiveness of the insulation, leading to the possibility of a fault in the form of a short circuit across the circuit conductors.

Resistivity

The resistance of a conductor material is normally very low and is expressed in microhms (μΩ). The specific resistance or **resistivity** of a material is expressed in the format μΩm. For example, the resistivity of copper is 0.0172 μΩm. This is the resistance of a one-metre cube of copper.

The specific resistance or resistivity of a material is represented by the Greek letter rho (ρ). Therefore the resistance of a particular conductor can be calculated as:

$$R = \frac{\rho l}{A}$$

where:

R = conductor resistance

ρ = cable resistivity in ohm-metres (Ωm)

l = cable length in metres (m)

A = cable cross-sectional area in square metres (m^2).

EXAMPLE

To calculate the resistance of 1000 m of 16 mm^2 annealed copper cable, where ρ = 0.0172 μΩm (or 0.0172 × 10^{-6} Ωm) and A = 16 mm^2 (or 16 × 10^{-6} m):

$$\text{use } R = \frac{\rho l}{A} = \frac{0.0172 \times 10^{-6} \times 1000}{16 \times 10^{-6}} = \frac{17.2 \times 10^{-6}}{16 \times 10^{-6}} = \frac{172.2}{16} = 1.075\,\Omega$$

Traditionally, in formulae the base SI units are always used (metres not millimetres, and ohms not microhms). Where the top line of a formula contains a 10^{-6} and the bottom line contains a 10^{-6} these values cancel each other out. This means that this formula is always an exception to the rule. As long as the value of resistivity used is microhm-metres and the cross-sectional area is in square millimetres, the two values can be input directly as the factors of 10^{-6} applied to both quantities always cancel out. So:

$$\frac{0.0172 \times 1000}{16} = \frac{17.2}{16} = 1.075\,\Omega$$

Effects of temperature

Generally, as the temperature of a material increases, so too the resistance of that material increases.

This variable is determined using the temperature coefficient of resistance, represented by the Greek letter alpha (α). For example, the coefficient for copper at 0 °C is 0.0043 Ω/°C and the resistance at 0 °C is represented by R_O. The coefficient for copper at 20 °C is 0.003 96 Ω/°C.

$$R_{t_1} = R_{t_2}(1 + \alpha T)$$

where:

R_{t_1} = resistance at the new temperature, t_1

R_{t_2} = resistance at a given temperature, t_2

α = temperature coefficient

T = temperature change.

EXAMPLE

If the resistance of 100 m of 2.5 mm^2 annealed copper cable is 0.5375 Ω at 20 °C and the coefficient α for copper at 20 °C is 0.00396 Ω/°C, its resistance at 50 °C is calculated as:

$$R_{50} = R_{20}(1 + \alpha T) = 0.5375(1 + 0.00396(50 - 20))\Omega$$

$$= 0.5375 \times 1.1188 \, \Omega$$

$$= 0.60 \, \Omega$$

INDUSTRY TIP

Not all insulators have a maximum temperature of 70 °C. Some are at 90 °C or even 105 °C.

In electrical installations, it is important to be able to calculate conductor resistance at the standard room temperature of 20 °C as well as at the conductor's working temperature, which may be 70 °C. In many publications used by electricians, tables used to calculate the resistance of circuits are all based on resistivity and factors used to adjust temperatures are based on temperature coefficients.

Many of these tables give factors for correcting resistance to a given temperature. On closer inspection of these tables, it can be seen that the resistance of a copper conductor changes by 2% for every five-degree change in temperature.

ACTIVITY

The IET On-Site Guide provides tables showing resistance values of conductors together with factors to adjust temperature. Identify the tables and study them.

If a conductor has a resistance of 0.5 Ω at 20 °C and the temperature is increased by 15 degrees, the resistance will increase by 6%. So the new resistance will be:

$$0.5 \, \Omega \times 1.06 = 0.53 \, \Omega$$

Multiplying the value of the resistance by 1.06 will increase it by 6%. To reduce the value of the resistance by 6%, it must be multiplied by 0.94.

INDUSTRY TIP

Conductor resistance increases as temperature increases but insulation resistance decreases as temperature increases.

EXAMPLE

A 10 mm^2 conductor is 27 m in length and is made of a material with a resistivity of 0.0172 $\mu\Omega$m at 20 °C.

Calculate the resistance of the conductor at 70 °C.

$$\text{Resistance at 20 °C} = \frac{\rho l}{A} = \frac{0.0172 \times 27}{10} = 0.046 \, \Omega$$

If the temperature increases to 70 °C, the change is 50 degrees. At 2% for every five degrees, this is a 20% increase, so apply a multiplier of 1.2.

$$0.046 \, \Omega \times 1.2 = 0.056 \, \Omega$$

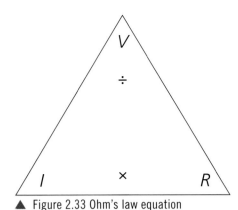

▲ Figure 2.33 Ohm's law equation

Applying Ohm's law

One of the most used electrical formulae is Ohm's law. It states that a potential difference across a conductor is proportional to the current passed through it.

This proportionality is equal to the resistance (R) of the conductor. Therefore $V = I \times R$, which is generally written as $V = IR$.

This gives the potential difference (V), given the current and resistance.

To calculate the resistance, divide both sides of the equation by I to get:

$$R = \frac{V}{I}$$

You can remember the three variations of the Ohm's law formula by using a triangle (Figure 2.33). Covering the value that needs to be found leaves the correct calculation visible. So if resistance is needed when current and voltage are known, cover resistance and V over I (V divided by I) is left. Likewise, to determine V, cover it and you are left with $I \times R$.

Series circuits

INDUSTRY TIP

Resistors in series are arranged one after the other in a string.

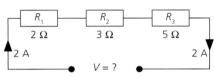

▲ Figure 2.34 Applying Ohm's law to circuits in series

Before considering Ohm's law to determine values of a series circuit, consider the characteristics of resistors in series. The total resistance of resistors in series is found by simply adding them together, so:

$$R_{total} = R_1 + R_2 + R_3 \dots$$

As the current has to flow through all the resistors in the series circuit, the current is the same through all, i.e. it is constant.

ACTIVITY

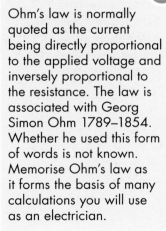

Ohm's law is normally quoted as the current being directly proportional to the applied voltage and inversely proportional to the resistance. The law is associated with Georg Simon Ohm 1789–1854. Whether he used this form of words is not known. Memorise Ohm's law as it forms the basis of many calculations you will use as an electrician.

EXAMPLE

If a current of 2 A is passing through each of the resistors in a series circuit, the voltage at the terminals of the supply is calculated as follows.

If $R_1 = 2\,\Omega$, $R_2 = 3\,\Omega$ and $R_3 = 5\,\Omega$

then:

$$R_{total} = R_1 + R_2 + R_3 = 2 + 3 + 5 = 10\,\Omega$$

To calculate the voltage across the supply terminals:

$$V = IR = 2\,A \times 10\,\Omega = 20\,V$$

As the current in a series circuit is constant, to calculate the potential difference across each resistor, assuming no resistance in any of the connections and no internal impedance in the supply, use Ohm's law on each resistor, where V_1 is the potential difference across R_1 and so on.

$$V_1 = IR_1 = 2 \times 2 = 4\,V$$
$$V_2 = IR_2 = 2 \times 3 = 6\,V$$
$$V_3 = IR_3 = 2 \times 5 = 10\,V$$

This adds up to the 20 V across the terminals, showing that the effect of resistors in series is to form potential dividers.

Try the following examples of resistance in series. Remember these key points.

- The total resistance is equal to all the resistances added together.
- The current is constant in a series circuit.
- The voltages across each resistance, when added together, will equal the supply voltage for the circuit.

ACTIVITY

1 Determine, for the circuit in Figure 2.35:

  a the total circuit resistance

  b the total circuit current (I_s).

2 Determine, for the circuit in Figure 2.36:

  a the total circuit resistance

  b the total circuit current (I_s)

  c the voltage drop across resistor 3 as indicated by the voltage meter V_3.

3 Determine, for the circuit in Figure 2.37:

  a the total circuit resistance

  b the resistance of R_2

  c the voltage drop across resistor 1 as indicated by the voltage meter V_1.

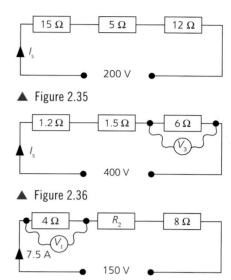

▲ Figure 2.35

▲ Figure 2.36

▲ Figure 2.37

Parallel circuits

With parallel circuits, the rules change. The total resistance is now found by using **reciprocals**, as shown.

$$\frac{1}{R_{total}} = \frac{1}{R_1} + \frac{1}{R_2} + \frac{1}{R_3} + \dots$$

In parallel circuits, the overall (total) resistance will be lower than the lowest resistance in the circuit. This is because the current can flow through the lowest resistance path as well as the other paths, effectively lowering the resistance.

Unlike series circuits, in parallel circuits voltage becomes the constant, with current varying across each resistor. The total circuit current is equal to the value of current flowing through each resistor, all added together.

EXAMPLE

If the same resistors ($R_1 = 2\ \Omega$, $R_2 = 3\ \Omega$ and $R_3 = 5\ \Omega$) are applied in parallel and the current flowing through the source is 2 A, calculate:

- the voltage at the terminals
- the voltage drop across each terminal of the resistors
- the current passing through each resistor.

→

ACTIVITY

In a parallel circuit, the greater the cross-sectional area, the lower the resistance. The reciprocal of a resistance, $1/R$, is the conductance (ability to conduct) and conductances can be added together. When the total conductance is found, the reciprocal of this will be the total resistance. The units of conductance are siemens (S).

Memorise the formula for the total resistance of parallel circuits and use it in calculations.

KEY TERM

Reciprocal: the reciprocal of any number is that number divided into 1. So the reciprocal of 10 is 1 divided by 10 which equals 0.1.

INDUSTRY TIP

Where there are two resistors in parallel, product over sum can be used. Where there are three resistors, find the parallel value of two and then use that answer with the third.

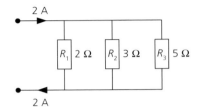

▲ Figure 2.38 Applying Ohm's law to circuits in parallel

KEY TERM

Common denominator: A denominator that can be divided exactly by all of the denominators in the question.

INDUSTRY TIP

Calculators do not all work in the same way. Therefore, any calculator key sequence suggested is only an example of what would be required on a standard scientific (non-programmable) calculator. If the key sequence does not give the expected result on your calculator, either ask your tutor for advice or refer to the manual for your calculator.

As the value of current through the source is known, calculate the voltage of the source by determining the total resistance value.

$$\frac{1}{R_{total}} = \frac{1}{R_1} + \frac{1}{R_2} + \frac{1}{R_3} \cdots$$

This can be approached in several ways. First, find a **common denominator**, in this case 30. Then work out how many times each of the original denominators divides into 30.

$$\frac{1}{R_{total}} = \frac{1}{2} + \frac{1}{3} + \frac{1}{5} = \frac{15 + 10 + 6}{30} = \frac{31}{30}$$

This gives the reciprocal of the total resistance, so turn both sides of the equations upside down.

$$R_{total} = \frac{30}{31} = 0.967\Omega$$

Alternatively, you can use the button marked on a calculator as ▪x⁻¹ and apply: ▪2 ▪x⁻¹ ▪+ ▪3 ▪x⁻¹ ▪+ ▪5 ▪x⁻¹ ▪= ▪x⁻¹ ▪= and the answer should be 0.967 Ω. Always remember to push the final ▪x⁻¹ to get the true value.

To calculate the voltage across the supply terminals, use:

$$V = IR = 2 \times 0.967 = 1.934 \text{ V (2 V)}$$

As the voltage in a parallel circuit is common, the voltage drop across each resistor is 2 V.

To calculate the current through each resistor, where I_1 is the current through resistor R_1 etc., apply the appropriate form of Ohm's law.

$$I_1 = \frac{V}{R_1} = \frac{1.934}{2} = 0.967 \text{ A}$$

$$I_2 = \frac{V}{R_2} = \frac{1.934}{3} = 0.645 \text{ A}$$

$$I_3 = \frac{V}{R_3} = \frac{1.934}{5} = 0.387 \text{ A}$$

The total current flowing through the source is 2 A.

The passage of an electric current represents a flow of power or energy. When current flows in a circuit, power loss occurs in the conductors due to the conductor resistance. This results in heat dissipation and voltage drop.

Power in basic electrical circuits

It is known that if a potential difference of 1 volt exists between two points, then 1 joule of energy is used in moving 1 coulomb of charge between the points. Therefore:

1 joule = 1 coulomb × 1 volt or $W = QV$

We also know that power is energy used over a period of time. In electrical circuits we also determine power using the amount of potential difference and current. So:

$$P = VI$$

Where:

P is electric power, in watts (W)
I is the current, in amperes (A)
V is the potential difference, in volts (V).

Also, as voltage can be determined using Ohm's law, by multiplying the voltage and resistance, then:

$$P = (IR)I = I^2R$$

Power loss or consumption is proportional to the square of the current flow, I^2. Thus, assuming a constant resistance in a circuit, if the current doubles (due to a corresponding increase in voltage) there will be a four-fold increase in power.

Also as:

$$I = \frac{V}{R} \text{ and } P = V \times I$$

Then:

$$P = \frac{V \times V}{R} = \frac{V^2}{R}$$

Note that these formulae are not complete for AC circuits as a power factor needs to be taken into account. Power factors will be covered in detail in Book 2, Chapter 2.

So, using the derived formula, the following calculations are possible.

INDUSTRY TIP

In a series circuit the current is the same in each resistor. In a parallel circuit the voltage is common to each resistor. This is an important principle that you should commit to memory.

EXAMPLES

1 If a resistor of 10 kΩ is connected to a 100 V DC supply, the power dissipated in the resistor is calculated using:

$$P = \frac{V^2}{R} = \frac{100 \times 100}{10\,000}$$

$$= \frac{10\,000}{10\,000} = 1\,W$$

2 Calculate the working (hot) resistance of a 60 W 230 V lamp, using:

$$P = \frac{V^2}{R}$$

ACTIVITY

Calculate the resistance of a 3 kW electric fire connected to a 230 V supply.

Therefore:

$$R = \frac{V^2}{P} = \frac{230 \times 230}{60}$$

$$= \frac{52\,900}{60} = 881.67\,\Omega$$

3 To determine the power dissipated in a resistor when a current of 100 A passes through and voltage 100 V is applied to the circuit:

$$P = VI = 100 \times 100 = 10\,\text{kW}$$

Calculate the value of the resistance in the same circuit:

$$P = \frac{V^2}{R}$$

Therefore:

$$R = \frac{V^2}{P} = \frac{100 \times 100}{10\,000} = \frac{10\,000}{10\,000} = 1\,\Omega$$

Voltage drop

The flow of electric current through a conductor results in a drop in electrical pressure (voltage), referred to as voltage drop or volt drop, due to the resistance to the current flow presented by the conductor resistance. This loss of voltage represents a loss of energy, which is reflected in the generation of heat.

The voltage drop is determined by the amount of current flowing in a conductor and the resistance of that conductor. It is calculated from the product of the current flowing, measured in amperes (A), and the conductor resistance, measured in ohms (Ω), which is simply Ohm's law.

Voltage drop	=	Current	$\times$	Resistance
V	=	I	$\times$	R

EXAMPLE

If a circuit had a total conductor resistance of 0.5 Ω and the load at the end of the circuit had a current demand of 15 A, the voltage drop in the circuit would be:

$$0.5\,\Omega \times 15\,\text{A} = 7.5\,\text{V}$$

So if the supply voltage at the origin of the circuit was 230 V, the voltage at the load would be:

$$230\,\text{V} - 7.5\,\text{V} = 222.5\,\text{V}$$

If voltage drop is excessive, there may not be enough voltage at the end of the circuit for the load to operate correctly. In order to reduce the amount of voltage drop, the cable resistance must be reduced. This can be done by increasing the cross-sectional area of the circuit conductor.

Effects of electric current

The three main effects of electrical current (similar to sources of electricity) are:

- thermal (heating)
- chemical
- magnetism.

In this section we deal with the thermal and chemical effects. Magnetism is covered in detail in the following section (page 106).

Thermal (heating)

When current flows in a wire, apart from the flow of electrons, there is a thermal effect; the wire starts to heat up. The amount it heats up depends on factors such as the cross-sectional area of the wire, the amount of current flowing and the material that the wire is made of. The heating effect of electricity is used in electric fires and other heaters. Variations of this heat effect are used to make light from light-bulb (lamp) filaments, which give off large amounts of light as they glow white hot as a result of the current passing through the thin filament.

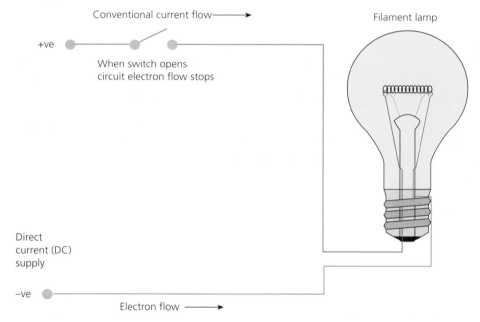

▲ Figure 2.39 An electric light circuit using a DC source such as a battery

The effect of current passing through a wire and producing heat is a major consideration when designing electrical installations and will be covered at length during your course. Current that produces heat can be useful in electrical installations, for example in:

- electric heating
- lighting
- cooking
- circuit or equipment protection devices such as fuses or circuit breakers
- monitoring equipment.

ACTIVITY

Early incandescent lamps used carbon filaments that were quite fragile. Unfortunately, carbon has a negative coefficient of resistance. This means that, as it gets hotter, the resistance goes down and so it will carry more current and become hotter still. This continues until the filament burns out. For this reason, carbon filaments were limited to small power ratings. What other material has a negative coefficient of resistance?

INDUSTRY TIP

Larger cables have a lower power loss than smaller cables carrying the same current. Although they save on power loss, larger cables cost more to buy and install.

There are also disadvantages, which include:

- circuit cables heating up, causing failure
- equipment getting too hot, causing danger
- energy loss.

A cable is designed to carry electricity from one place to another and is not supposed to heat up by any large amount. If it does heat up, it is using energy to do so which means less energy is available where it is required.

Chemical

KEY TERM

Electrolyte: a chemical solution that contains many ions. Examples include salty water and lemon juice. In major battery production, these may be alkaline or acid solutions, or gels.

When electric current is passed through an **electrolyte**, this causes basic chemical changes as ions are allowed to move to the positive electrode (anode), creating the process of electroplating or electrolysis. This process is used to coat material as, for example, in copper cladding on steel. If a copper-based solution were used, as shown in the diagram, the steel forming the negative electrode (cathode) would become coated with copper.

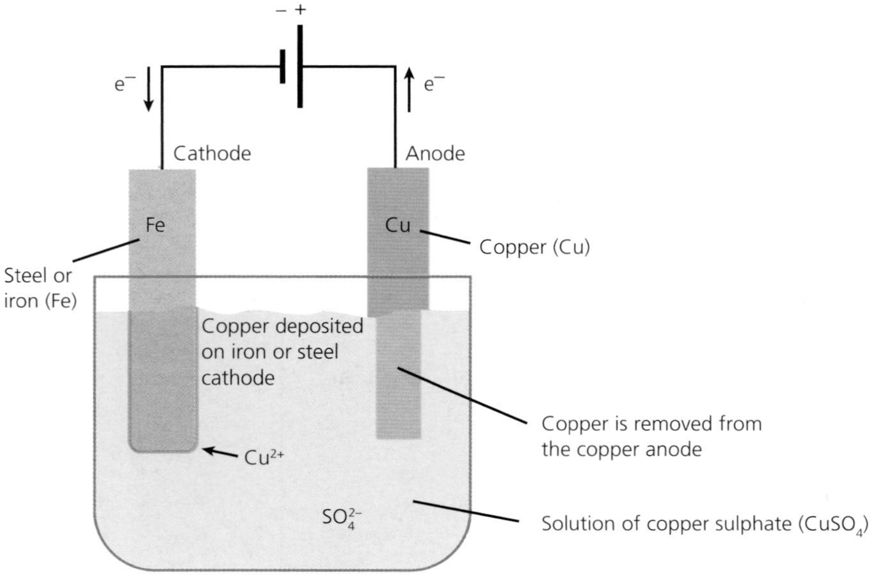

▲ Figure 2.40 The chemical process of electroplating copper onto an iron or steel object

5 MAGNETISM AND ELECTRICITY

Magnetism

Electromagnets are often thought of as large magnets, found in vehicle salvage yards, lifting scrap metal from one place to another. Although this is one example, electromagnets are also used in lots of everyday items. For example, simply pressing a doorbell, or the release unit on a controlled access door, activates or de-activates an electromagnet. Without electromagnets many everyday tasks would be far more difficult.

First, consider what a magnet is.

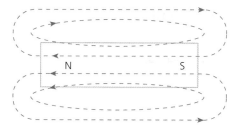

▲ Figure 2.41 The pattern of magnetic flux lines that pass through a magnet from south to north

The bar magnet in Figure 2.41 shows a magnet and its north and south poles. It also shows the pattern of the lines of magnetic flux that pass through the magnet from south to north, and also outside the magnet from north to south. These flux patterns can be seen when a piece of paper is put over a bar magnet and iron filings are sprinkled over the paper. When the paper is tapped, the iron filings form a pattern because they are drawn into the flux lines.

> **INDUSTRY TIP**
>
> Remember: like poles repel, unlike poles attract.

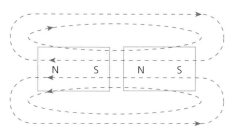

▲ Figure 2.42 Two magnets, showing that opposite charges attract

When two magnets are put together, with a north pole facing a south pole, the lines of flux move together in the same direction. This causes the magnets to attract, pulling together and forming one larger magnet. Opposites attract.

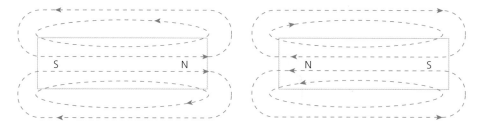

▲ Figure 2.43 Two magnets, showing that like charges repel

When two magnets are placed with the same poles together, the flux paths move against each other. The force of the magnetic flux causes the magnets to repel and move away from each other.

The planet we live on is a giant magnet with a magnetic field. People navigate around the world using this magnetic field by placing a small piece of iron on a pivot. Like the iron filings, this small piece of iron follows the flux direction. It is called a compass.

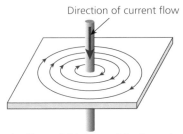

▲ Figure 2.44 Concentric rings of magnetic flux centre around the conductor

Direction of current flow

Magnetic flux patterns of current-carrying conductors

Experimentation with a compass needle or iron filings on a sheet of paper with a conductor passing through it shows that a magnetic field is created around a conductor when current flows through it. If the current is removed, the effect on the compass or iron filings disappears.

This effect occurs throughout the length of a conductor. However, the effect on iron filings on a sheet of paper shows a 'slice' of the field in the plane where the paper is at right angles to the conductor (Figure 2.44).

Current and field convention

It is usual to indicate current flow in a conductor because there is a three-dimensional relationship between current flow and magnetic field. Current flowing away from the viewer is shown with a cross, rather like an arrow or dart passing through a tube. Current flowing towards the viewer is shown as a large dot, like an arrow or dart point emerging from a hollow tube (Figure 2.45).

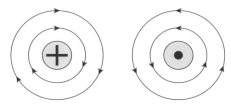

▲ Figure 2.45 Direction of magnetic field around a conductor (shown in cross-section): current flowing away from view (left) and current flowing into view (right)

The direction of the magnetic field (field rotation) of the concentric rings can be checked with a compass needle. When current flows away from the viewer, the magnetic field rotates clockwise. When current flows towards the viewer, the magnetic field rotates anticlockwise. The magnetic field rotates in the same way as a screw: clockwise to tighten the screw (forcing it away), anticlockwise to undo it (drawing it closer).

The strength of the magnetic flux is proportional to the current flowing through the conductor. The more current flowing, the stronger the magnetic field will be.

Placing two conductors together changes the effects. If two conductors are placed together in a conduit, for example, with the current flowing in opposite directions, there is a cancelling effect between the opposing magnetic fields, as long as the magnetic fields are of equal strength. This arrangement is therefore adopted in electrical installations. Magnetic fields can cause problems in electrical installations and therefore need to be cancelled and minimised as far as is reasonably practicable.

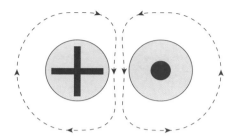

▲ Figure 2.46 Cancellation effect of opposing conductors

Where conductors are placed together, with the current flowing in the same direction, there is an additional effect. This is undesirable in electrical installations as the increase in the magnetic field will cause additional losses in the circuit and possibly electromagnetic compatibility issues.

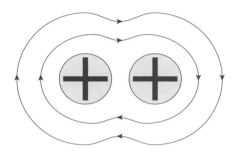

▲ Figure 2.47 Totalling effect of magnetic fields

Solenoids

The strength of a magnetic field is proportional to the current flowing through the conductor. Even with high currents passing through the conductors, the field produced is relatively weak, in terms of useful magnetism. To obtain a stronger magnetic field a number of conductors can be added by turning or winding the cable.

The most common form of this is the solenoid, which consists of one long insulated conductor wound to form a coil. The winding of the coil causes the magnetic fields to merge into a stronger field similar to that of a permanent bar magnet. The strength of the field depends on the current and the number of turns.

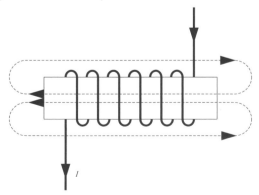

▲ Figure 2.48 A cable wound around a tube: the current at the top moves away from the viewer and the current at the bottom moves towards the viewer

INDUSTRY TIP

Remember that parallel conductors with currents flowing in opposite directions will push away from each other. Currents flowing in the same direction will cause the conductors to pull towards each other.

INDUSTRY TIP

Winding the conductor on an iron core will considerably increase the strength of the magnetic field.

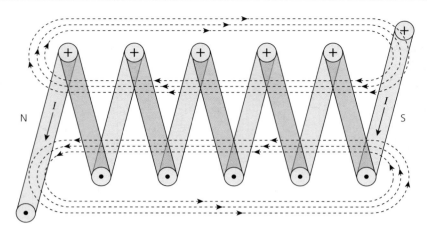

▲ Figure 2.49 Coiling produces a bar magnet effect

As a solenoid is the electrically powered equivalent of a bar magnet, its field strength is dependent on the current passing through it. The magnetic field can be switched on or off.

The polarity of a solenoid is determined by the current direction. Using the NS rule, the letter N and/or S can be drawn, following the current direction.

The arrow heads on the letters, as shown in this diagram, indicate the direction of the current flow.

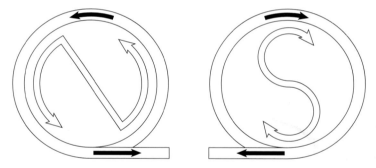

▲ Figure 2.50 Tracing the letters shows the direction of the magnetic field rotation

The polarity of a solenoid can also be determined using the right-hand grip method. If the fingers of the right hand follow the current flow direction, the thumb points to the north pole.

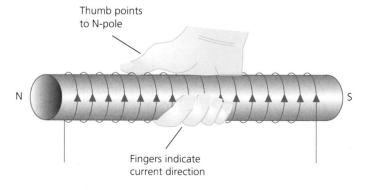

▲ Figure 2.51 Holding a solenoid in a right-hand grip indicates the direction of magnetic field

Units of magnetic flux

The unit of magnetic flux is the weber (pronounced 'veyber'), abbreviated to Wb. It is represented by the Greek letter phi (Φ). Magnetic flux is a measure of the quantity of magnetic flux, not a density.

Flux density (the amount of flux in a given area) is represented by the symbol β, which is measured in webers per square metre (Wb/m^2), called teslas (T). One weber of flux spread evenly across a square metre of area gives a flux density of 1 tesla. Therefore:

$$\beta = \frac{\Phi}{A}$$

where:

β = magnetic flux density in teslas (T), or webers per square metre (Wb/m^2)
Φ = magnetic flux, in webers (Wb)
A = the cross-sectional area of flux path, in square metres (m^2).

INDUSTRY TIP

The unit of magnetic flux is called the weber (Wb) and is named after Wilhelm Eduard Weber 1804–1891. The unit of magnetic flux density is called the tesla (T) and is named after Nikola Tesla 1856–1943. You should remember the units, what they represent and their symbols.

Force on a current-carrying conductor

If a current-carrying conductor is suspended in a magnetic field, the field induced by the current in the conductor reacts with the main magnetic field. This reaction of the two fields creates a force which moves the suspended conductor. The amount of force acting on the conductor is determined by:

$$F = \beta L I$$

where:

F = force acting on the conductor, in newtons (N)
β = magnetic flux density, in teslas (T), of the main field
L = the length of the conductor within the main magnetic field (m)
I = the current passing through the conductor (A).

EXAMPLE

A conductor 0.5 m in length is placed in a magnetic field having a density of 0.5 teslas. If 10 A is passed through the cable, determine the force acting on the cable.

$$F = \beta L I$$

So:

$$F = 0.5 \times 0.5 \times 10 = 2.5 \text{ N}$$

How alternators produce sinusoidal waveform outputs

Alternators use rotating magnets to generate electricity. When the magnetic field cuts through the conductor, which is wound on iron cores in the stator, an electromotive force (e.m.f.) is produced and current is induced in the conductor.

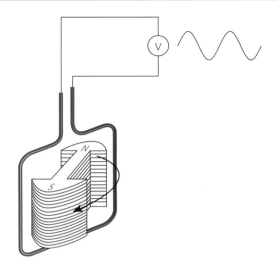

▲ Figure 2.52 Simple alternator action

Calculating magnitudes of a generated e.m.f.

If you were to experiment with the electromagnetic induction apparatus shown in Figure 2.52, you would find that when the magnet is changed for a stronger one, the flux increases, and a greater e.m.f. is generated, indicated by greater deflection on the voltmeter. Furthermore, if the conductor can be doubled up by winding it into a coil, the deflection on the meter will be twice that for a single conductor. Speeding up the crossing of the magnetic field by the conductor also causes a greater deflection. The following formula can be deduced from these three effects:

$$E = \beta l v$$

where:

E = induced e.m.f., in volts (V)
β = flux density of the magnetic field, in teslas (T)
l = the length of the conductor in the magnetic field, in metres (m)
v = velocity, in metres per second (m/s).

EXAMPLE

If a conductor of length 0.1 m cuts a magnetic field of 1.2 T at a velocity of 15 m/s, the generated e.m.f. is calculated as follows:

$$E = \beta l v = 1.2 \times 0.1 \times 15 = 1.8 \text{ V}$$

If the conductor is twice the length (0.2 m), the e.m.f. is:

$$E = \beta l v = 1.2 \times 0.2 \times 15 = 3.6 \text{ V}$$

Static induction

Static induction is the induction of a current in a circuit where no physical movement has taken place.

The induction is caused by the rising or collapsing of the magnetic field effectively 'cutting' the conductor. The value of the statically induced e.m.f. depends on the change of total magnetic flux.

$$E = \frac{\Phi}{t}$$

and

$$\Phi = Et$$

where:

$E =$ induced e.m.f. in volts (V)
$\Phi =$ total magnetic flux change, in webers (Wb)
$t =$ time for the flux change, in seconds (s).

EXAMPLE

If a coil induces an e.m.f. of 250 V and it takes 10 ms for the current to fall to zero, the flux change is calculated as follows:

$$\Phi = Et = 250 \times 10 \times 10^{-3} = 2.5 \text{ Wb}$$

If the same coil takes twice as long (20 ms or 20×10^{-3} seconds), the change in the magnetic flux is:

$$\Phi = Et = 250 \times 20 \times 10^{-3} = 5 \text{ Wb}$$

KEY FACT

Remember, all formulae need to be in base SI units, so 10 milliseconds is equal to 10×10^{-3} seconds.

Sources of electromotive force

When electric current flows, energy is dissipated because it cannot be created or destroyed. As energy cannot be created, electrical energy has to be converted from an existing form of energy. The form of energy converted may be chemical, as in a battery, it may be mechanical, as in a generator, or a combination of materials reacting to a source of energy such as a solar photovoltaic (PV) cell reacting to sunlight.

In the early days of electrical research, electricity was believed to be a fluid, which circulated as a result of an applied force. The term 'electromotive force' (e.m.f.) (E) was, and still is, used.

In determining units of electricity, e.m.f. is defined as the number of **joules** (J) of work required to move 1 C of charge around a circuit. This unit of joules per coulomb is referred to as a volt (V).

$$1 \text{ volt(V)} = \frac{1 \text{ joule (J)}}{1 \text{ coulomb (C)}}$$

INDUSTRY TIP

Electricity can be produced by magnetic, chemical or heating methods. Very little is produced by the latter method and chemical sources (cells and batteries) have a small capacity compared to generators.

KEY TERM

Joule: The unit of measurement for energy (W), defined as the capacity to do work over a period of time.

EXAMPLE

If a battery of 12 V gives a current of 5 A for 10 minutes, the amount of energy provided over the 10-minute period is calculated as follows.

To find total energy: $W = Q \times V$

Total charge transferred:

$$Q = It = 5 \times (10 \times 60) \text{ C}$$

$$Q = 3000 \text{ C}$$

$$W = Q \times V = 3000 \times 12$$

$$W = 36\,000 \text{ J or 36 kJ}$$

Electromotive force can be produced through:
- a chemical source
- heat
- electromagnetic induction (see page 128).

Chemical sources

When two different metals are placed in an electrolyte, ions are drawn towards one metal and electrons to the other. This is called a cell and it produces electricity. A set of several cells joined together is called a battery.

Heat

Simply wrapping a copper wire around a nail and heating one side with a flame can produce electricity, although in very small amounts. This is known as thermoelectric generation. Because the two metals react to the differences in temperature on the heated side and the cool side, a magnetic effect occurs, which creates a current and e.m.f. This process is sometimes called the Seebeck effect. The principle is used in thermocouples, which are used to sense temperature. The amount of electricity generated is in proportion to the temperature.

In some waste disposal plants where waste is burned, this effect is used to generate electricity.

Waveform

The characteristic of AC is its **oscillating** waveform, referred to as a 'sine wave'. This oscillating current arises through the changing position of the winding relative to the magnetic field within the alternator generating the current (Figure 2.53).

In practice, the AC supply at a workplace is unlikely to have a waveform that is a pure sine wave because of harmonics (contamination) arising from the connected loads. This will result in distortion of the waveform shape. The shape of the waveform of an AC supply may be displayed on an oscilloscope.

Because of its oscillating nature, alternating current and its associated voltage have additional factors that must be considered in describing the nature of the

ACTIVITY

Find a lemon, a zinc-coated nail and a piece of copper. Place the nail and copper into the lemon at opposite ends, ensuring there is a good gap between them inside the lemon. Use a sensitive voltmeter to measure the voltage between the two metals. You can see that a cell has been produced.

KEY TERM

Oscillating: moving back and forth in a regular pattern.

supply. These include the frequency of oscillation and the 'effective' current and voltage, whose values are continuously changing with time.

The output of an AC system, when measured and tracked, is usually referred to as a waveform.

As the rotating machine induces voltage, the value rises to a peak, falls to zero, then falls to a peak negative value, and then rises back to zero.

This single waveform represents one full turn of the alternator.

In Figure 2.53 the sinusoidal waveform is produced as the rotating conductor cuts the magnetic field set up by the permanent magnet (left). As the conductor is rotated at a constant speed and the number of conductors is fixed, the only variable is the amount of magnetic flux being cut, which can change the value of e.m.f. induced. This is represented by the waveform (right).

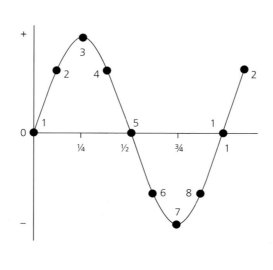

▲ Figure 2.53 The e.m.f. generated per rotation: the time taken for the cycle to return to its starting point is periodic time (*t*), shown on the waveform from the first position 1 to the second position 1

As the conductor in the diagram above starts at position 1, the direction of motion is in line with the field. Therefore no flux is crossed and no e.m.f. is induced.

As the conductor rotates, the number of lines of flux increases together with the angle at which the conductor cuts the flux lines, until the maximum value of e.m.f. is produced at position 3. As the conductor moves from position 3 to 5, the e.m.f. decreases to zero, at which point the conductor cuts no magnetic flux lines because it moves in the same direction as the flux. As the conductor rotates from position 5 to 7, the induced e.m.f. increases, reaching its peak value because the conductor is cutting the maximum number of flux lines at right angles. This time the polarity is reversed, and reaches the negative peak value. As the conductor rotates back to position 1, the number of flux lines cut is reduced, so that the output decreases to zero.

This is the complete cycle, producing a sinusoidal waveform (or sine wave) with zero at the start, the mid-point and the end.

This waveform is a sine wave in which the instantaneous voltage v (the voltage at any one point) can therefore be calculated as:

$$v = V_{max} \times \sin \theta$$

where:

V_{max} = the maximum induced voltage in the coil

$\sin \theta$ = the sine of the angle at that point.

The value v can be plotted for any angle, as shown in the table.

ACTIVITY

As the conductor rotates one full circle, it rotates 360°, so the sine wave is a representation of a circle over a period of time. Plot a sine wave where V_{max} is 325 V.

Coil angle $\theta(°)$	0	30	45	60	90	135	180	225	270	315	360
$v = V_{max} \sin \theta$ where V_{max} = 100 V	0	+50	+70.71	+86.67	+100	+70.71	0	−70.71	−100	−70.71	0

The values can be plotted for any angle to give a value for the instantaneous value based on the value of V_{max}.

Frequency and periodic time

Using UK frequencies, AC values change at a rate of 50 Hz, which means that the cycle repeats 50 times every second. This means that the voltage and current waveforms are rising to maximum positive and maximum negative and back to zero every 0.02 seconds. Because of this, the calculation to determine values is slightly more complex than that for DC calculations.

The time taken for the cycle to return to its starting position is the periodic time (t). The number of cycles per second is called the frequency (f), measured in hertz (Hz). Frequency is defined as:

ACTIVITY

The standard AC frequency in the UK is 50 Hz. Find out the frequencies used in other countries and the reasons for any differences that you find.

$$f = \frac{1}{t}$$

KEY TERM

Periodic time: the amount of time to complete one cycle.

where:

t = **periodic time**, in seconds (s)
f = frequency, in hertz (Hz)(cycles per second).

In the UK, the electricity supply frequency is 50 Hz. Therefore the time (t) to complete one cycle is:

$$t = \frac{1}{f} = \frac{1}{50} = 0.02 \text{ s}$$

As the values change, different values of voltage and current will be obtained, depending on the actual time and the position on the waveform the instant that the value is measured. The symbols for these values are represented by small letters: instantaneous voltage (v) and instantaneous current (i).

Maximum or peak voltage and current

The **mean** voltage (V_{av}) or mean current (I_{av}) of the waveform is zero because half a cycle is positive and the other half is negative, cancelling each other. However, to obtain the **mean voltage** or current of any half cycle it is necessary to multiply the peak voltage (V_{max}) or peak current (I_{max}) by 0.637. This is normally calculated by dividing a half waveform into enough equal divisions to be accurate but not too cumbersome. The values are added up and divided by the number of values taken to give the mean value (i.e. 0.637 peak for a sine wave).

The **root mean square** (RMS) value of a waveform is the equivalent value of AC that provides the same heat or work as a DC output over the same time period. To find the RMS value, each individual value is squared, these squares are added together and then the total is divided by the number of values to give the mean square. The square root of this value gives the RMS, which for a sine wave is 0.707 of the peak value.

Values of AC are taken to be RMS values unless otherwise specified because that is the effective current and the same relationship applies to the voltage associated with the AC that has the same waveform. When we consider the supply voltage to a house in the UK as 230 V, this is the RMS value. The actual peak value is 325 V. Peak values are relevant in some circumstances – for example, concerning the specification of cable insulation which must be capable of withstanding the peak voltage.

Peak-to-peak value

The **peak-to-peak value** is the measurement between the positive peak and the negative peak on a cycle. This equates to twice the peak value of any half cycle, assuming that the centre of the waveform is based at zero.

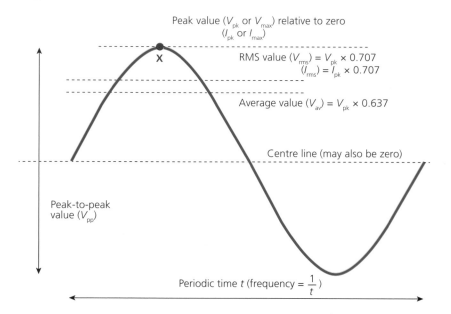

▲ Figure 2.54 Different values of a sine wave

KEY TERMS

Mean: The mean is the average of the numbers; a calculated 'central' value of a set of numbers. To calculate the mean, just add up all the numbers of the set, then divide by how many numbers there are in the set.

Mean voltage: the peak value multiplied by 0.637.

Root mean square (RMS): The square root of the mean of the squares of the value. It is the value we specify as the nominal voltage. In the UK, the supply voltage is 230 V AC, which actually peaks at 325 V.

INDUSTRY TIP

'Amplitude' is also used to describe peak value.

ACTIVITY

What is the peak voltage if the RMS voltage is:
- 110 V
- 240 V
- 400 V?

KEY TERM

Peak-to-peak value: the value of voltage or current between the positive peak and negative peak.

How AC generators work

For simplicity, the description below relates to one phase. However, it is important to remember that there are three phases, displaced at 120° from each other (which will be covered in detail in Book 2, Chapter 2).

KEY TERM

Rotor: the moving part of a generator or motor, which rotates through the magnetic field.

As each pair of conductors passes through the strongest part of the magnetic field at right angles, the maximum electromotive force (e.m.f.) is induced into that particular phase. At that point, the other two pairs are in a weaker part of the field and a lower voltage is induced. The moving **rotor** is connected to the stationary external connections by brushes and slip rings, which keep each phase in constant contact.

The output of an AC system, when measured and tracked, is usually referred to as a waveform. This is because, as the rotating machine induces an e.m.f., the value rises to a peak, falls to zero, then to a negative peak value and then rises back to zero.

INDUSTRY TIP

Rotor bars are also formed by casting aluminium into the rotor.

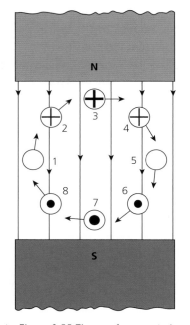
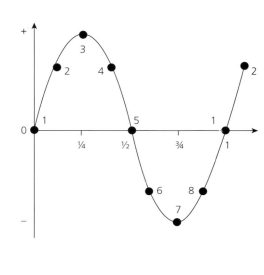

▲ Figure 2.55 The e.m.f. generated per phase per rotation

The time taken for the cycle to return to its starting position (from position 1 back to 1 in the example above) is the periodic time *t*. This process can be described in terms of Faraday's law because the rotation of the coil continually changes the magnetic flux through the coil and therefore generates an e.m.f.

Power distribution

In the UK, large amounts of electricity are generated at high voltage in power stations. This is typically between 23 and 25 kV and is transformed to EHV (extra high voltage) 275 kV or 400 kV systems through **step-up transformers**. Once the electricity is transmitted to its region, it is transformed down to a more manageable voltage through **step-down transformers.** (See pages 124–9 for more on transformers.) These distribution systems then deliver electricity at the correct voltage for the load, usually ending with an 11 000 V or 400 V transformer to supply both three- and single-phase installations at a local level of 230 V or 400 V.

A network of circuits, overhead lines, underground cables and substations link the power stations and allow large amounts of electricity to be transmitted around the country to meet the demand. Alongside the seven local distribution networks operating at 132 kV, 66 kV, 33 kV and 11 kV, there are also four high-voltage transmission networks operating at 400 kV (super grid) and 275 kV (the grid) in the UK. The 400 kV network was installed in the 1960s to strengthen the 275 kV system which began operating in 1953. The 400 kV network has three times the power carrying capacity of one 275 kV line and eighteen times the capacity of a 132 kV line.

Primary distribution by the Distribution Network Operators (DNOs) is usually carried out at 132 kV, using double circuit steel tower lines feeding primary substations, which in turn feed supplies at either 66 kV or 33 kV. The purpose of these primary substations is to supply larger industrial installations and the secondary distribution networks in urban and rural areas. Secondary distribution networks carry supplies from the primary substations via overhead lines on wooden poles or underground cables. Customers are connected at low voltage 230–400 V single or three phase.

Interconnected capacity

Interconnectors between Europe and the UK provide a pooling of capacity and diversity of supplies between the UK and Europe. This is provided by High Voltage Direct Current (HVAC) which is a way of conveying electricity over very long distances with fewer transmission losses than an equivalent HVAC solution. It also provides greater control over the transmission of electricity, including ability to change size and direction of power flow. Direct current supplies are usually obtained from AC mains supplies, first by using a transformer to change to the required voltage and secondly by using a rectifier to convert the AC supply to DC. Unfortunately, the rectification process is not perfect and some superimposed ripple is likely to appear on the DC output.

The current interconnectors in operation are:

- 2 GW between England and France
- 1 GW between England and the Netherlands
- 500 MW between England and Northern Ireland.

Other methods of generation

The majority of electricity generation is produced by the conversion of heat or thermal energy (steam) to some form of mechanical energy (by turning a turbine) which in turn forces a generator to turn.

KEY TERMS

Step-up transformer: a transformer that has a proportionally higher number of turns on the secondary (output stage) than on the primary (input stage).

Step-down transformer: a transformer that has a proportionally higher number of turns on the primary than on the secondary.

INDUSTRY TIP

A small number of power stations burn household refuse, which to a limited extent also addresses the landfill problem.

Standby supplies have diesel generators for use when the public supplies are not available.

ACTIVITY

Why are hydroelectric systems normally confined to Scotland and Wales?

The burning of **fossil fuels** (coal, gas, petroleum/diesel) and **nuclear fission** have been the main sources of heat to produce the steam to drive turbines.

Steam to drive turbines can also be produced by the following:

- **biomass** – such as wood, palm oil and willow
- **solar thermal** – Sun's heat energy is transferred to fluids
- **geothermal** – either directly from steam in the ground or via heat transfer.

Other means of turning a turbine are described below.

- **Water** – hydroelectric, pumped storage or micro hydro systems where the water turns turbine blades. Unlike large-scale hydroelectric schemes, such as Three Gorges Dam in China, micro hydro systems use small rivers or streams to produce up to 100 kW of power.

- **Wind** – the turbines used in a wind farm for commercial electricity generation are usually three bladed and can have a wing tip speed of 200 mph.

ACTIVITY

Dinorwig (see Figure 2.56) is an example of a pumped storage electricity scheme. Find out how and when Dinorwig produces electricity.

▲ Figure 2.56 Dinorwig (North Wales) pumped storage electricity generation scheme

▲ Figure 2.57 A typical off-shore wind farm installation

A small proportion of the UK's electricity is already generated from renewable sources but, with concerns about the depletion of fossil fuels, this is expected to grow significantly in the next few years. The Government is targeting a 34% reduction in **carbon emissions** by 2020, and 80% by 2050.

Other ways that electricity can be generated are:

- **solar photovoltaic cells (PV)** – these convert sunlight into electricity and although the PV concept is associated with small PV panels on domestic and commercial premises, many large PV power stations have been installed with hundreds of MW being produced
- **combined heat and power (CHP)** including micro CHP – in conventional power stations the waste heat is normally discarded via large cooling towers, whereas in a CHP generation system the waste heat or thermal energy is captured and used for heating schemes or production processes
- **batteries and cells** – when two different metals are placed in a chemical solution, or electrolyte, ions are drawn towards one metal and electrons to the other. This is called a cell and produces electricity. Many cells joined together are called batteries.

Source and arrangements of supply

A number of different electricity supply systems may be used in work premises, catering for specific requirements. These may be DC or AC operating at different voltages, and in the case of AC the supply may be single- or three-phase.

KEY TERM

Carbon emission: the polluting gas given off from the burning of fossil fuels such as gas, oil or coal.

Direct current is not used for public electricity supplies (with the exception of the links between England, Netherlands, Ireland and France) but has some work applications, such as battery-operated works plant (fork lift trucks, etc.). Certain parts of the UK railway system, in particular the London Underground and services in Southern England, also use DC for traction supplies.

Alternating current is the distribution system of choice for electricity suppliers all over the world. The main reason is its versatility. Using an AC supply permits wider scope concerning circuit arrangements and supply voltages, through the use of transformers, which enable the supply voltage to be changed up or down.

These different types of AC supply systems will now be explained.

Single-phase and neutral AC supplies

As with DC the simplest AC supply arrangement is a two-wire system, known as a single-phase supply. This is the arrangement provided for domestic premises, as well as for many supplies within work premises (lighting, socket outlets, etc.). The two conductors are referred to as the line conductor (L) and the neutral conductor (N). The neutral conductor is connected to earth at every distribution substation on the public supply system and therefore the voltage of the neutral conductor, with reference to earth, at any point should be no more than a few volts. The voltage between the L and N conductors corresponds to the nominal supply voltage (230 V).

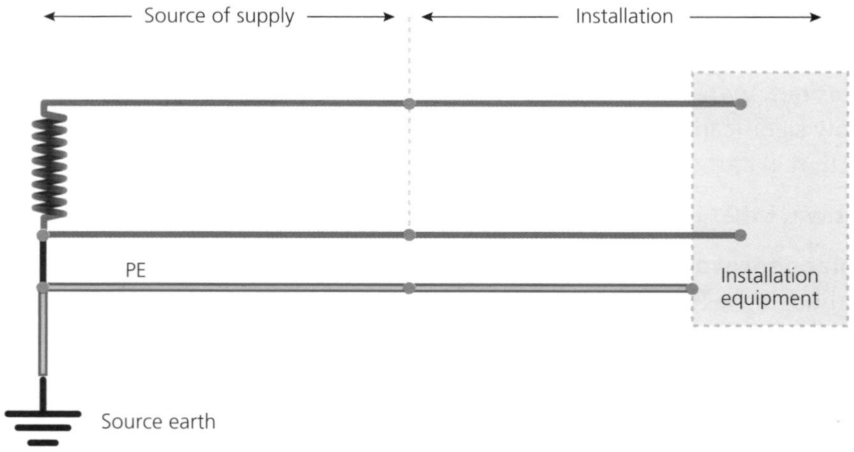

◄─── Source of supply ───► ◄─── Installation ───►

PE

Installation equipment

Source earth

▲ Figure 2.58 Single-phase AC supply

Three-phase and neutral AC supplies (star connected)

While single-phase AC supplies are adequate for domestic premises, the much higher loads typical of industrial and commercial premises would result in the need to use very large conductors to carry the high currents involved. The high currents would also give rise to a large volt drop.

However, it is possible to use a multi-phase arrangement, which effectively combines several single-phase supplies. If three coils spaced 120° apart are rotated in a uniform magnetic field, we have an elementary system which will provide a symmetrical three-phase supply.

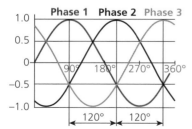

▲ Figure 2.59 Three-phase sine wave where each waveform is 120° apart

The usual arrangement is a three-phase system employing four conductors – three separate line conductors and a common neutral conductor, as shown in Figure 2.60. Most substation transformers in the distribution system that delivers power to houses are wound in a **delta**-to-star configuration. A neutral point is created on the star side of the transformer.

The three line conductors (L_1, L_2 and L_3) were previously distinguished by standard colour markings: red, yellow and blue. However, European harmonisation has now resulted in these conductor colours being changed to brown, black and grey.

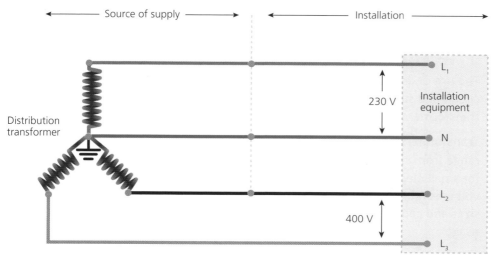

▲ Figure 2.60 Three-phase AC supply

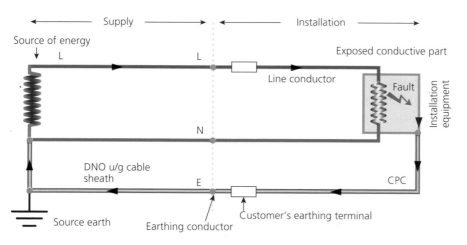

◄ Earth fault loop path

▲ Figure 2.61 Earth fault loop

KEY TERM

Delta: where the windings of the transformer are arranged in a triangular formation with the start of one winding connected to the end of another, meaning the voltage across each winding is the same as the line voltage. This will be covered in greater detail in Book 2, Chapter 2.

ACTIVITY

Why is the star point of the transformer connected to earth?

The earthing system adopted will determine the earth fault loop impedance, and this will determine the method of protection against electric shock (Figure 2.61).

- TN-C-S systems tend to have low earth fault loop impedances external to the installation, of the order of $0.35\,\Omega$.
- TN-S systems tend to have higher earth fault loop impedances compared to TN-C-S systems. The typical maximum declared value is $0.8\,\Omega$.
- When TT systems are adopted, the resistance of the installation earth will be high (of the order of $100\,\Omega$). This means that residual current devices (RCDs) will need to be adopted for protection against electric shock as they operate at much lower earth fault currents than standard protective devices.

The operating principle of transformers

The use of alternating current, rather than direct current, gives more scope for circuit and supply voltages as they can be changed up or down by the use of a transformer.

Transformers are fundamental to the safe and efficient use of electricity. They range from step-down transformers, which provide an extra-low voltage supply for small electrical appliances, to large step-up transformers that produce voltages of up to 400 kV for power transmission purposes. The wide range of sizes and capabilities corresponds to the everyday requirements of electricity usage.

Although there are many different types and sizes of transformer, they all operate on the same principle. The basic principle is that in two independent coils (windings), a change in the magnetic flux of one coil can induce a magnetic change in the second coil. This is known as static inductance. It is further enhanced when the two electrically separated coils are wound onto a common magnetic core, which creates a common magnetic circuit.

If an AC supply is applied to one coil (known as the primary), the magnetic flux produced by the first coil rotates and cuts through the second coil (known as the secondary), producing an e.m.f. in the secondary. This configuration works with AC or chopped DC. The field in the first coil rises and falls rapidly, causing inductance in the adjacent coil.

The ease with which voltages can be changed up or down, coupled with the fact that it is more economical to transmit electricity over long distances at high voltage, is an important reason why AC has been adopted for mains electricity supplies.

Transformers are very efficient devices, particularly when working on full load and, since there are no moving parts, require little maintenance.

KEY TERM

Transformer: an item of electrical equipment which is generally used to change the voltage and current from one value to another. When a step-up transformer is used, the output voltage is higher than the input voltage but the current decreases.

INDUSTRY TIP

When a current flows in the secondary, it produces a secondary flux which opposes the mutual flux and weakens it, allowing more primary current to flow.

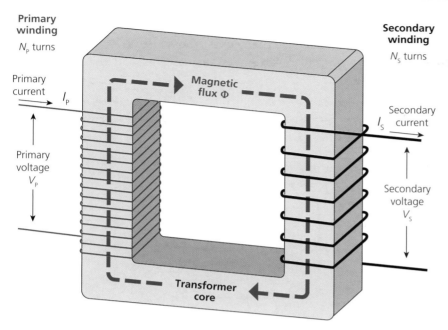

▲ Figure 2.62 Simple transformer arrangement

Types of transformers

There are a wide range of electrical transformers all designed for different uses and purposes. The designs may differ but the fundamental principles are the same.

Power transformers

Power transformers are the most common type of transformer and are used as:

- step-up transformers in the transmission of electrical energy from power stations (25 kV–400 kV) and step down from 400 kV to 132 kV
- step-down transformers in distribution substations from 132 kV–11 kV, and 11 kV to 400 V for use in commercial and domestic situations
- step-down transformers to convert mains voltage 230 V to extra-low voltage to power electronic equipment.

The construction of insulated laminations minimises the eddy current produced during the transformation process.

For large power transformers used at generating and distribution substations, oil is used as a coolant and insulating medium. For small-rating transformers, the oil is circulated through ducts in the coil assembly and through cooling fins fixed to the body of the transformer tank. In higher ratings, the oil circulates through separate air-cooled radiators that may use pumps and fans to aid the cooling process.

Cast resin power transformers, where the windings are encased in epoxy resin, are often used where there is a fire risk in indoor situations.

Current transformers

Current transformers (CTs) have many uses. Those that involve metering require extremely accurate current transformers, such as Class X CTs.

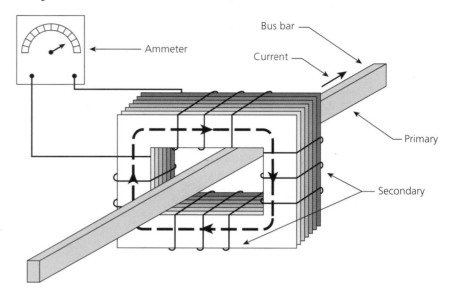

▲ Figure 2.63 A current transformer with secondary winding around the primary conductor

A current transformer has a primary winding, a magnetic core and a secondary winding. The core and secondary winding surround the primary winding, which is a simple conductor giving one single turn.

The AC flowing in the primary conductor produces a magnetic field in the core, which then induces a current in the secondary winding circuit.

It is extremely important when positioning the current transformer to ensure that the primary and secondary circuits are efficiently coupled to give an accurate reading. It is also very important to ensure that the current transformer is always connected on the secondary side, either by a measuring instrument or a shorting link. If the CT is left open circuit, a large voltage discharge will occur.

Isolation transformers

Isolation transformers are available for step-up, step-down or straightforward 1:1 isolation purposes. Their uses are quite diverse but ultimately they are intended to ensure electrical separation from the primary supply, usually for safety purposes.

These transformers have additional insulation and electrostatic shielding. In some cases, one or both sides of the transformer remains separated from earth by a resistance or they have no connection at all.

A common use for isolation transformers is the domestic shaver socket arrangement, which is isolated from earth to minimise any shock hazards from touching live parts. Because the parts are isolated and the current cannot flow back to its origin (the secondary side of the transformer) via the earth path, the shaver socket cannot deliver a shock under first-fault conditions. Isolation

ACTIVITY

A typical current transformer (CT) would have a ratio of 800:5 so for every 800 A on the primary there would be 5 A on the secondary, allowing a 5 A meter to be used if the maximum current on the primary was 800 A. What size meter would be needed if the CT had a ratio of 1360:5 and the maximum current on the primary was still 800 A?

transformers are also used in many medical situations in order to prevent first-fault failures. These systems are also known as medical isolated power supplies (IPS).

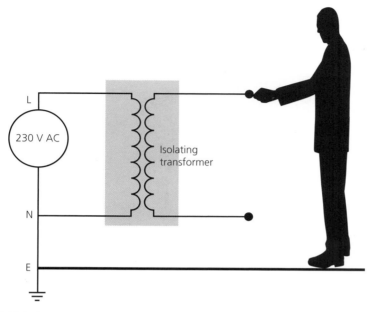

▲ Figure 2.64 An isolated supply, such as in a BS EN 61558-2-5 shaver socket outlet, minimises the risk of shock

Voltage transformers

A voltage or potential transformer has two windings wound around a common core. See the diagram of the simple transformer arrangement on page 125.

Transformer cores

The core of a transformer is generally one of two types: core or shell.

The shell-type transformer is regarded as being more efficient as the magnetic flux is able to circulate through two paths around the core. The core-type transformer only channels flux through one path, meaning that some of the flux is lost at the core corners (leakage flux).

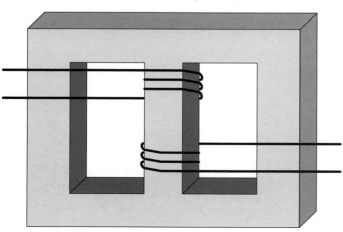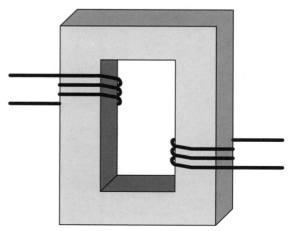

▲ Figure 2.65 Shell-type core (left) and core-type core (right)

HEALTH AND SAFETY
Remember, you will still get a shock between the two transformer output terminals.

INDUSTRY TIP
The lower voltage winding is normally wound closest to the core for safety reasons. Some windings are placed on top of each other; other types use a sandwich arrangement.

ACTIVITY
There is a third type of core. Use your research skills to find out what it is.

Transformer laminations

Transformer cores are made up of laminations. Laminations are thin, electrically insulated slices of metal that, when stacked on top of one another, form a large core. These thin laminations are used to reduce eddy currents in the transformer core.

▲ Figure 2.66 Laminations used to form a transformer core

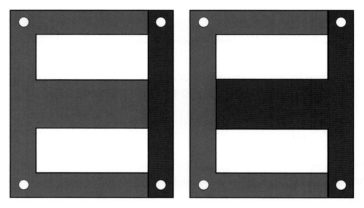

▲ Figure 2.67 Laminations are stacked to form a shell-type core

Eddy currents are products of **induction**. Small rotating currents, like swirling currents seen in rivers, rotate around a transformer core. If allowed to, these currents produce heat that ends up as a loss of energy. Making a transformer core with thin, insulated laminations prevents eddy currents from flowing, thus reducing any loss. Eddy current loss is a form of iron loss – that is, losses in the core of a transformer.

Another type of loss is hysteresis loss. This is due to the transformer core being magnetised in one direction, then re-magnetised in the other direction as the supply alternates. If the material requires energy to re-magnetise it, this also becomes a loss. Careful consideration of the material used to construct the core reduces hysteresis loss.

Further losses in transformers are copper losses. This is a loss due to the heating effect of current passing through the windings. The windings are resistances and as in any resistance, the power dissipated, or lost, is usually determined as I^2R.

This is because power is determined by:

$$P = V \times I$$

and voltage is determined by:

$$V = I \times R$$

so power can also be expressed as:

$$P = I \times I \times R \text{ or } I^2R$$

Inductance

There are two types of inductance in relation to transformers: self-inductance and mutual-inductance.

Self-inductance is where a transformer winding induces a magnetic field that rotates around the core and induces a current back into the winding, limiting current flow. This principle is used in choke/ballast units in fluorescent luminaires.

Mutual-inductance is where a primary winding induces a magnetic field that induces a current into a second winding, as in current and voltage transformers.

Relationship of e.m.f. produced and number of turns

The proportion of e.m.f. produced is related to the number of turns on the primary coil with respect to the secondary coil. This relationship can be determined by:

$$\frac{E_1}{E_2} = \frac{N_1}{N_2}$$

where:
E_1 = e.m.f. induced in the primary (V_1)
E_2 = e.m.f. induced in the secondary (V_2)
N_1 = number of primary turns
N_2 = number of secondary turns.

As power transformers have very low impedance, there is negligible error in assuming that the voltage on the primary is the same as the e.m.f. induced on the primary and likewise for the voltage and e.m.f. on the secondary.

INDUSTRY TIP

When working out transformer ratios, use the winding to winding values – that is, phase voltage to phase voltage ($V_p - V_s$).

A reasonable assumption is that:

$$\frac{E_1}{E_2} = \frac{V_1}{V_2}$$

Therefore:

$$\frac{V_1}{V_2} = \frac{N_1}{N_2}$$

The ratio of primary turns to secondary is referred to as the turns ratio, represented by:

$$\frac{N_1}{N_2}$$

Current is also affected by the turns ratio of a transformer, but in reverse to voltage, so:

$$\frac{V_1}{V_2} = \frac{N_1}{N_2} = \frac{I_2}{I_1}$$

INDUSTRY TIP

Notice the current is the opposite way round to the voltage and number of turns.

EXAMPLE

A transformer is wound with 560 turns on the primary and 20 turns on the secondary. Suppose that 230 V AC is applied to the primary winding. Calculate the output voltage.

$$\frac{V_1}{V_2} = \frac{N_1}{N_2}$$

Then:

$$\frac{230}{V_2} = \frac{560}{20}$$

So:

$$V_2 = \frac{230 \times 20}{560} = 8.21\,V$$

If the secondary current is 4 A, calculate the input primary current.

$$\frac{N_1}{N_2} = \frac{I_2}{I_1}$$

ACTIVITY

Try these calculations with different combinations of primary and secondary turns or voltages.

Then:

$$\frac{560}{20} = \frac{4}{I_1}$$

So:

$$I_1 = \frac{4 \times 20}{560} = 0.14 \text{ A}$$

Transformer power ratings

As a transformer is not a load, simply a method of changing voltage and current, it is rated in kVA. This is because certain loads may include a power factor, which can increase the current demand by the load. As a result, the rating of the transformer is given in volt amperes (VA) instead of kilowatts (kW). So when selecting a transformer for a particular load, the actual current drawn by the load must be multiplied by the voltage.

⑥ ELECTRONIC COMPONENTS USED IN ELECTROTECHNICAL SYSTEMS AND EQUIPMENT

The function of electronic components in electrical systems

Most electronic systems use many different electrical components in their power supply and in their operational systems.

Security alarm systems use full-wave rectification and smoothing through capacitors to supply a 12 V operating system. In addition, the closed-loop system uses transistor and similar technology to convert low-level signals from components such as passive infrared (PIR) detectors into an alarm output signal to components that operate the alarm.

Thyristors or SCRs are used extensively in motor speed-control circuits for heating and other applications, where motors and pumps require variable output. The ability to control an output waveform is essential in controlling the motor speed.

Heating control systems use a number of components. The most important element of any form of heating control is probably the ability to sense temperature in the airspace or water systems that are being heated.

A thermistor is used to determine accurately the temperature of the space or heating medium. The heating control system then uses feedback from the sensor to determine how much heat needs to be passed. The temperature is controlled via valves and/or variable speed pumps.

▲ Figure 2.68 A range of different alarm detection devices using different technologies

▲ Figure 2.69 Typical thermistor-based space temperature detector

Diacs and triacs are used in lighting control circuits. The ability to trigger the device through a separate voltage allows dimming to be provided via proprietary dimming systems (Figure 2.70).

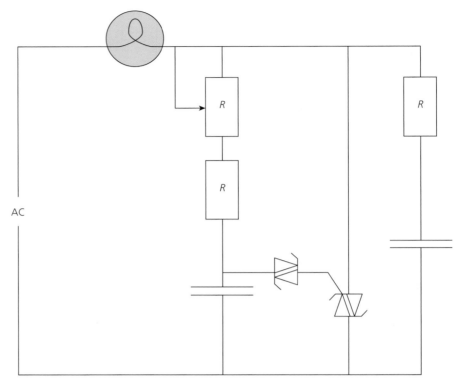

▲ Figure 2.70 Typical diac and triac controlled dimming arrangement

How electronic components work

In today's world of micro-components, it is becoming increasingly common to replace a whole circuit board rather than replacing single components. Nevertheless, it is far easier to diagnose faults in electrical systems if you understand how particular electronic components function.

Diodes

A diode is a silicon P-N junction, which allows current flow in one direction, but not the other. When current flows through a diode, it is called 'forward bias'. When current is restricted, it is called 'reverse bias'. There are several types of diode, from the simple one just described used for rectification or signalling, to:

- a zener diode, which only allows current flow when a set voltage is reached
- a light-emitting diode (LED), which emits light when current flows through it
- a photo diode, which allows forward bias current flow when it detects light.

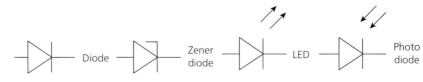

▲ Figure 2.71 Symbols for different types of diode

ACTIVITY

How could a capacitor, zener diode and resistor be used to smooth the output of a full-wave bridge rectifier?

Diacs

A diac is a junction of two zener diodes, with two terminals. It works on AC circuits, hence the name **di**ode for **AC**. A diac will not allow current flow unless a pre-set voltage is reached. Once this voltage is reached, current can flow in both directions. Current will continue to flow until the voltage falls below the level set, at which point the diac restricts current flow.

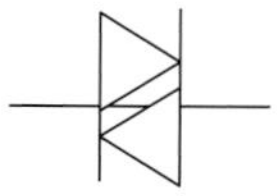

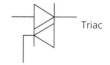

▲ Figure 2.72 Symbol for a diac

Thyristors

A thyristor is a solid-state switch that allows current flow between two of its terminals if a small current is sensed on the third. There are two types: silicon-controlled rectifiers (SCR) and triacs.

SCRs

The SCR is similar to the diode, in that current can only flow between the anode and cathode in one direction. However, it also has a gate terminal, which activates the switch when a small current is sensed on that terminal. Essentially, it allows a large current to be controlled by a small current. The SCR will continue to allow current flow between anode and cathode until the gate current is stopped. It does not require a constant gate terminal current, except when allowing the main current to pass.

ACTIVITY

What effect would varying the trigger angle have on a triac?

Triacs

A triac has three terminals, one called the gate. If the gate senses a very small control current, AC is allowed to flow between the other two main terminals (known as MT_1 and MT_2). If the gate current is removed, the device will stop current flow when the alternating cycle reaches 0 V.

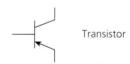

▲ Figure 2.73 Symbol for a triac

Transistors

The transistor is the fundamental building block of modern electronics and the reason why electronic systems are now so affordable.

The three terminals on a bipolar transistor are known as base (B), collector (C) and emitter (E).

▲ Figure 2.74 Symbol for a transistor

A transistor may be used either as a switch or an amplifier. When the base of an NPN transistor is grounded (0 V), no current flows between emitter and collector, so the transistor is off. If the base voltage is increased above 0.6 V, a current will flow from emitter to collector and the transistor is on. If the base current varies in value, the emitter to collector current will follow this pattern of variation with a larger or smaller current flow; in this situation, the transistor acts as an amplifier.

A PNP transistor operates in the same way as an NPN transistor but with current flow allowed in the reverse direction.

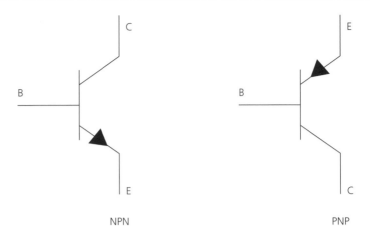

NPN   PNP

▲ Figure 2.75 An NPN and a PNP transistor, showing the polarity of each device

Another type of transistor is the field effect transistor (FET), which has terminals marked gate, source and drain. The FET is much cheaper to produce as it requires less silicon. It also has the major advantage of operating at virtually no current on the gate terminal as long as a voltage above 0.6 V is present.

Resistors

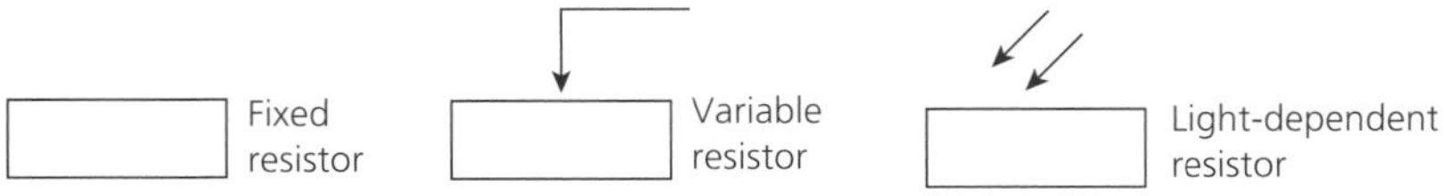

Fixed resistor	Variable resistor	Light-dependent resistor

▲ Figure 2.76 Symbols for different types of resistor

Resistors are used to control or reduce current flow in electronic circuits. With a sufficiently high resistance, they can also be used as voltage dividers on certain circuits to allow a fixed voltage, less than the input voltage, to be obtained. Fixed-value resistors are either made from carbon film with an insulated coating, as shown below, or are wire wound for larger power applications.

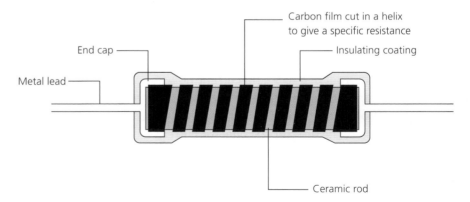

Carbon film cut in a helix to give a specific resistance

End cap

Insulating coating

Metal lead

Ceramic rod

▲ Figure 2.77 Section through a carbon-film resistor

Carbon-film resistors are colour-coded to indicate their value and tolerance as shown below:

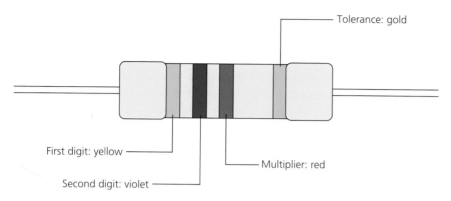

Tolerance: gold

First digit: yellow

Second digit: violet

Multiplier: red

▲ Figure 2.78 Resistor colour-coding system

▼ Resistor colour values

Colour	Digit	Multiplier	Tolerance
Black	0	1	
Brown	1	10	1.0%
Red	2	100	2.0%
Orange	3	1 000	
Yellow	4	10 000	
Green	5	100 000	0.5%
Blue	6	1 000 000	0.25%
Violet	7	10 000 000	0.1%
Grey	8		0.05%
White	9		
Gold		0.10	5.0%
Silver		0.01	10.0%

ACTIVITY

A resistor is identified by the colour bands red, violet, orange and gold. What is its value?

Wire-wound resistors are normally coded in order to establish the value. For example, a 2R resistor is 2 Ω, whereas 2R2 is 2.2 Ω.

Thermistors

A thermistor is a type of resistor in which the resistance varies significantly with temperature. This variation is so well defined that there is a definite temperature-related use for them. Thermistors typically achieve high precision within a limited temperature range, typically −90 °C to 130 °C.

Thermistors are widely used as temperature sensors, self-resetting overcurrent protectors for self-regulating heating elements and current inrush limiters.

Photoresistors

These are resistors that vary in resistance, depending on the amount of light falling on them. They are often referred to as photocells and are used to control lighting as day/night switches.

Thermistor

▲ Figure 2.79 Symbol for a thermistor

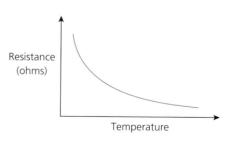

Resistance (ohms)

Temperature

▲ Figure 2.80 Effect of temperature on resistance in a thermistor

Variable resistors or potentiometers

These resistors are used to vary resistance in a circuit manually. Their applications are wide, including use as sound volume controllers and speed controllers.

Capacitors

▲ Figure 2.81 Symbol for a capacitor

Capacitors are widely used in electrical circuits in many common electrical devices. A capacitor is a passive two-terminal electrical component used to store energy electrostatically in an electric field, rather than by chemical reaction, as in a battery. (Originally capacitors were known as condensers, but the original term has now been widely superceded.)

Capacitors vary widely, but all contain at least two electrical conductors separated by a dielectric (insulating layer), which acts as an insulator between the conducting plates. The plates are usually made from foils. The capacitance is varied by the area of the plates and the size of gap between the plates. The narrow gaps that are used require a very high dielectric strength.

ACTIVITY

Name five different types of capacitor.

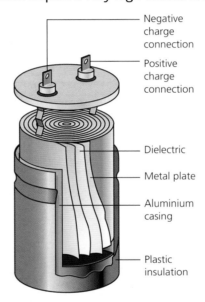

▲ Figure 2.82 Construction of a typical capacitor

Rectifiers

A rectifier is an electrical device that uses diodes to convert alternating current (AC), which periodically reverses direction as it cycles, to direct current (DC), which flows in only one direction.

Half-wave and full-wave rectifiers are available.

Half-wave rectification

In half-wave rectification of a single-phase supply, either the positive or negative half of the AC wave is passed, while the other half is blocked. Half-wave rectification requires a single diode in a single-phase supply, or three in a three-phase supply.

INDUSTRY TIP

Three-phase rectification requires three diodes for half-wave and six diodes for full-wave rectification.

As only one half of the input waveform reaches the output, the mean voltage is lower than full-wave rectification.

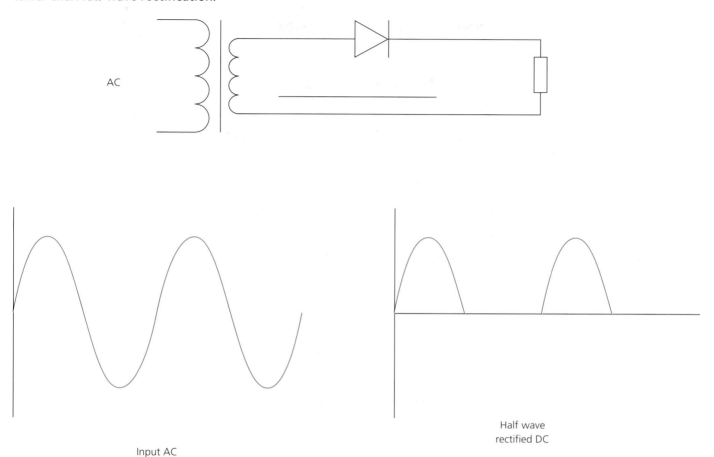

Input AC

Half wave rectified DC

▲ Figure 2.83 Half-wave rectification

Full-wave rectification

A full-wave rectifier converts the whole of the input, both positive and negative components of the waveform, to one of constant polarity at its output. Full-wave rectification output gives a pulsating DC waveform with a higher average output voltage than its half-wave counterpart.

The unit works with two diodes and a centre-tapped transformer, or four diodes in a bridge configuration, as shown in Figure 2.84 below.

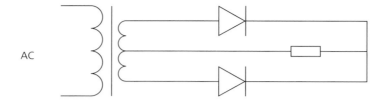

Two-diode method with centre-tapped transformer

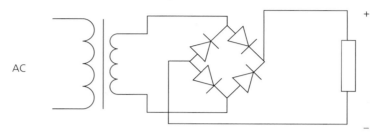

Four-diode or bridge rectifier

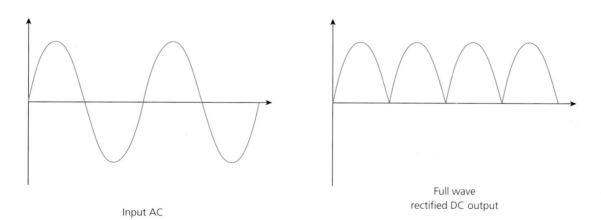

Input AC

Full wave
rectified DC output

▲ Figure 2.84 Full-wave rectification using two diode and bridge arrangements

Invertors

An invertor works in the opposite way to a rectifier by converting DC to AC. Invertors that use electronic components to convert DC to AC are known as static invertors as they do not move (unlike coupling an AC generator to a DC motor which does move).

By using SCRs in a DC circuit connected to a transformer, the SCRs can switch a large current, using a small signal current in opposite directions which is the same as AC.

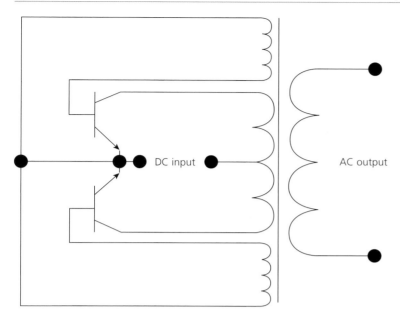

▲ Figure 2.85 A simple invertor circuit using SCRs

Invertors are commonly used to convert, or invert, the DC produced by photovoltaic cells on the roof of a building to AC, which can be fed into the building's electricity supply.

▲ Figure 2.86 Photovoltaic cells on the roof of a building

139

Test your knowledge

1 What number is represented by 2.3×10^{-3}?

A 0.23

B 0.023

C 0.0023

D 0.00023

2 Which two values of a right-angled triangle are used to determine the sine of an angle?

A Opposite side and hypotenuse.

B Adjacent side and hypotenuse.

C Right angle and opposite side.

D Adjacent and opposite sides.

3 What is the SI unit of measurement for charge?

A Farad.

B Coulomb.

C Henry.

D Newton.

4 What would be the force acting on the ground below a mass of 10 kg due to gravity?

A 0.981 N

B 9.81 N

C 98.1 N

D 981 N

5 How many electrons does a copper atom have?

A 10

B 17

C 23

D 29

6 A series circuit has three 12 Ω resistors connected to a 200 V DC supply. What is the total circuit current?

A 50 A

B 32 A

C 16.6 A

D 5.56 A

7 What happens to the strength of the magnetic field around a conductor when another conductor, carrying equal current in the opposite direction, is placed next to it?

A Doubles in strength.

B Halves in strength.

C Remains the same strength.

D Cancels out to no strength.

8 What does the thumb indicate when using the right hand 'grip rule'?

  A North pole.

  B South pole.

  C Current flow.

  D Voltage flow.

9 What are transformer cores made from to reduce eddy current loss?

  A Copper bars.

  B Plastic tubes.

  C Laminated metal.

  D Glass sheeting.

10 What electronic component allows current to flow in one direction only?

  A Resistor.

  B Diac.

  C Capacitor.

  D Diode.

11 A 1.5 mm^2 copper conductor has a resistance of 0.8 Ω at 20 °C. Determine the conductor length given the resistivity of copper is 0.0172 μΩm.

12 Three suppliers are quoting the following prices for a specialised luminaire.

Supplier	Price £	Discount on price %
Lights and Shades	145.00	20%
Square Wholesaler	120.00	6%
Dark No More	132.00	15%

Determine which supplier is the cheapest overall.

13 Determine, for the circuit below, the:

  a total resistance

  b supply current (I_s).

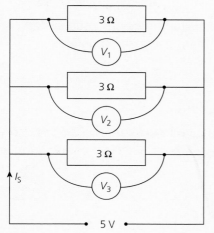

14 State **each** of the values A, B, C and D marked on the image.

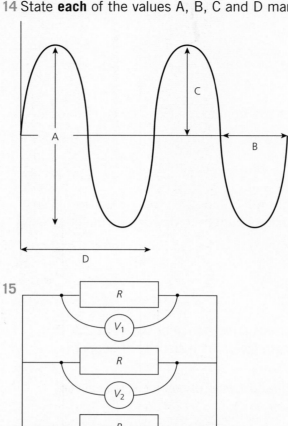

15

a   State the value of voltage V_1.

b   The supply current (I) is 25 A and two of the resistors are 10 Ω. Determine the resistance of the third resistor.

Practical task

Ask your tutor for a selection of resistors. Using the colour code system, work out the resistances and then check them using an ohmmeter.

CHAPTER 3

ELECTRICAL TECHNOLOGY

INTRODUCTION

Large amounts of information need to be absorbed and understood by everyone within the construction industry. Most of this information is guidance or commentary on Regulations. However, there are also laws covering electrical work and the way it is carried out that need to be obeyed.

The law will be satisfied if the installer complies with various documents, including **BS 7671:2018** The IET Wiring Regulations, 18th Edition, the Institute of Engineering and Technology (IET) Guidance Notes and other related guidance. The first part of this chapter will help you to identify the relevant duties set out in **statute** law and commercial contract law (**civil law**) and to know how to carry out those duties.

The rest of the chapter deals with technical considerations of working in the electrical industry.

INDUSTRY TIP

You can access information relating to the IET Wiring Regulations at: https://electrical.theiet.org/wiring-regulations/index.cfm?IET%20Wiring%20Regulations

Access to the Institute of Engineering and Technology (IET) Guidance Notes is via: https://electrical-standards.theiet.org/

KEY TERMS

Statute: a law made by Parliament as an Act of Parliament.

Civil law: law that deals with disputes between individuals and/or organisations, in which liability is decided and compensation is awarded to the victim.

Learning objectives

This table shows how the topics in this chapter meet the outcomes of the different qualifications.

Topic	Electrotechnical Qualification (installation) or (maintenance) 5357-004	Level 2 Diploma in Electrical Installations (Buildings and Structures) 2365 Unit 203	Electrical Installation 8202 Unit 204
1 Regulations related to electrical activities	1.1	1.1; 1.2; 1.3; 1.4	5.1; 5.2
2 Technical information	1.1	2.1; 2.2; 2.3; 2.4	5.3; 5.4; 5.5
3 How electricity is supplied		5.1; 5.2; 5.3; 6.1; 6.2; 6.3	1.1
4 Intake and earthing arrangements		4.1; 5.4	1.2; 1.3; 1.4
5 Consumers' installations	2.1; 2.2; 2.3; 2.4; 2.5; 6.1; 6.2; 6.3; 6.4; 7.1	3.1; 3.2; 3.3; 3.4; 3.5; 3.6; 4.2; 4.3; 4.4	3.1; 2.2; 2.3; 3.1; 3.2; 3.3; 3.4; 3.5; 4.1; 4.2; 4.3

1 REGULATIONS RELATED TO ELECTRICAL ACTIVITIES

In Chapter 1, we looked at the many Statutory and Non-statutory Regulations, Guidance material and Codes of Practice relating to working in the construction industry. Below is a reminder of some of those Regulations and other documents relating specifically to electrical installation work.

Statutory regulations

A Statutory Regulation is easily distinguishable from other regulations because its full title usually contains the word 'Regulations' and always has an SI (Statutory Instrument) number. Statutes (Acts of Parliament) are known as primary legislation. They lay down the general framework of the law, which is added to over the years through numerous Regulations. Statutory Regulations are known as secondary legislation.

Health and Safety at Work etc. Act 1974

The Health and Safety at Work etc. Act 1974 (HSW Act) is the primary legislation covering occupational health and safety in Great Britain. This Act has been amended over the years by various pieces of legislation to enforce health and safety law in Great Britain.

Principles of the HSW Act

All workers have a right to work in places where risks to their health and safety are properly controlled. Health and safety is about stopping anyone getting hurt at work or becoming ill through work. It sets down employers' responsibilities for health and safety, but also requires employees to help keep themselves and others safe. The HSW Act requires employers to do the following.

- Assess risks in the workplace and to put in place measures to prevent harm and to inform all concerned about who is responsible for the removal or reduction of the risks.
- Consult with workers and health and safety representatives to protect employees from harm in the workplace.
- Provide free health and safety training to enable employees to do their jobs safely.
- Provide, free of charge, any equipment or personal protective equipment (PPE) required for employees to do their jobs, and ensure it is properly maintained and remains functional.
- Provide adequate welfare facilities such as toilets, washing facilities and drinking water.
- Provide adequate first aid facilities, as appropriate for the type of work being carried out.
- Report deaths and major injuries at work to the Health and Safety Executive (HSE) Incident Contact Centre.
- Report other injuries, diseases and dangerous incidents under the Reporting of Injuries, Diseases and Dangerous Occurrences Regulations 1995 (RIDDOR).

INDUSTRY TIP

You can access the Health and Safety at Work etc. Act 1974 (HSW Act) at: www.legislation.gov.uk/ukpga/1974/37

The Health and Safety at Work (Northern Ireland) Order 1978 applies in Northern Ireland and can be accessed at: www.legislation.gov.uk/nisi/1978/1039

- Have insurance that covers employees in case they get hurt at work or become ill through work.
- Work with any other employers or contractors sharing the workplace or providing employees (such as agency workers), so that everyone's health and safety is protected.

The HSW Act requires employees to:

- follow the training received when using any equipment supplied by the employer
- take reasonable care of their own and other people's health and safety
- co-operate with the employer on health and safety matters
- tell someone (the employer, supervisor or health and safety representative) if they think the work or inadequate precautions are putting anyone's health and safety at serious risk.

If an employee has concerns about health and safety in the workplace:

- They should talk to their employer, supervisor or health and safety representative.
- They should seek advice from the general information about health and safety at work on the HSE website at www.hse.gov.uk.

If, after talking with their employer, they are still worried, an employee can contact their local enforcing authority for health and safety and the Employment Medical Advisory Service (EMAS) via the HSE website.

Electricity at Work Regulations 1989

The Electricity at Work Regulations 1989 (EAWR) are made under the HSW Act. This means that breaches in the EAWR will result in action being taken under the HSW Act.

In addition to the duties imposed by the HSW Act, the EAWR impose duties on **duty holders** in respect of systems, electrical equipment and conductors, and work activities on or near electrical equipment under their control. Managers of mines and quarries are also included, despite mines and quarries having other special regulations.

The EAWR cover the principles of electrical safety that apply to work activities and systems, including anything that influences them, equipment, isolation and safety systems, and the **competency** of those working with electricity.

The EAWR apply beyond those situations that we traditionally associate with the dangers of

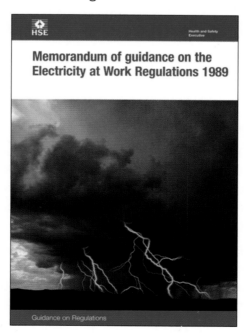

▲ Figure 3.1 Memorandum of guidance (HSR25)

electricity, including voltages outside the scope of **BS 7671:2018** The IET Wiring Regulations, 18th Edition (see below). The EAWR cover battery operated systems to extra high voltage transmission supplies.

The HSE produces a memorandum of guidance (HSR25) (Figure 3.1), which is free and provides guidance on how each Regulation should be interpreted. Each Regulation covers a specific topic, as follows:

- Regulation 4, Systems, work activities and protective equipment
- Regulation 5, Strength and capability of electrical equipment
- Regulation 6, Adverse or hazardous environments
- Regulation 7, Insulation, protection and placing of conductors
- Regulation 8, Earthing or other suitable precautions
- Regulation 9, Integrity of referenced conductors
- Regulation 10, Connections
- Regulation 11, Means for protecting from overcurrent
- Regulation 12, Means for cutting off the supply and for isolation
- Regulation 13, Precautions for work on equipment made dead
- Regulation 14, Work on or near live conductors
- Regulation 15, Working space, access and lighting
- Regulation 16, Persons to be competent to prevent danger and injury.

Electricity Safety, Quality and Continuity Regulations 2002

The Electricity Safety, Quality and Continuity Regulations 2002 (ESQCR) impose requirements regarding the installation and use of electric lines and the apparatus of electricity suppliers, including provisions for connection with earth. The safety aspects of these regulations are administered by the HSE. All other aspects are administered by Government and the industry.

The ESQCR may impose requirements, usually on supply companies or those associated with the supply of electricity, in addition to those of the EAWR. Designers of installations have a responsibility to ensure they meet the ESQCR.

Provision and Use of Work Equipment Regulations 1998

The Provision and Use of Work Equipment Regulations 1998 (PUWER) ensure that work equipment, such as electric drills and cutters, as constructed or adapted, is suitable for the task and safe to use, taking into account all the risks of using it in a specific work environment. This applies to any equipment, machinery, appliance, apparatus or tool for use at work.

For example, in order to be fit for purpose, construction site equipment needs to cope with a wide range of weather conditions including wet weather. Equipment also needs to be robust enough and sufficiently protected to withstand mechanical impact or abrasion.

It should help to reduce, to as low a level as possible, the risk of electric shock through external influences, including from reduced low-voltage systems such as the 110 V system used on UK construction sites. This may mean that additional protection must be provided.

INDUSTRY TIP

The HSE publication HSR25 (Figure 3.1) interprets the regulations and gives guidance on how to meet the requirements. It is freely available at: www.hse.gov.uk/pUbns/priced/hsr25.pdf

INDUSTRY TIP

The ESQCR can be accessed here: www.legislation.gov.uk/uksi/2002/2665/pdfs/uksi_20022665_en.pdf

INDUSTRY TIP

A copy of Provision and Use of Work Equipment Regulations can be downloaded free of charge from the HSE website: www.hse.gov.uk/pubns/priced/l22.pdf

Control of Substances Hazardous to Health (COSHH) Regulations 2002

The law requires employers to control substances that are hazardous to health and to protect employees and other persons from the hazards of substances used at work, by risk assessment, control of exposure, health surveillance and incident planning. The COSHH regulations contain very detailed information relating to each type of harmful substance, depending on its physical state and how it will harm individuals or groups of people.

The COSHH regulations cover:

- chemicals and products containing chemicals
- dusts, fumes, vapours and mists
- **nanotechnology**
- gases and **asphyxiating gases**
- biological agents (germs) and germs that cause diseases.

The COSHH regulations do not cover lead, asbestos or radioactive substances, which are dealt with under separate Regulations.

Duties of employers

In order to reduce the exposure of workers to hazardous substances, employers are required to do the following:

- **Identify the health hazards** – using information about any chemicals or substances used in the workplace that might be a hazard.
- **Decide how to prevent harm to health** – using risk assessments to determine who may be harmed by a chemical or substance and how someone might be harmed.
- **Provide control measures to reduce harm to health** – looking at the potential for harm and introducing protection measures, such as local exhaust ventilation (LEV), respiratory protective equipment (RPE) or personal protective equipment (PPE).
- **Make sure protective measures are used** – including ensuring that workers are informed of the risks associated with a hazard, what the protective measures are and how to comply with them, and the health and safety and disciplinary consequences of failing to use the measures.
- **Keep all control measures in good working order** – all measures must function correctly and any failing or potential failing must be reported and rectified. Equipment that is not in good working order must be taken out of service until a repair has taken place.
- **Provide information, instruction and training for employees and others** – this is vitally important to ensure that those who work with or are potentially exposed to harmful substances understand what they are being exposed to, know what to do to prevent a situation getting out of control and becoming an emergency, and know how to deal with particular situations or emergencies.

INDUSTRY TIP

You can access the Control of Substances Hazardous to Health Regulations at: www.legislation.gov.uk/uksi/2002/2677/pdfs/uksi_20022677_en.pdf

KEY TERM

Nanotechnology: where a material or substance is created by changing matter at an atomic or molecular level. In this way, science may create new materials or substances that are stronger or more suited to a particular application. Nanotechnology is widely used in medical applications but may also be found in everyday products such as sunscreen, cosmetics and paints, finishes and surface coatings on everyday products.

KEY TERM

Asphyxiating gas: a gas that can cause suffocation. If someone is suffering asphyxia, they are unable to breathe.

- **Provide health monitoring in appropriate cases** – this applies to people who regularly work with or who may have been accidentally exposed to a harmful substance. If adverse effects on health become apparent, the relevant healthcare can then be undertaken.
- **Plan for emergencies** – providing an emergency contingency plan so that workers and managers know exactly what to do should the need arise.

Company name		Department/site		Assessment discussed with employees (include date)		
Potential hazard	Potential harm and those affected	Existing measures	Additional measures required	Who	When	Check
Oil-based fluid and fine particle in lathe sumps	Dermatitis Metal workers and turners	Fluid stops when machine stops 0.4 mm nitrile gloves	Improve washing facilities Apply barrier cream before and after work			
		Check oil temperature and particle concentration	Check appearance daily Fit sump thermometer			
		Change and launder overalls once a week	Keep oily rags out of pockets			

Also	Action taken	Action needed			
Thorough examination and test for COSHH compliance					
Supervision	Yes				
Instruction and training	Yes				
Emergency plans	Spillage clearance	Practice			
Health surveillance		Improve record-keeping			
Monitoring					
Review date:	Review observations/actions Any significant changes:				

▲ Figure 3.2 Typical COSHH risk assessment form

Published Guidance and Codes of Practice

Guidance on the legal requirements of certain Statutory Regulations and how to interpret them, and Approved Codes of Practice (ACoPs), are published or approved by the HSE and the relevant Secretary of State. These documents often contain a copy of the legislation they are giving guidance on. They have a special legal status in law. Where a duty holder does not follow the Guidance or Code of Practice, they must show that their course of action or omission is no less safe.

INDUSTRY TIP

The HSE website contains many Guidance documents, which are free to download here www.hse.gov.uk/guidance/index.htm

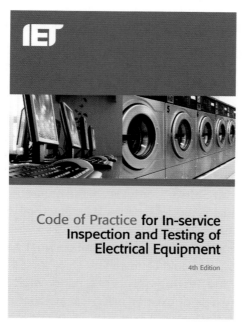

▲ Figure 3.3 IET Code of Practice for In-service Inspection and Testing of Electrical Equipment

Non-statutory Regulations

Not all documents using the term 'regulations' are Statutory Instruments. For example, **BS 7671:2018** The IET Wiring Regulations, 18th Edition, commonly called the IET Wiring Regulations, are non-statutory. They form the **National Standard** in the UK for low-voltage electrical installations. They deal with the design, selection, erection, inspection and testing of electrical installations operating at a voltage up to 1000 V AC. Work undertaken in accordance with BS 7671:2018 is almost certain to meet the requirements of the Electricity at Work Regulations 1989. BS 7671:2018 is supported by a number of Guidance Notes published by the IET.

Other non-statutory rules include the rules and requirements of regulating bodies such as the NICEIC, the National Association of Professional Inspectors and Testers (NAPIT), the Electrical Contractors' Association (ECA) and trades unions such as Unite. Individuals and enterprises that want to belong to these organisations must comply with the relevant organisation's rules. In certain circumstances, compliance

KEY TERMS

National Standard: based on International Standards produced by the International Electrotechnical Commission (IEC), member nations create their own versions specific to their needs. Other **CENELEC** countries use the term 'rules' rather than 'regulations'. For example, the national wiring standard in the Republic of Ireland is the National Rules for Electrical Installations (ET101).

CENELEC: the Comité Européen de Normalisation Electrotechnique, which is the European Committee for Electrotechnical Standardization.

with such rules, such as those for competent person schemes (CPSs), assists in compliance with specific statutory regulations or requirements.

BS 7671:2018 The IET Wiring Regulations, 18th Edition

BS 7671:2018 The IET Wiring Regulations, 18th Edition is the basis for all other guidance documentation, such as the IET On-Site Guide or the eight IET Guidance Notes (GNs). BS 7671:2018 is broken down into seven parts with further appendices providing information relating to the seven parts. The parts are in a logical order relating to the design and installation of an electrical installation.

- **Part 1: Scope, Object and Fundamental Principles** – sets out what the standard covers (scope), why it is setting out (object) and what risks need addressing (fundamental principles).
- **Part 2: Definitions** – provides detailed descriptions of technical words and phrases, including abbreviations and symbols, used in the document.
- **Part 3: Assessment of General Characteristics** – details what a designer must assess prior to designing an electrical installation and before any equipment is installed.
- **Part 4: Protection for Safety** – provides detail on methods that may or must be used for protecting against particular risks electricity may pose, such as electric shock, fire or overcurrent.
- **Part 5: Selection and Erection** – gives detail on what particular equipment can be used and how it should be installed.
- **Part 6: Inspection and Testing** – gives criteria for what should be inspected and tested for every new or existing electrical installation.
- **Part 7: Special Installations or Locations** – sets out requirements where a particular location provides greater risk. In these locations, all relevant previous Regulations still apply but, because of the risk, some changes are needed to overcome the risk.

IET On-Site Guide

The purpose of the IET On-Site Guide is to provide guidance on the design, installation and inspection and testing of standard installations such as those in dwellings, shops and small offices. By following the guidance given, the need for complex design calculations is reduced.

IET Students' Guide to the wiring regulations

While this Guide is intended for students learning about electrical installations, the title should not put off more experienced electricians as it simplifies a lot of the technical aspects of **BS 7671:2018** The IET Wiring Regulations, 18th Edition and provides good guidance on the electrical industry, basic design procedures, good working practices and common calculations.

IET Guidance Notes

The IET Guidance Notes (GN) are intended to provide more detailed guidance on aspects of BS 7671:2018 but the numbering of the eight GNs does not necessarily reflect the Parts of BS 7671:2018. The eight Guidance Notes are:

 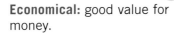

- GN1: Selection and Erection
- GN2: Isolation and Switching
- GN3: Inspection and Testing
- GN4: Protection Against Fire
- GN5: Protection Against Electric Shock
- GN6: Protection Against Overcurrent
- GN7: Special Locations
- GN8 Earthing and Bonding

The IET also provide specific Guidance documents on aspects of electrical installation work such as:

- Guide to Consumer Units
- Considerations for DC Installations
- Installation of Electric Vehicle Charging installations
- In-service Inspection and Testing of Electrical Equipment
- Guide to the Building Regulations
- Guide to Emergency Lighting Installations

Applying Statutory Regulations

Statutory Regulations cover the whole range of areas dealt with by the law. Each set of regulations lays down the detailed requirements that have to be met to comply with a specific Act of Parliament. In construction-related industries, most legislation focuses on health and safety.

Statutory Regulations place specific requirements on duty holders to do something, provide something or declare something.

HSE inspections

HSE inspectors routinely inspect places of work. Although it is unusual to call unannounced, they have the right to enter and inspect any workplace without giving notice.

On a routine visit, rather than responding to a complaint or particular incident, inspectors inspect the workplace, work activities, and methods and procedures used for the management of health and safety, and check compliance with health and safety law relevant to the specific workplace. The inspector is empowered to talk to employees and their representatives, and take photographs and samples.

Where breaches in requirements are observed, the inspector can serve the following, depending on the severity of the breach.

- **Informal warning** – where the breach is minor, the inspector can explain on site what is required to the duty holder (usually the employer), following up with written advice explaining best practice and the legal requirements.
- **Improvement notice** – where the breach is more serious, the notice will state what has to be done, why, and when the remedial action has to be completed by. The inspector can take further legal action if the notice is not complied with within the time period specified. The duty holder has the right to appeal to an industrial tribunal.

> **INDUSTRY TIP**
>
> The IET also publish a comprehensive suite of Codes of Practice known as IET Standards. A full list of these can be found at: http://electrical.theiet.org/books/standards/index.cfm

> **HEALTH AND SAFETY**
>
> In the case of health and safety law, non-compliance can impact on an organisation's reputation. For example, some clients monitor HSE incidents before appointing contractors and sub-contractors. More significant failures can result in fines and, for the most serious cases, prison sentences.

- **Prohibition notice** – where there is a risk of serious personal injury, the inspector can stop the activity immediately or after a specific period. The activity cannot be resumed until remedial action has taken place. The duty holder has the right of appeal to an industrial tribunal.
- **Prosecution** – in certain circumstances, the inspector may consider prosecution necessary to punish offenders and deter other potential offenders.

In addition, particular pieces of legislation may carry specific penalties, such as unlimited fines and custodial sentences. These measures are generally taken to court by an enforcing authority. In cases of health and safety legislation violation, this is done by the HSE.

When cases go to court, specialist lawyers are called on to work through the complexities of the relevant area of law. Often the evidence of expert witnesses is sought to obtain professional opinions on the subject. The outcome will be an acquittal or a sentence with remedial or corrective action.

In general, employers are responsible for the actions of their employees and, where these result in breaches in legislation or accidents, it is the employer that is punished. However, where a breach is entirely due to an employee, the employer has the right to impose its own disciplinary processes, which, at worst, may result in a finding of gross misconduct and dismissal.

Breaches of Non-statutory Regulations

Breaches of Non-statutory Regulations, such as British Standards (BSs) and supporting guidance, usually amount to a breach of contract and often revolve around quality. Contractual requirements for the provision of services or standards usually exceed any legal requirements. When requirements have been breached, financial compensation for damages is often sought.

In Figure 3.5, showing the hierarchy of work-related legislation and guidance, it might appear that British Standards and Guidance documents are less important than Statutory Regulations. However, the best practices contained in these examples of non-statutory documents are actually used as benchmarks in legal proceedings. Failure to meet them can often result in a breach of legal requirements.

This will lead to the statutory enforcement system and criminal courts. For example, a breach of the non-statutory IET Wiring Regulations may have a significant effect on safety, ultimately breaching statutory legislation.

Even if a breach of non-statutory regulations is not a legal issue, the consequences can be far-reaching. The industry may lose faith in the offending organisation's ability to perform, resulting in loss of customers and tenders. In the case of the rules of trade and similar organisations, breaches may result in **expulsion** from the relevant organisation.

2 TECHNICAL INFORMATION

There are many different sources of technical information available, including reference books, textbooks and commentaries, to help you interpret different elements of electrical engineering. They are aimed at different levels, from the trainee to the experienced engineer. It is important to recognise the different types and levels of information available so that it can be used to best effect.

In addition to industry information, guidance and documentation, technical information needed to carry out installation work can also be contained in the form of drawings and specifications.

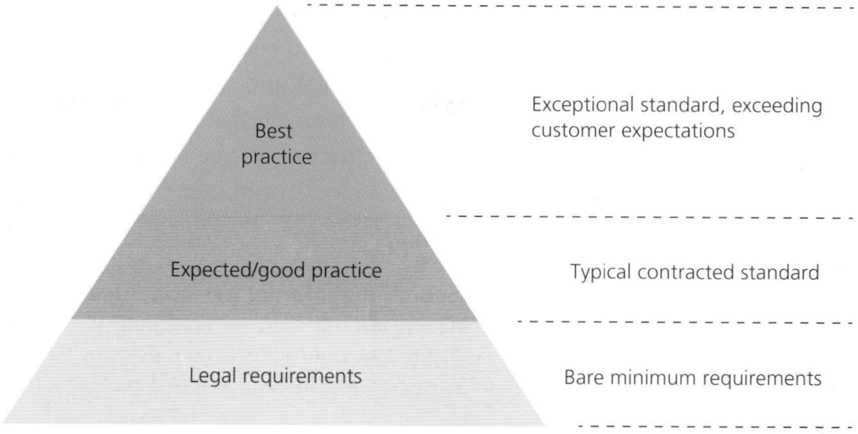

▲ Figure 3.4 Standards required by Statutory and Non-statutory Regulations

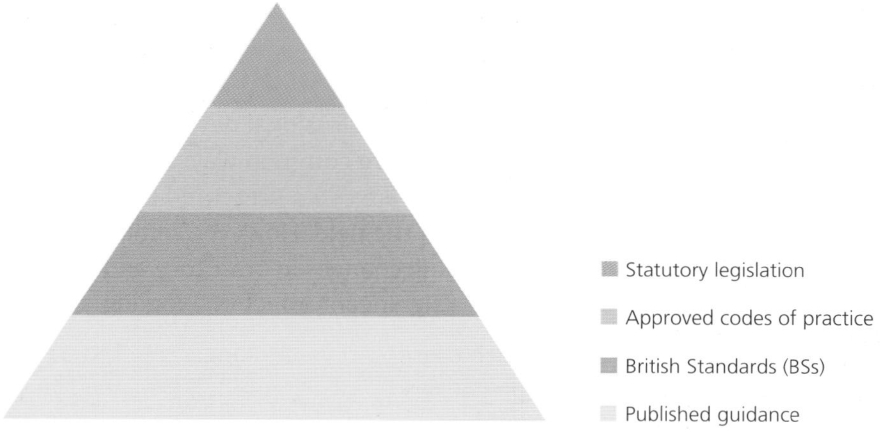

Best practice — Exceptional standard, exceeding customer expectations

Expected/good practice — Typical contracted standard

Legal requirements — Bare minimum requirements

Sources of technical information

▲ Figure 3.5 Hierarchy of work-related legislation and guidance

■ Statutory legislation
■ Approved codes of practice
■ British Standards (BSs)
■ Published guidance

Most electrotechnical information is found in book or electronic (ebook) format, although some can be delivered verbally. It can provide, for example, support and calculations to solve particular problems or commentary on compliance with legal requirements. Sources include:

- technical specifications
- continuous professional development (CPD) presentations
- Guidance and Approved Codes of Practice (ACoPs)
- Government websites
- technical standards
- technical information and guidance from professional institutions.

Technical specifications

As part of the manufacturing process, multi-national manufacturers employ engineering specialists to carry out testing and development on their products, giving them a level of expertise that is not normally found in such depth anywhere else in the industry. As a result, they are well equipped to provide the technical specifications in addition to the installation, operation, maintenance and safety information they supply with their products. Examples could be lighting manufacturers or manufacturers of protective devices.

> **HEALTH AND SAFETY**
> Under health and safety legislation, the employing organisation is obliged to provide sufficient information and training to protect those accessing their premises from harm. For example, a specialist chemical manufacturing company must advise or train all those who access their site (including visitors) on what to do if they are exposed to specific chemicals. Simply washing with a seemingly harmless substance such as water after exposure to a chemical may, for instance, be at least inadequate and at worst even dangerous.

Their technical information is usually available as free downloads from the internet and they often provide free **seminars** and training to the electrotechnical industry based around the solutions that their products provide. The information will obviously be biased towards their own products, but nevertheless can be very useful. Many of these seminars are CPD-accredited with trade institutions and training organisations.

ACTIVITY

Search the internet for manufacturers of electrical equipment such as cable management systems and see what courses they offer. These are sometimes listed as continuing professional development (CPD) seminars. See what other support and literature these manufacturers offer.

The employing organisation can often hold significant information relating to a particular site or installation.

Continuing professional development presentations

Continuing professional development (CPD) seminars at a workplace or institution are common ways of giving information and know-how to the end user. They are often presented by an industry specialist and acknowledged expert in the field. However, seminars can also present updates to standards and upcoming changes in Statutory and Non-statutory Regulations, all of which can have a significant effect on working methods.

Such presentations no longer need to be attended in person. They can also be delivered live or recorded for a web-based audience via a webinar. After a presentation, viewers can send queries to the organiser, which are then dealt with by the speaker in a short question-and-answer session. Webinars are often left posted on the internet so they can be referred to later.

Guidance and ACoPs

These types of documents are provided to help interpret the law. Approved Codes of Practice, more commonly known as **ACoPs**, have a special status in law and may contain a copy of the Statutory Regulations they refer to. They have been approved by the relevant Secretary of State and, in the case of health and safety issues, by the Health and Safety Executive (HSE).

INDUSTRY TIP

You can find a wide range of electrotechnical webinars available on IET TV at www.theiet.org

KEY TERM

ACoPs: Approved Codes of Practice issued by the Health and Safety Executive.

INDUSTRY TIP

Electrical Safety First is the campaigning name of the Electrical Safety Council and is the UK's leading charity on electrical safety. They work closely with industry bodies publishing all types of electrical safety information for both consumers and industry experts. See what they do at www.electricalsafetyfirst.org.uk

INDUSTRY TIP

The HSE and Electrical Safety First also publish many guidance documents and leaflets to help employers and employees understand their duties under specific Statutory Regulations and achieve best practice in their area. Check the Electrical Safety First Best Practice Guides at www.electricalsafetyfirst.org.uk/electrical-professionals/best-practice-guides

Technical standards

British Standards (BSs) are documents published to provide industry specific guidance and, in some instances, information on Regulations. Each British

Standard gives information and the minimum best practice for a particular industry, put together by representatives and experts from the given industry.

British Standard European Norms (BS ENs) are similar documents but contain information and guidance from corresponding European Norms, based on good practice agreed across Europe. The UK is a member of CENELEC (the European Committee for Electrotechnical Standardization) and can therefore influence the content of electrotechnical ENs. However, UK legislation remains more important. For example, statutory requirements in the Electricity Safety, Quality and Continuity Regulations 2002 override certain continental European practices. Other UK documents such as **BS 7671:2018** The IET Wiring Regulations, 18th Edition are based on **harmonised documents** (HDs) from International Standards.

Other technical standards include the National Engineering Specification (NES), which was developed from former Government department standards. These standards are now maintained by private companies and licensed to the relevant industry.

Technical information and guidance from professional institutions

Most professional institutions provide support and publish technical information for their members. This type of information has usually been accumulated over many years by practitioners in the industry and then put together by respected members of the institution, so it is full of useful technical advice.

Some institutions give general advice on design or installations – for example, the Plumbing Engineering Services Design Guide from the Chartered Institute of Plumbing and Heating Engineering, or the wide range of Chartered Institution of Building Services Engineers (CIBSE) guides.

Other institutions focus on particular elements of electrotechnical engineering, providing technical solutions to particular tasks or advice on developing specific systems of work. They include application manuals and guidance publications – for example, the IET series of guidance notes, which provide commentary on different areas of the Wiring Regulations. In addition to the IET, there are many other authors and publishers who provide commentary and guidance on **BS 7671:2018** The IET Wiring Regulations, 18th Edition.

Types of drawings

Numerous sets of drawings are produced to communicate information about individual systems or collections of systems within a project. They include:

- plans or layout drawings
- schematic (block) diagrams
- wiring diagrams
- circuit diagrams.

Drawings usually use British Standard symbols. However, in order to represent the wide variety of materials and equipment available, non-standard symbols may also be used. A legend or key explains them.

KEY TERM

Harmonised document: agreed by member nations, such as members of CENELEC, who jointly agree what rules or regulations should be in place for a particular risk or situation. For example, HD 60364-4-41 Protection Against Electric Shock is what Chapter 41 of BS 7671:2018 is based on.

Plans or layout drawings

Plans or layout drawings are used to locate individual systems within the overall project and give an indication of the scale of the project. In addition, there may be drawings to show specific fixing, assembly and/or completion details. This is often the case where complex construction, lifting or use of a crane is required. These details may be provided in elevation, plan or both. They will then become part of the contractor's method statement when the project goes into the construction phase. Figure 3.6 is an example.

Schematic diagrams

Schematic diagrams, sometimes referred to as block diagrams, can serve many purposes, but are primarily provided to show the overall functionality of a system, including interfaces and operational requirements. This means they are intended to show a sequence of control and connection but not necessarily scale or actual positions in the wiring. For example, if a three-plate lighting system were drawn as a wiring diagram, the light would be wired before the switch even though the switch controls the light. A schematic would show the switch before the light as a sequence of control. Schematic or block diagrams are typically used to show distribution systems and circuits in large electrical installations having many remote distribution boards.

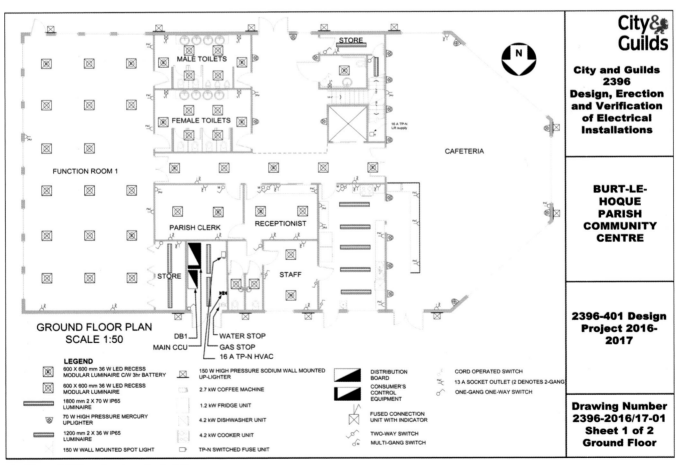

▲ Figure 3.6 Plan drawing of a building

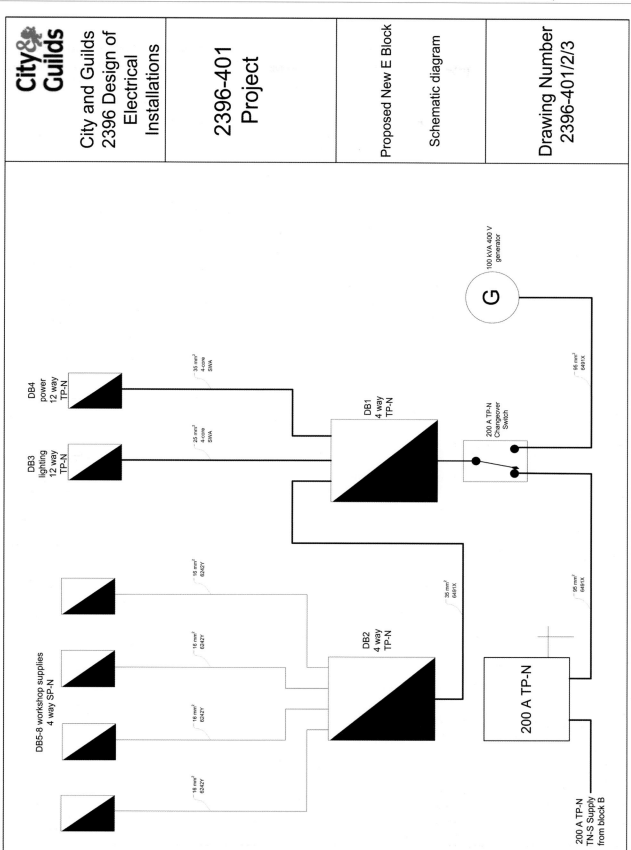

City & Guilds

City and Guilds
2396 Design of
Electrical
Installations

2396-401
Project

Proposed New E Block

Schematic diagram

Drawing Number
2396-401/2/3

DB4
power
12 way
TP-N

35 mm²
4-core
SWA

DB3
lighting
12 way
TP-N

25 mm²
4-core
SWA

DB5-8 workshop supplies
4 way SP-N

16 mm²
6242Y

16 mm²
6242Y

16 mm²
6242Y

16 mm²
6242Y

DB1
4 way
TP-N

DB2
4 way
TP-N

35 mm²
6491X

200 A TP-N
Changeover
Switch

100 kVA 400 V
generator

G

95 mm²
6491X

95 mm²
6491X

200 A TP-N

200 A TP-N
TN-S Supply
from block B

▲ Figure 3.7 Typical schematic drawing

157

Wiring diagrams

Wiring diagrams are generally provided to show in detail how a system or collection of systems is put together. This type of drawing shows locations, routing, the length of run and types of systems cabling. These types of diagrams are sometimes mistakenly referred to when it is actually a circuit diagram that is required.

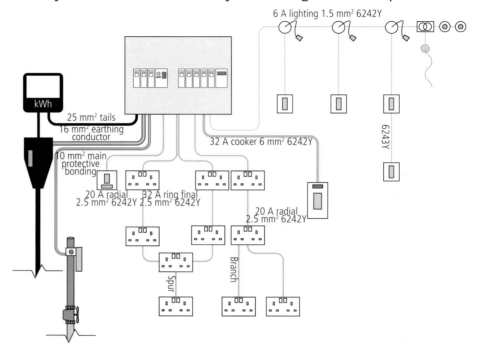

▲ Figure 3.8 Typical wiring diagram

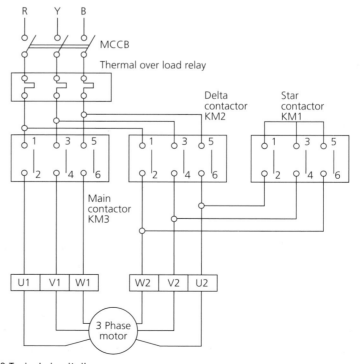

▲ Figure 3.9 Typical circuit diagram

Circuit diagrams

Circuit diagrams contain information on how circuits and systems operate. They can be provided as detailed layouts, although some information such as length of run may be omitted for clarity. In most instances, these diagrams are used for **diagnostic** purposes so that designers, installers and maintainers understand how their actions may influence a particular component or arrangement.

At different stages of work, the information on the drawings and diagrams is given in different levels of detail. The information becomes more detailed and accurate as the project develops.

Levels of drawing

There are five main levels of drawing:

- sketch drawings
- design and tender drawings
- contract drawings
- working or construction drawings
- as fitted or installed drawings.

Sketch drawings

Sketch drawings are often hand-drawn sketches to help the designer demonstrate ideas and interpretations. They are usually drawn in the early stages of a project when the scheme and designs are still **fluid**. As a result, they can be subject to multiple developments as every stakeholder is adding input to and potentially changing the scheme.

Sketch drawings are also often used at later stages of a project. Once the contractors have produced their working drawings, the designer may issue new sketch drawings for particular design details. These drawings become a **contractual variation** for that portion of the works and are recorded as a variation by the contract administrator.

Design and tender drawings

Design and tender drawings are produced by a designer to communicate the main design intent of the works. Although they are generally intended to be read in conjunction with the specification and other written documents, it is often easier to show the geographical scope of works by hatching out or blanking out areas not covered on a drawing.

Design drawings are usually first exchanged between designers to ensure co-ordination, and then to communicate items for consideration under the Construction (Design and Management) Regulations 2015 (CDM 2015) with other designers and stakeholders, including the client.

At the design stage of a project, the designer has a duty to **eradicate** hazards (whether notifiable under the CDM Regulations or not) where possible. A number of hazards may be unavoidable and these need to be pointed out to contractors. Residual hazards should be identified on the drawings or schedules.

KEY TERM

Diagnostic: concerned with identifying problems.

KEY TERM

Fluid: open to change or development.

KEY TERM

Contractual variation: where changes to the original agreed contract take place and are documented using drawings and variation orders.

KEY TERM

Eradicate: remove or get rid of.

The Health and Safety Executive (HSE) prefers residual hazards or hazard reduction strategies to be indicated on drawings.

The design drawings are then put together with any stakeholder comments and issued with the specifications and schedules as tender drawings. Details of the potential hazards and hazard reduction strategies on the drawings help contractors (tenderers) to understand any potential difficulties or additional requirements for completing the task safely. Allowances can then be incorporated into the pre-construction health and safety plan so that construction is carried out in a safe manner.

Contract drawings

Following tender award, the tender drawings usually form part of the contract on which the scope and detail of work has been priced. These drawings are usually re-issued, containing any contractual variations issued during the tender period. They are then referred to as contract drawings.

Construction or working drawings

Construction or working drawings are usually produced by the building services sub-contractor. They are usually drawn at a very large scale and contain additional information not given on the previous drawings (e.g. on fixings, expansion measures, etc.). This level of detail allows installation exactly to the drawing.

As fitted or installed drawings

In building services, as fitted or installed drawings are normally produced by the sub-contractor or installer.

These drawings should represent what has been installed, but also contain information on any hazards and risk reduction strategies employed, and instructions on how to clean, maintain or deconstruct the works if necessary. They become part of a wider set of operation and maintenance manuals (O&M).

Using symbols in drawings

It is important that drawings are easy to understand, and that includes the symbols used. British Standard symbols should be used wherever possible. Where no standard symbol is available, a specialist symbol can be used which is created by the person making the drawings. Whether the symbols are British Standard or not, they should be explained in a key or legend. In particular areas, such as lighting design, a more detailed legend specifying fitting type, manufacturer, source, output and catalogue reference number of products might be required to assist the reader to understand the type and quality of the component as well as how many are needed and where they are to be located.

Unless a project is very small and positions are obvious, the actual positions of items indicated by symbols have to be interpreted by the installer from scale working drawings. If the scale of a drawing is 1:100, for example, a 4 mm symbol if scaled up would be 400 mm across. This is obviously not what is intended and where four or five different symbols come together it would be difficult to position them correctly even at a larger scale of 1:50. Working drawings are normally provided at 1:50, often with dimensioned values, to allow the installer to interpret them more easily.

One-way switch, indicating the number of switches at one point		Key-operated switch	
Two-pole one-way switch		Socket outlet	
Cord-operated single-pole one-way switch		Switched socket outlet	
Two-way switch		Double socket outlet	
Dimmer switch		Ceiling-mounted lighting point	
Period-limiting switch		Emergency lighting point	
Time switch		Wall-mounted lighting point	
Thermostat		Lighting point with integral switch	
Push button		Single fluorescent lamp	
Luminous push button		Three fluorescent lamps	

▲ Figure 3.10 Examples of British Standard symbols for electrical installation drawings

Understanding scale in drawings

Most drawings are produced at a scale that allows the information to be readable. In order to determine an appropriate scale, it is essential to know the size of the actual object, room or building. Next, it is important to decide on the size of paper for the drawing. Very little detail can be communicated on smaller sheets of paper such as A4 or A3, but A2, A1 and A0 drawings need to be printed and copied on expensive specialist equipment.

Drawings can be created with **computer-aided design (CAD) software** and stored on networks, so they can also be communicated electronically. Electronic media is not, however, always practical on site and paper documents are still important.

KEY TERM

Computer-aided design (CAD) software: specialist software for producing drawings, which has a library of symbols and features, such as walls and doors, that can be dragged and dropped into the drawing. In the days before CAD, a draughtsperson would have to painstakingly draw every detail on every drawing by hand.

INDUSTRY TIP

It is tempting to view drawings using tablets or smartphones. Just remember, though, that viewing electronic drawings doesn't necessarily show them at the proper scale. This is where printed drawings are particularly useful, and potentially worth the significant cost.

IMPROVE YOUR MATHS

Drawing scales are represented as a ratio. The first number of the ratio represents the measurement of the object on the drawing, while the second number represents its actual size. For example, on a scale of 1:50, a line on a drawing measuring 40 mm (4 cm) represents a wall 50 times bigger in reality, so the wall is actually:

40 mm × 50 = 2000 mm or 2 m

Or to look at it the other way round, a 4 m (4000 mm) wall in reality is 50 times smaller on a drawing so the line length would be: $\frac{4000 \text{ mm}}{50} = 80 \text{mm}$

Typical drawing scales frequently used are:

- 1:20 – where 50 mm on a drawing is 1000 mm in reality
- 1:50 – where 20 mm on a drawing is 1000 mm in reality
- 1:100 – where 10 mm on a drawing is 1000 mm in reality
- 1:500 – where 2 mm on a drawing is 1000 mm in reality

INDUSTRY TIP

Drawings may have to be reduced to fit on sheets of A3 until the full versions have been printed out, as many sites have easy access to an A3 printer but not to larger ones.

ACTIVITY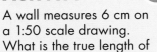

A wall measures 6 cm on a 1:50 scale drawing. What is the true length of the wall?

Paper sizes

In order to understand drawings at different scales and reductions, it is helpful to understand paper sizes.

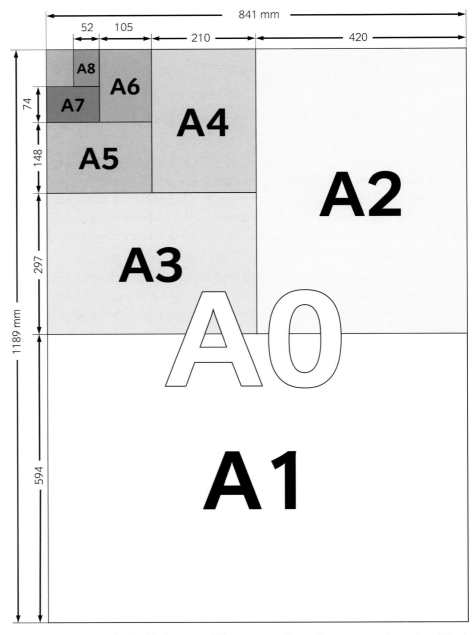

▲ Figure 3.11 The relationship between different paper sizes; all measurements are in millimetres

Paper sizes are based on proportions referred to as the golden rectangle. Size A0 is twice the size of A1, A1 is twice the size of A2, and so on through the range of paper sizes. The numbers increase as the paper size gets smaller.

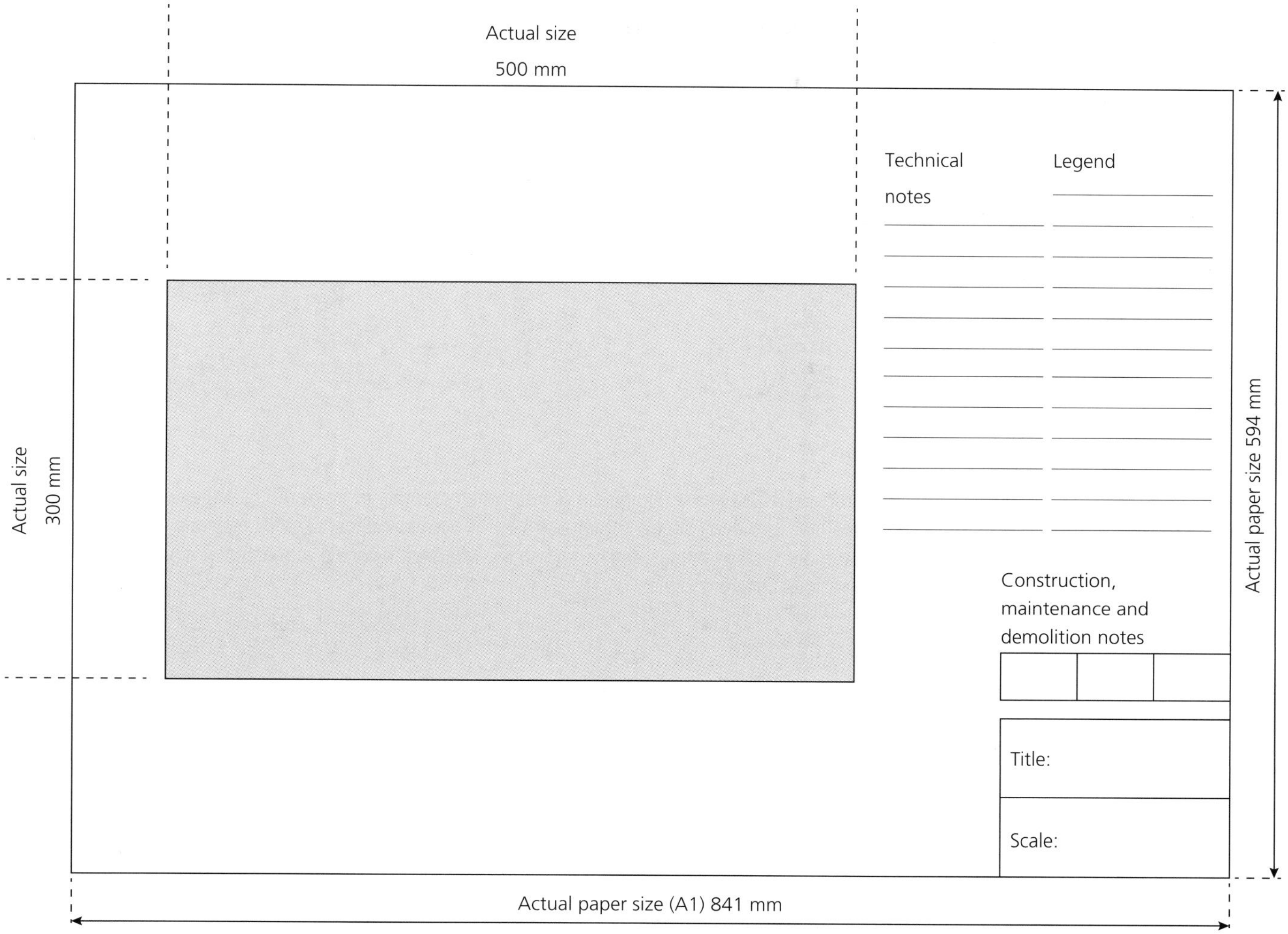

Actual size
500 mm

Technical notes

Legend

Actual size
300 mm

Actual paper size 594 mm

Construction, maintenance and demolition notes

Title:

Scale:

Actual paper size (A1) 841 mm

▲ Figure 3.12 The relationship between the scale of a drawing and the actual paper size

IMPROVE YOUR MATHS

As the size of paper reduces, the width of one size becomes the length of the next size down. This is a 71% reduction in size. Reducing by two paper sizes – for example, from A0 to A2 – produces an overall 50% (70.07% of 70.07%) reduction in size.

It is important to select a scale that is clear at the paper size you intend to use. For example, drawings of a whole building of 30 × 50 m, at a scale of 1:100, will fit onto A1 reasonably well, leaving enough space for title blocks and additional notes. The layout of the building will be 500 × 300 mm on paper.

In this example, it is possible to fill the paper further using a scale of 1:75.

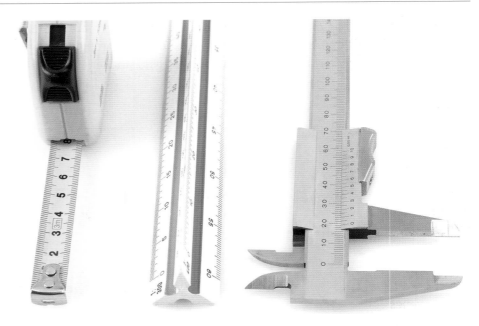

▲ Figure 3.13 Measuring equipment is vital when creating or interpreting scale drawings. The tape measure (left) shows actual size 1:1, the vernier callipers (right) measure accurate sizes of materials such as cables, pipes, and so on, whereas the scale rule (middle) measures distances to various different scales

Reduction

When drawings can be stored and viewed on screen, they are often only printed out for use by installers on site. Often paper copies are reduced in size for storage on A3 sheets. Then the user can quickly identify the drawing on paper before calling up the relevant drawing on screen. The readability on A3 will depend on the size and scale of the original. For example:

- 1:50 at an original paper size of A1 will become 1:100 on A3
- 1:50 at an original paper size of A0 will become 1:100 at A2 or 1:141 at A3.

The greater the reduction, the less readable the drawings become.

③ HOW ELECTRICITY IS SUPPLIED

Electricity is supplied throughout Great Britain by National Grid, which was formerly the Central Electricity Generating Board (CEGB). National Grid owns and maintains the high-voltage electricity transmission system in England and Wales, as well as operating the system across Great Britain. As electricity cannot be stored in significant amounts, National Grid is responsible for balancing **supply with demand**.

Electricity generating methods

Electricity in Great Britain is generated in a number of ways:

- nuclear power
- coal-fired power

INDUSTRY TIP

You can access more information about Nation Grid Electricity via their website: www.nationalgrid.com/uk

KEY TERM

Supply with demand: producing electricity (the supply) at the rate it is being used (demand).

- gas-fired power
- renewable energy
- other technologies.

Nuclear power

The nuclear industry is responsible for generating a large proportion of the **base load** of electricity needed in Great Britain. Nuclear reactors are used to provide the base load because they cannot easily be turned on or off. The nuclear industry currently has 16 reactors.

Nuclear fission, which takes place in the reactors, creates heat. The heat is used to create high-pressure steam, which turns turbines coupled to generators to create electricity.

Coal-fired power

Coal is used as a fuel to heat up water to create high-pressure steam. This steam turns turbines coupled to generators to create electricity. Traditionally, coal-fired power stations were the primary source of power generation in Britain. With the decline of coal production, rising costs of importing coal and changes in legislation relating to emissions, this source of energy has been squeezed by other producers. In addition, coal is a fossil fuel, which cannot be replaced.

As a result of all these factors, large-scale providers of coal-fired electricity are now looking at more carbon-efficient schemes such as large-scale biomass to reduce their **carbon footprint**.

▲ Figure 3.14 The coal-fired Drax Power Station in North Yorkshire, England

Gas-fired power

Gas is used to create steam that can be used to generate electricity in the same way as coal.

KEY TERM

Base load: basic amount of electricity.

KEY TERM

Carbon footprint: a measure of carbon dioxide emissions from burning fossil fuels during the production and transportation process. This may be offset by the amount of carbon released naturally by decay. For example, biomass fuel is made from plant material that would release carbon dioxide as it rots naturally if left – so burning it efficiently as a fuel, and gaining use from it as a fuel, means the carbon footprint is reduced making it carbon efficient.

ACTIVITY

What is the power loss in a cable having 0.2 Ω resistance when:

- 10 A is drawn through the cable
- 20 A is drawn through the cable?

KEY TERM

I^2R **losses**: power losses caused by the current heating the cable and the heat energy being lost.

Demand for gas-fired power in Britain has increased as the demand for coal-fired power has declined. Gas is a fossil fuel, like coal, but it is now the largest producer of electricity in Britain.

Renewable energy

Renewable technologies account for an ever-growing share of British electricity production. These include wind turbines, solar photovoltaic panels, and so on. Building energy targets and targets for reducing atmospheric carbon dioxide levels mean that the renewable and green energy sectors are set to increase in importance. These are covered in more detail in Book 2.

Other technologies

Other technologies for producing electricity, including small-scale generation and hydro-electricity, represent less than 3% of electricity production in Britain.

Pumped storage systems allow hydro-electric power to meet demand at peak times. The system uses off-peak electricity to pump water from a low level to a higher level, where it can be stored. At peak demand times, the stored water is released through turbines to create electricity. The cycle can then begin again.

The table shows the proportion of fuels used in the UK to generate electricity, as of April 2017.

Fuel	Percentage of overall UK electricity production
Coal	2%
Oil	3%
Gas	41%
Nuclear	24%
Renewables	30%

Transmission voltages

Large-scale electricity generation is carried out by producing superheated steam, which is then released through turbines to drive large, high-voltage alternators, which turn the kinetic energy into electrical energy.

The energy output from the alternators is fed to transformers that raise the voltage from 20–30 kV up to 275 kV (275 000 V) or 400 kV (400 000 V). Transmission voltages need to be this high in order to minimise power losses through the cable (I^2R **losses**) and keep cable sizes manageable.

Pylons are large metal structures that keep high-voltage and extra-high-voltage cables away from people for safety and convenience. They also require inspection and maintenance because they are exposed to all weather conditions and a great deal of wear and tear.

HEALTH AND SAFETY

Electrical arcs are extremely dangerous to work around. Anyone approaching extra high-voltage cables could be electrocuted without touching the conductor. Intruders in high-voltage substations have died without actually touching the equipment, due to electrocution caused by a lightning-bolt-type shock from the equipment.

At high voltage, electricity can be conducted through the air for very short distances. The higher the voltage, the greater the **arc** length can be.

In sites where there is high-voltage equipment, appropriate insulators are used to prevent short circuits between the site structure and the conductors that carry the current. The higher the voltage being handled, the greater the length of the insulators required.

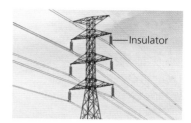

▲ Figure 3.15 An electricity pylon with insulators

KEY TERM

Arc: in this instance, an electrical breakdown of a gas that produces an ongoing plasma discharge, resulting from a current through a normally non-conductive medium such as air.

▲ Figure 3.16 Some insulators have to be very large

IMPROVE YOUR MATHS

Power losses in a cable

A transmission cable has a given resistance due to its cross-sectional area (csa) and length (R). We could say that the cable's resistance is 20 Ω. If a locality served by the cable has an overall consumption of 1 MW (1 000 000 W) and the supply cable voltage is 400 V, the current drawn through the cable would be:

$$\frac{1\,000\,000}{400} = 2500 \text{ A}$$

So the loss would be:

$$2500^2 \times 20 = 125\,000\,000 \text{ W}$$

which means the loss is greater than the consumption.

If the voltage is 400 kV (400 000 V), then:

$$\frac{1\,000\,000}{400\,000} = 2.5 \text{ A}$$

So the loss would be:

$$2.5^2 \times 20 = 125 \text{ W}$$

which is much more acceptable as a standing loss.

ACTIVITY

What would be the losses for a cable having a 10 Ω resistance used to supply a community consuming 20 000 kW?

Balancing the load

Electricity is not easily stored, so National Grid has to assess power supplies to cope with peak demand times as well as off-peak times. It is a complex task to predict **load profiles**, negotiate on- and off-grid time for generators and manage any export of electricity to or import from France and/or Ireland.

At a local level, substation transformers need to keep loads balanced as closely as possible across the three phases. If this isn't achieved, the transformer becomes less efficient. This is why three phase distribution boards should have their loads balanced as closely as possible over the three phases. Consumers may be fined by the distribution network operator (DNO) if their loading isn't well balanced.

Distribution voltages and networks

Once the electricity has been transmitted to the particular region in which it is required, National Grid depends on historical infrastructure to transform it from 400 kV or 275 kV to a distribution voltage of 132 kV.

At this point, distribution is handed to one of the 14 regional distribution network operators (DNOs) to manage and maintain. As the regions are quite large, high voltages such as 132 kV are required to minimise losses. At these supply voltages, cables are still generally on pylons, although some of the infrastructure is buried underground.

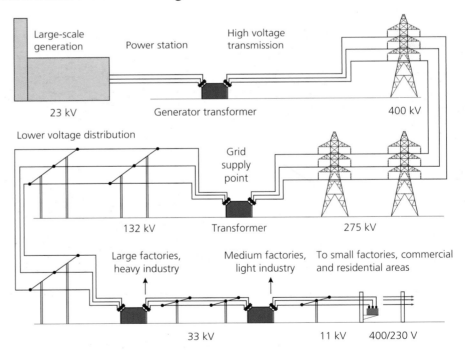

▲ Figure 3.17 Transmission and distribution voltages

Because the switchgear and transformer areas are very large, voltages of 132 kV are generally difficult to manage and maintain close to towns and cities. In order to make localised distribution more manageable, the voltage is reduced to 33 kV.

In rural areas, supplies are distributed through cables on wooden, or similar, poles, the height of which can be as little as 6–7 m. These power lines run alongside roads and across fields. In urban areas the cables are usually buried underground or can be supplied on small overhead pylons.

These supply cables terminate at primary substations. These substations are large and unsightly due to the size of the switchgear they accommodate, so they are normally located in areas around the edges of cities that are not commercially sensitive, although some are integrated into city centres. Primary substations transform the electricity to lower voltages for distribution. Some industrial users want electricity at 33 kV, while other specialist plants such as chemical works use 132 kV.

Primary substations also distribute 11 kV or 6.6 kV supplies to 'domestic' transformers located closer to the consumers. These supplies are usually distributed via cables buried underground in towns and cities or on wooden poles in more rural areas. Many public or large buildings take their electricity at high voltage (usually 11 kV) due to the size of their demand, which usually has to be in excess of 500 kVA.

▲ Figure 3.18 A primary substation

Local transformers, which are usually 1000 kVA (1 MW) or even 500 kVA, serve individual buildings or residential areas. These transformers can be smaller, especially when pole-mounted, where they are limited to around 200 kVA. From this point, electricity is supplied to the consumer either as a single-phase 230 V or a three-phase 400 V system.

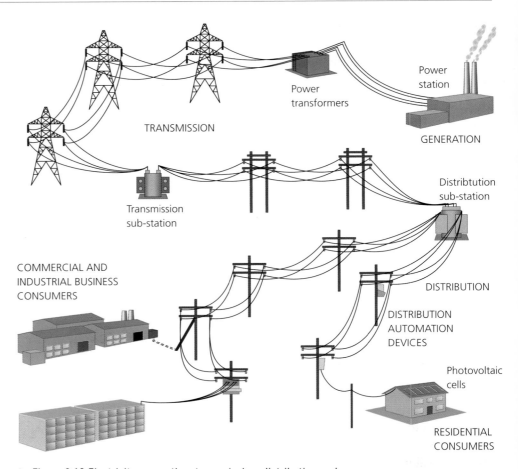

▲ Figure 3.19 Electricity generation, transmission, distribution and use

4 INTAKE AND EARTHING ARRANGEMENTS

The type and rating of electricity supply to an existing installation, given by the distribution network operator (DNO), has a great bearing on the design of an installation. The type and rating of equipment, for example, and the methods of protection against electric shock are areas greatly affected by the supply to an installation. For new buildings, those factors will influence the type of supply requested.

Three factors relating to the supply are:

- the number of live conductors
- the maximum current rating
- the earthing system used.

Live conductors

Supplies to small domestic-type installations are normally single-phase 230 V, meaning there are two live conductors (line and neutral). Where supply has two live conductors, it is described as two-wire.

Larger installations, which need more than 100 A, will normally be given a three-phase supply, allowing the larger load to be spread over the three phases at the substation transformer and keeping the system more efficient. This will be explained further in Book 2, Chapter 2.

The table shows the live conductor arrangement within the UK.

Type	Arrangement	Where
Single-phase 230 V L–N	Two-wire	Dwellings, small shops or small commercial buildings
Three-phase 400 V L–L	Three-wire (no neutral)	Very uncommon as supplies to installations, as all circuits must be three-phase balanced loads
Three-phase and neutral 230 V L–N : 400 V L–L	Four-wire	Larger retail outlets, larger commercial or industrial buildings

Current

Another factor that affects the supply is the current rating known as **maximum demand**. Most normal-sized dwellings in the UK will have a maximum demand less than 100 A and many houses in a road or estate will be supplied from a single substation transformer. Most commonly, house supplies are rated at 80 A.

Larger installations, which have a maximum demand over 100 A, will likely be supplied using three phases in order to spread the load over the supplier's equipment. Where a building or facility needs a supply over 400–500 A per phase, it will likely have its own substation transformer installed on site.

The maximum demand for an electrical installation will be limited by the rating of the DNO's service fuse so where an installation is to be added to, a check must always ensure that the supply has enough capacity for the extra load.

Where a new building is to be given a new supply, the DNO must be told what the proposed maximum demand for the electrical installation is so they can provide an adequate supply. This would be the job of the installation designer to provide this using **diversity**.

The DNO's service fuse will be fitted inside the electrical service head (cut-out) which is the enclosure the supply cable terminates into before the tails go to the meter. Most DNO fuses will be in a fuse carrier that is 'sealed' to the service head, meaning it can only be removed or replaced by a person approved by the DNO or the DNO themselves. The seal is a small wire that, with a special sealing crimp tool, locks the fuse to the service cut-out so it has to be broken if the fuse is removed. Connections to meters will also be sealed in this way. If you need to remove a fuse from a service head, you must always contact the DNO.

It is always important to know the supply rating, but it is not always easy to see what the DNO fuse rating is. In this situation, you should always contact the DNO who can tell you all the supply details such as rating.

KEY TERMS

Maximum demand: the full load current of an installation assuming all circuits are drawing their design current at one time. As most installations do not actually do this, designers apply diversity.

Diversity: allows the designer to use skill and experience to calculate what the full load may be, given that some circuits operate at certain times (e.g. lighting) while others operate for very short times (e.g. hand driers in toilet facilities). The result of applying diversity to the maximum demand is usually used to determine the supply characteristics.

INDUSTRY TIP

Typical current ratings of DNO service fuses are 63 A, 80 A, 100 A, 125 A, 150 A, 200 A, 300 A, 400 A and 500 A.

Earthing arrangement

Five types of **earthing system** are defined in **BS 7671:2018** The IET Wiring Regulations, 18th Edition:

- TN-S
- TN-C
- TN-C-S
- TT
- IT

TN systems

TN systems are configurations where the distribution network operator (DNO) provides the means of earthing. The two types commonly used in the UK are TN-S and TN-C-S.

TN-S systems

A TN-S earthing configuration has only one neutral-to-earth connection, which is as near as practicable to the source (supply transformer). In low-voltage supplies, the consumer's earth terminal is connected to the metallic sheath of the supply cable.

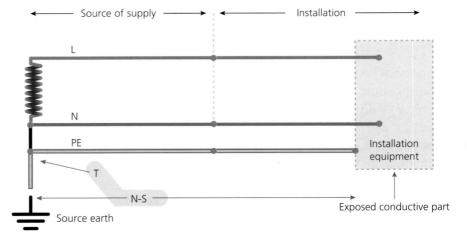

▲ Figure 3.20 TN-S system arrangement, showing the neutral-to-earth connection, separate supply cables throughout their length and the earth-to-neutral connection close to the supply

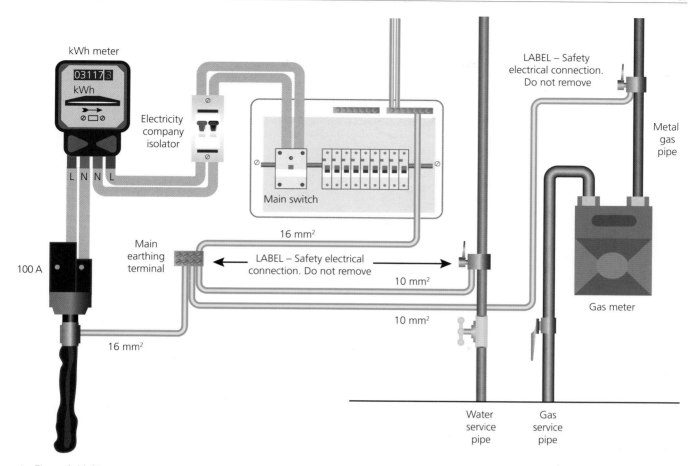

▲ Figure 3.21 Single phase TN-S system, showing intake equipment (the supply company switch is not always installed)

Figure 3.20 shows that the earthing conductor is connected to the DNO's supply cable metallic sheath. This sheathing provides a separate route back to the substation transformer. Because the return path is usually a material such as steel, the DNO will normally declare a maximum external earth fault loop impedance (Z_e) of 0.8 Ω.

TN-C-S systems

This configuration is now very common throughout the UK as it allows the DNO to provide a low-voltage supply with a reliable earthing arrangement to many installations across the country.

The TN-C-S arrangement is also known as protective multiple earthing (PME). It relies on the neutral being earthed close to the source of supply and at points throughout the distribution system. There is also a neutral-to-earth connection at the

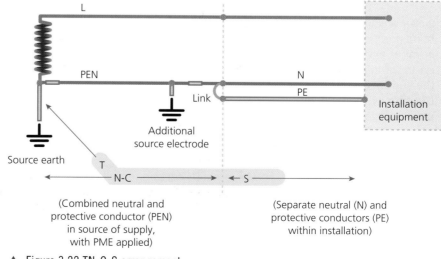

▲ Figure 3.22 TN-C-S arrangement

intake of the installation. As the DNO uses the combined neutral and earth return path (known as a protective earthed neutral or PEN), the maximum external earth fault loop impedance declared by the DNO is 0.35 Ω.

There may be a number of consumers using the supply cable. A rise in current flow will create a voltage rise in the PEN, which needs multiple connections to the general mass of earth along the supply route.

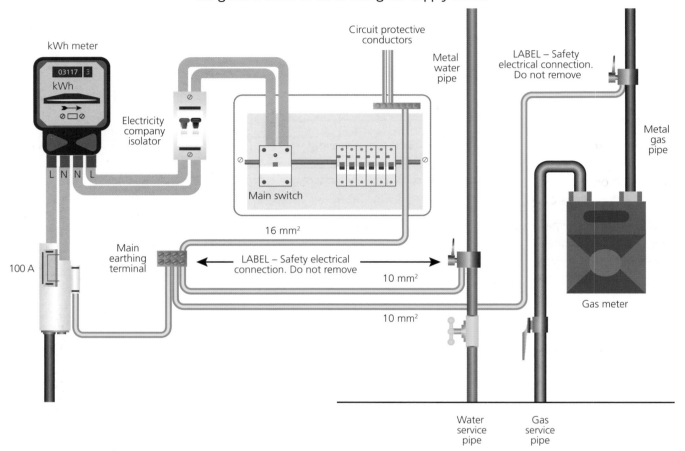

▲ Figure 3.23 Single phase TN-C-S (PME) arrangement

Although TN-C-S, or PME, is the most common type of earthing configuration in the UK, there are risks associated with it. If the PEN conductor were to become open circuit in the supply, current flowing through the installation would not have a path back to the substation through the supplier's cable. Instead, current could try to follow an alternative path through the earthing system of the installation, which may include service pipework or the general mass of earth.

Unfortunately, it could be people who make that link between the earthed metallic equipment and the earth.

As a result, certain installations such as petrol filling stations and some construction sites and caravan parks cannot be supplied by TN-C-S arrangements. Even in domestic or commercial installations, some restrictions may apply such as certain outbuilding supplies. Where this occurs, TT systems are preferred.

TT systems

This type of system is configured in very much the same way as a TN-S system in terms of the earthing of the supply source. However, a TT system does not provide the consumer with an earth connection. Instead, the earth for the consumer's installation has to be supplied by the consumer, usually by driving earth rods or burying metallic plates or strips into the ground so as to provide a path of low enough impedance through the ground to give protection.

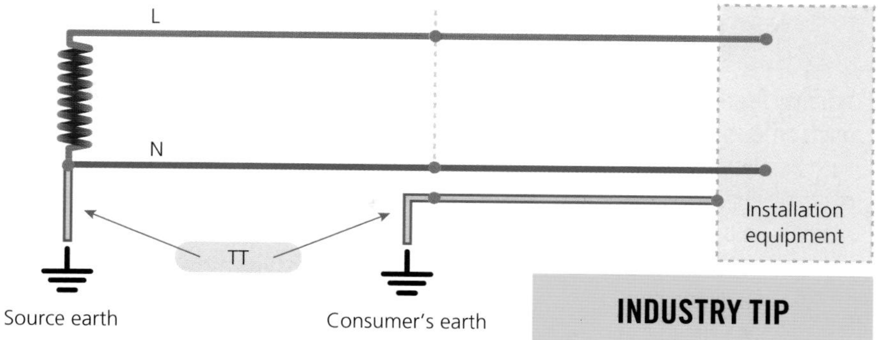

▲ Figure 3.24 TT arrangement

TT systems are usually installed either where a TN-C-S arrangement is not permitted (e.g. in a petrol filling station or in rural installations where the supply is provided on overhead poles) or where there is little or no opportunity to provide other types of system.

As the earth return path uses the general mass of earth, external earth fault loop impedance values (Z_e) may be very high where different soil types exist, meaning further shock protection measures such as RCDs may be required to provide automatic disconnection of supply (ADS).

INDUSTRY TIP

In some situations, the external earth fault loop impedance is too high for a conventional test instrument to measure. In this situation, the consumer's earth electrode resistance is measured, which includes the resistivity of the soil around it. As this method does not provide a measurement of the earth return path (Z_e), it is recorded as R_A. R_A is the resistance of the earth electrode.

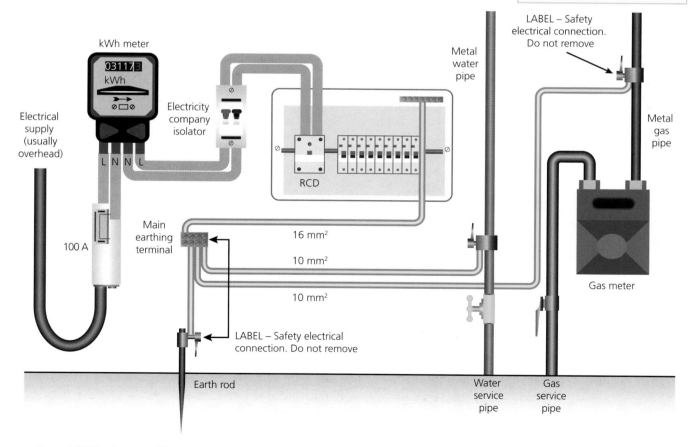

▲ Figure 3.25 Single-phase TT arrangement, including earth electrode and RCD

5 CONSUMERS' INSTALLATIONS

An installation is made up of many parts and in certain ways to make them safe and functional. These include:

- Consumers' Control Unit (CCU) or distribution board (DB)
- circuits
- earthing, bonding and protection
- isolation and switching.

Consumers' Control Unit (CCU)

Generally, the first unit in an installation, where the supply is split into circuits, is called the Consumers' Control Unit (CCU). This may sometimes be referred to as a distribution board or DB and sometimes simply as a consumer unit (CU). The CCU will have protective devices inside. They control and protect the circuits and could include:

- circuit breakers (CB)
- residual current breakers with overload (RCBO)
- fuses.

The CCU will also have a main switch, which can be used to isolate the entire installation. In a domestic dwelling, the main switch must be double-pole, meaning it isolates both line and neutral. In some three-phase installations that are part of a TN earthing arrangement, the main switch may be three-pole, switching the three-line conductors only. In this situation, the neutral can be isolated by a link in the CCU.

Three-phase installations having a TT earthing arrangement must have four-pole switches isolating all live conductors.

The main switch in a CCU within a house must be the type that is rated to switch the full load current. Some isolator devices which act as main switches are not rated for full load current and, if they are used regularly to switch full load, they will eventually fail.

Modules and ways

When you install a distribution board or CCU, the numbers of circuit breakers or main switches that can be installed are described differently:

- *Modules* – describes the number of components that can be fitted into the unit. For example,

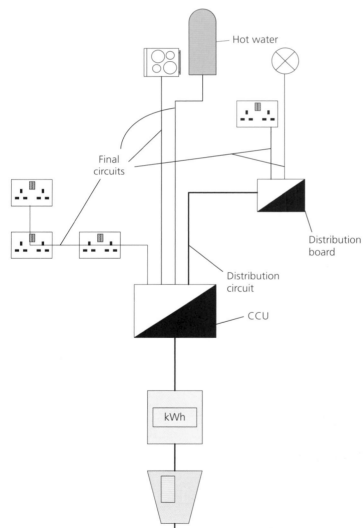

Figure 3.26 Circuit arrangements in a typical installation

a main switch controls both line and neutral so is classed as a two-module device. A circuit breaker is single-pole, so it is one module. If you needed a CCU to control six circuits and have a main switch, you would need an eight-module board.

- *Ways* – describes the number of circuits that can be controlled. A six-way single-phase board could house six circuit breakers as well as the main switch.

A six-way three-phase board can house six three-phase circuit breakers or 18 single-phase circuit breakers. It could also house a combination of the two such as 2 × three-phase (six ways) and 12 × single-phase (12 ways).

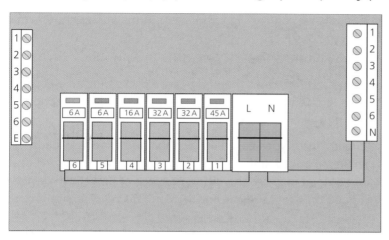

▲ Figure 3.27 A six-way distribution board with eight modules

Split-way

Some CCUs are known as split-way boards. This means that some of the modules can be protected by one RCD or switch and the others by another RCD or switch. Split-way boards should always have one main switch that can isolate all circuits. The advantage of split-way boards is that any fault on one circuit will only trip part of the installation protected by one of the RCDs.

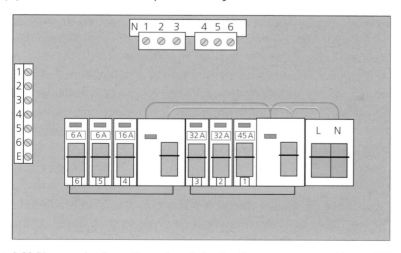

▲ Figure 3.28 Diagram showing split-way board showing three ways protected by one RCD and three ways protected by another. In total, the DB is 12 module

Different types of circuit

The **BS 7671:2018** The IET Wiring Regulations, 18th Edition state:

> The number and type of circuits required for lighting, heating, power, control, signalling, communication and information technology, etc. shall be determined from knowledge of:
>
> (i) location of points of power demand
>
> (ii) loads to be expected on the various circuits
>
> (iii) daily and yearly variations in demand
>
> (iv) any special conditions, such as harmonics
>
> (v) requirements for control, signalling, communication and information technology, etc
>
> (vi) anticipated future demand if specified.

This requirement demonstrates the importance of dividing an electrical installation into different circuit types to meet the different functions of a complete working electrical installation.

By dividing circuits into types based on location and expected loads, system reliability can be enhanced and cable kept to sensible sizes. This is because different types of load may require different types of protective device. The designer will need to assess the current demand, taking diversity into account.

Circuits can be categorised in the following way:

- **Lighting circuits** – provided in all installations. They are low-power circuits with an almost constant load when lights are operating. For more detail, see the section on lighting circuits below.
- **Power circuits** – provided for socket outlet circuits. They require larger protective devices due to the demand for current and, in most instances, require RCD protection due to the additional risk presented by the portable equipment connected to the socket outlets. For more detail, see the section on power circuits, starting on page 181.
- **Alarm and emergency systems** – require additional protective measures, which include separating them from other circuits so that they are not affected by any faults on those other circuits. Further protection by wiring the circuit in fireproof cabling, may be considered.
- **Data and control circuits** – normally wired in different types of cable to a low-voltage electrical system. However, the electrical supplies need more rigorous cabling and earthing arrangements to reduce interference between the different wiring systems and to meet electromagnetic compatibility (EMC) requirements.

Lighting circuits

Lighting circuits are generally rated at 6 A but can, in some installations, be rated at 10 A or 16 A. Consideration must be given to the type of lighting point used for higher rated circuits due to requirements of Section 559, Luminaire and Lighting Systems in **BS 7671:2018** The IET Wiring Regulations, 18th Edition.

Depending on the type of wiring system, lighting circuits may be wired:

- **three-plate** – for circuits wired in **composite cable** such as thermosetting insulated and sheathed flat profile cable, commonly found in domestic dwellings
- **conduit method** – for circuits wired in single-core cables within a suitable containment system.

Three-plate

Because composite cables limit the freedom to take a single conductor between one point and another, lighting circuits wired in composite cables require more joints within the circuit. These additional joints in the cable are called loop-ins and are usually in the form of a third live terminal within a light point such as a ceiling rose. As the rose contains a third terminal, the system gets the name 'three-plate'. Although it is traditional to install the three-plate at the light point, this is not necessarily the case in all situations. Some installers will install the three-plate at a switch, normally when the wiring in the ceiling is difficult to access such as in multi-storey flats or flat roofs.

Figure 3.29 shows both the connections, and that a two-core with circuit protective conductor (cpc) cable is used to connect the various points. Other light points can be installed on the circuit by looping off the ceiling rose. The circuit can be modified to allow two-way switching. Two-way switching is where one light point or luminaire can be controlled by two switches.

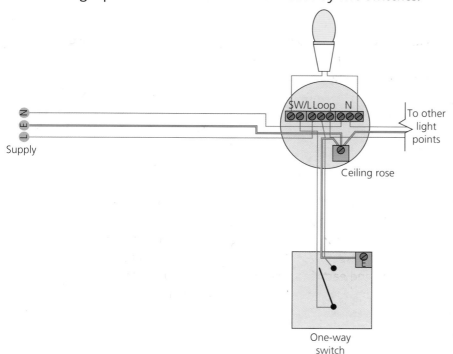

One-way switch

Figure 3.29 A three-plate one-way lighting circuit

Linking the two switches using a three-core and cpc cable allows the circuit to be opened or closed from either switch. Two-way switches do not actually open and close, but divert the current from the common terminal (C) to either terminals L1 or L2. The two cables that link terminals L1 and L2 between the switches are called strappers. To allow the light point or luminaire to

KEY TERM

Composite cables: multi-cored cables, in which the cores are surrounded by a sheath providing mechanical protection.

Conduit method: the common term used when the circuit is wired in single-core cable. This could be in trunking or conduit, but it means that some cables do not need to be broken at the light point and can be wired directly to the point where they need to be terminated.

be controlled from one or more further switches, the strappers need to be interrupted by one or more intermediate switches as shown.

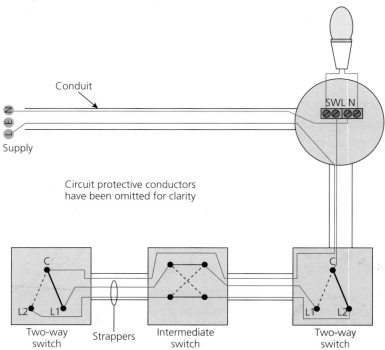

▲ Figure 3.30 A three-plate circuit controlled via two-way switches

With three-plate systems, it can be seen from the diagrams that cable cores need jointing to enable them to be extended to a further point. One example of this is the common connection between the two-way lighting that requires jointing in the intermediate switch.

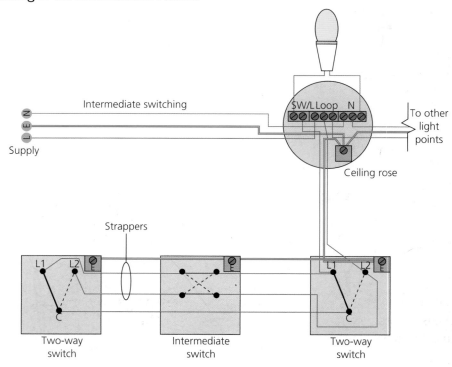

▲ Figure 3.31 Three-plate with intermediate switching

More and more, three-plate wiring is being carried out at the switch. This allows a neutral at the switch, which could be required where home automation devices are used. This method also has the advantage of leaving fewer connections at the light point where specific light fittings are installed instead of ceiling roses and these light fittings do not have the provision for a permanently live connection.

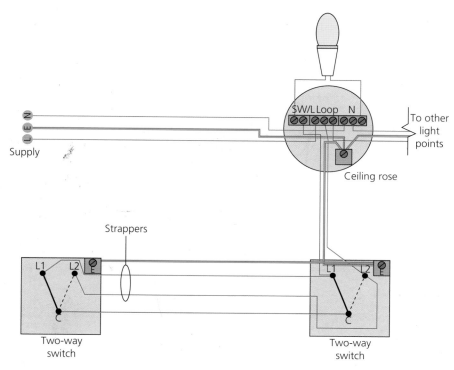

▲ Figure 3.32 Two-plate intermediate circuit using single-core cables

Power circuits

Power circuits generally supply socket outlets but may also supply individual appliances. Power circuits may be wired in two ways:

- ring-final
- radial.

Ring-final circuits

Ring-final circuits are the traditional means of wiring socket outlet circuits within the UK. The reason for their use was to provide a high number of conveniently placed outlets adjacent to the loads. The circuit load is shared by two conductors in parallel. This also assists in improving voltage drop as the conductors being in parallel reduce the overall resistance.

Ring-final circuits are able to supply an unlimited number of outlets serving a maximum floor area of 100 m². Permanent loads, such as immersion heaters, should not be connected to a ring-final circuit. Ring-final circuits supplying kitchens should be arranged to have the loads equally distributed around the circuit. The plugs and fused spur units are normally fitted with 3 A or 13 A fuses.

With technology reducing the consumption of appliances and equipment, the need for a ring-final circuit is being questioned as they do have several disadvantages. If a ring-final circuit becomes open circuit, this may not necessarily be detected by the user as power will still be distributed to all socket outlets. Circuit conductors may then become overloaded leading to the possibility of fire. Earth fault loop impedances and circuit resistance values between live conductors may also increase due to this, leading to increased disconnection times. Testing continuity of ring-final circuits is also more time-consuming than testing the continuity of radial circuits. Designers today are very likely to specify radial socket outlet circuits with a reduced nominal rating of protective device due to these reasons. This also reduces the inconvenience of many socket outlets being lost due to a single circuit failure.

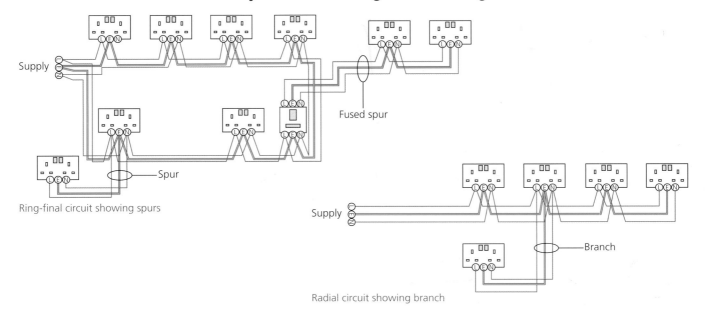

▲ Figure 3.33 Ring-final circuit showing spurs and radial circuit showing branch

Radial circuits

Radial circuits may be selected to supply multiple socket outlets for the reasons given above, and they are also chosen as a means of supplying individual appliances or dedicated fixed appliance circuits.

Wiring systems for different environments

Different types of wiring system are available because not all systems are suitable for all environments or compatible with all types of construction.

Domestic installations

▲ Figure 3.34 Twin and earth cable

These installations tend to use twin and earth cables (referenced as 6242Y by wholesalers) or similar types of sheathed cable. This type of cable is ideal because it is compatible with the building products used and the finishes expected by the end user. It is also easy to install because it is flexible and can be bent and shaped into tight locations. Although other types of cables could be used, the relatively low cost of this type is also important in the domestic market, which is very cost sensitive.

> ## INDUSTRY TIP
>
> Holes for cables in joists should be drilled as level as possible, to allow ease of drawing in. Holes drilled at sharp angles will abrade the cables and damage the outer sheath.

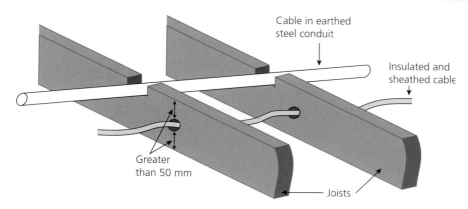

Cable in earthed steel conduit

Insulated and sheathed cable

Greater than 50 mm

Joists

▲ Figure 3.35 Installation method when crossing joists

There are various types of containment systems suitable for PVC/PVC 6242Y wiring systems. However, these tend to be used in non-domestic circumstances (see pages 186–187).

Commercial installation

There are so many different types of commercial building that this category probably includes the widest range of electrical installations. Commercial installations can be found in large converted houses, listed buildings, purpose-built portal frame buildings, multi-storey offices and shopping complexes with pre-tensioned concrete structures. In all these scenarios the installations need to be compatible with the structure and the finished layouts need to be either as cost effective or as sympathetic to the aesthetics of the building as possible.

Any, or all, of the following systems may need to be considered in commercial installations.

Supply distribution systems are usually found in the 'back-of-house' areas or common structural cores of commercial buildings where special finishes to the building fabric are not required. In such functional areas it would not be unusual to see steel wire armoured (SWA) cable with low smoke and fume sheathing, which is usually described as XLPE SWA LSF cabling.

Fire alarm and similar systems need to be wired in fire-resistant cabling such as mineral-insulated, copper-clad cables (MICC), a soft-skinned fire-resistant cable such as FP200 or, where enhanced cabling is required by Building Regulations, FP200 plus or similar products.

▲ Figure 3.36 Prysmian FP200 fire-resistant cable

▲ Figure 3.37b The Kitemark is used to identify that products, for example fire alarms, conform to a particular specification or standard (source: BSI)

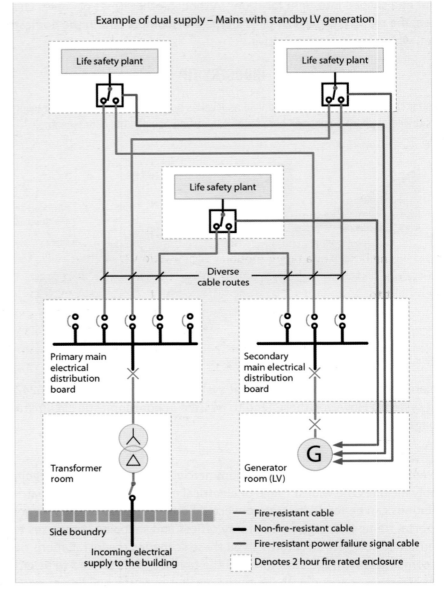

▲ Figure 3.37a Typical fire-fighter lift arrangement using 2-hour fire-resistant cabling (source: BSI)

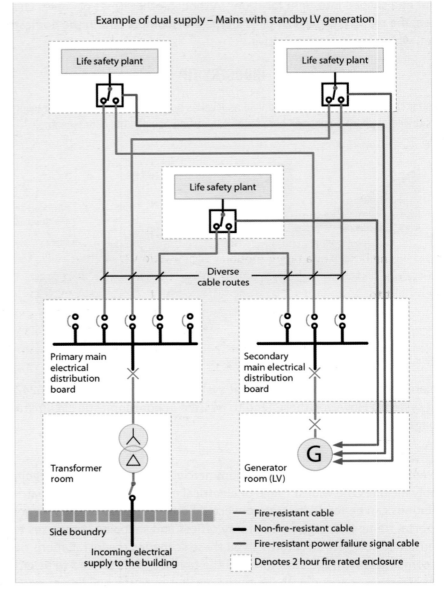

▲ Figure 3.38 Prysmian FP600 fire-resistant armoured cables

Safety of life systems are fire resistant, usually for 120 minutes. They include FP600 ranges of cabling, which are tried and tested for fire-fighting equipment, sprinkler pumps, lifts, and so on. These systems are usually separated from other systems so that they are not affected by faults on any of the other systems.

For systems in public areas, it is likely that the installation will be in recessed conduits and that single cables will be concealed in the structure. Office developments often have raised floors and power is provided through floor boxes plugged into underfloor 63 A busbar systems. With the rise in prefabrication of site services, modular wiring (often known as 'plug n play' systems) may even be used.

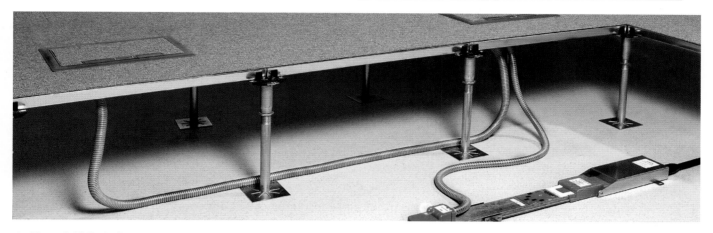

▲ Figure 3.39 Underfloor busbar and plug-in floor box arrangement

Modular wiring is part of a larger modular construction manufactured off site. It includes pipework, ductwork and cabling. The modular construction is then brought to site and installed at an appropriate time in the construction programme. This reduces the number of skilled operatives needed to install the electrical system, as well as labour time generally. This often reduces costs. Modular wiring cannot be used for fire-fighting or safety of life systems as these must be kept separate from other systems.

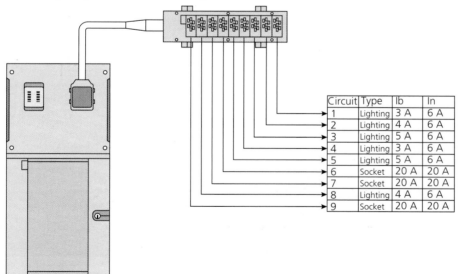

Circuit	Type	Ib	In
1	Lighting	3 A	6 A
2	Lighting	4 A	6 A
3	Lighting	5 A	6 A
4	Lighting	3 A	6 A
5	Lighting	5 A	6 A
6	Socket	20 A	20 A
7	Socket	20 A	20 A
8	Lighting	4 A	6 A
9	Socket	20 A	20 A

▲ Figure 3.40 Modular wiring diagram

Agricultural installations

These installations require the wiring system to be resistant to the elements, livestock and vermin. As there is an increased risk of damage, armoured cables, conduit and sealed trunking systems are used. PVC conduit is more suitable as it will not corrode.

ACTIVITY

Which section of BS 7671:2018 specifically deals with agricultural installations?

▲ Figure 3.41 Steel wire armoured (SWA) cable

Industrial installations

These installations are often located in harsh environments. Large amounts of SWA cable are installed on ladder racking along main routes from switch rooms. Where individual cables drop to serve distribution boards or other equipment, they are usually supported in baskets or trays.

Local wiring from distribution boards is normally fed to the relevant outlet through thermoplastic singles in conduit and trunking systems or through thermoplastic SWA cabling on ladder and basket arrangements.

Single insulated cables

Wiring systems using single insulated cables are common in industrial and commercial environments. The insulation is generally thermoplastic or LSF XLPE (low smoke and fume cross-linked polyethylene cable). These types of wiring systems are vulnerable to mechanical damage, so they must be contained in either complete trunking or a conduit system.

Installations in hazardous environments

Hazardous environments require robust cabling systems to prevent any possibility of ignition or triggering an explosion. The majority of large supply cables are SWA cables on ladder racking.

▲ Figure 3.42 Ladder racking containing SWA cable

Where fire resistance is required, however, mineral-insulated, copper-clad (MICC) cables are used to provide strength, integrity and fire resistance.

INDUSTRY TIP

Cable basket is a lightweight system that normally clips together.

▲ Figure 3.43 An oil refinery is obviously a hazardous environment with the risk of explosion

▲ Figure 3.44 A petrol forecourt, although familiar, is still a hazardous environment due to the risk of ignition

Containment systems

Most containment systems are not exclusive to a particular type of wiring system or installation.

Cable baskets can be used to contain many types of cable that have more than basic insulation, including PVC/PVC 6242Y cables, MICC, soft-skinned fireproof cables and, on a smaller scale, steel wire armoured cables. Cable baskets are often used for Category 5A and Category 6 TCP/IP data cabling.

▲ Figure 3.45 IT cables on basket

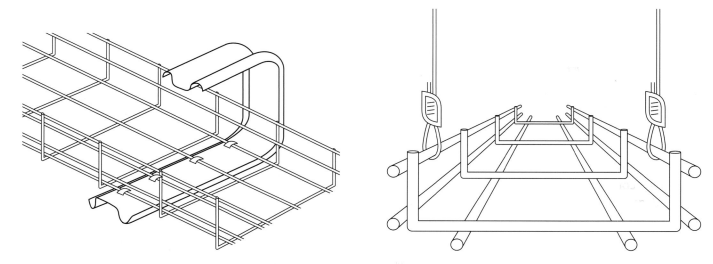

▲ Figure 3.46 Methods of fixing baskets

Ladder racking and tray work containment systems are predominantly for industrial installations, although tray work can also be used for smaller armoured cables and MICC and other soft-skinned cables. There are various types and grades of tray work to suit different cables and installations.

INDUSTRY TIP

Trays can be bent, using a tray bender. Ladder is too heavy to be bent and manufactured fittings are used.

▲ Figure 3.47 Returned flange traywork

Ladder racking is used almost exclusively for armoured cables and typically where large amounts of heavy cables are installed because it is strong enough to withstand the weight.

▲ Figure 3.48 Ladder racking

Conduit systems

In the UK, conduit systems are used in a number of ways. They include complete systems for single thermoplastic or XLPE cables with just basic insulation. This gives mechanical protection to prevent damage to the insulation and reduces the risks of shock or fault hazard.

Conduit systems can be made of plastic, galvanised metal, stainless steel or black enamelled steel. Where there is risk of damp or corrosion, PVC or galvanised conduit should be selected. Where strong mechanical protection is required, galvanised conduit is normally used.

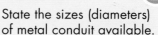

ACTIVITY

State the sizes (diameters) of metal conduit available.

INDUSTRY TIP

Always try to draw all cables into conduit simultaneously, to prevent damage. Trunking systems allow easy access, to enable the addition of extra cables at a later date.

▲ Figure 3.49 Black enamel finished metal conduit and coupling

Conduit systems are also used in part for protection of wiring system 'drops' to electrical equipment, so that walls or sections of building work can be completed and the cables pulled in later.

Couplings are used to connect lengths of conduit. Where a conduit enters a box that does not have a threaded entry, it is terminated by a coupling and a brass bush to complete the mechanical connection.

▲ Figure 3.50 This stop end box has been converted into a back entry box using a conduit coupling and male bush arrangement. In proprietary assemblies, the box would have a manufactured thread into which a conduit with a male thread could be screwed

▲ Figure 3.51 Systems of joining conduits including back entry boxes, which can be manufactured or fabricated

Where two systems cannot be screwed together, a running coupling might be fabricated on site using two couplings, a nipple, a locknut and a length of conduit with an extra long thread. However running couplings create a potential weakness in the conduit and are best avoided.

Spacer-bar saddles are the most commonly used method of securing conduit. Distance saddles are used where conduit is fixed to uneven or damp surfaces. Where hygiene is important, hospital saddles are used to allow easy cleaning of the shaped saddle and around the conduit.

Trunking systems

In the UK, most of the wiring contained in trunking systems is single insulated cable. However, trunking systems can be used to carry many different types of

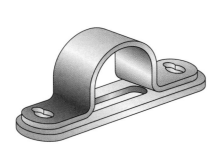

▲ Figure 3.52 A spacer-bar saddle

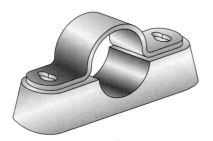

▲ Figure 3.53 A hospital saddle

cables, depending only on whether they will fit. The systems are available in a range of shapes with different numbers and profiles of compartments. Standard earth fault loop impedance values are often less on site than calculated. This is because conduit and trunking systems, and their support systems, can work as parallel conductors in real life, but cannot realistically be calculated.

▲ Figure 3.54 A floor trunking system with a bend and jointing section

Dado trunking systems are often found in offices where there are large numbers of data and power outlets. They provide a multi-compartment system allowing data, telephone and power cables to be kept separate and distributed around the work space. They also offer flexibility in that the positions of outlets can easily be changed, to a degree, any time after installation to meet changing requirements.

▲ Figure 3.55 Dado trunking in cross section

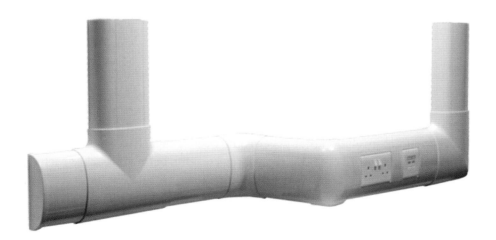

▲ Figure 3.56 Dado trunking profile

▲ Figure 3.58 Anchor bolt fixing

▲ Figure 3.59 Anchor bracket

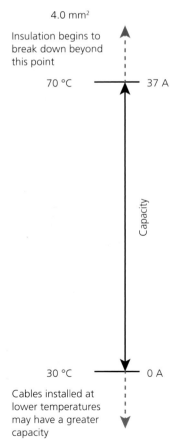

4.0 mm²

Insulation begins to break down beyond this point

70 °C ——————— 37 A

Capacity

30 °C ——————— 0 A

Cables installed at lower temperatures may have a greater capacity

▲ Figure 3.60 PVC/PVC flat twin, 70 °C thermoplastic, 30 ° C ambient

Fixings

The choice of fixings depends very much on the combined weight of the containment system and the cables in it. Often, many kilograms of cable need to be supported. For example, larger cable such as four-core 240 mm² steel wire armoured cable can weigh in the region of 13 kg/m. So a simple fixing with a round headed wood screw and wall plug is often not enough.

▲ Figure 3.57 Simple wood screws and plugs are often not strong enough to fix containment systems and the cables in them

It is quite likely that specialist advice from proprietary fixings manufacturers will be required, along with confirmation from the structural engineer that the floor above can cope with the loading from below. Where the load is considerable, substantial fixings such as anchor bolts or anchor brackets will be required.

Determining minimum current-carrying capacity

Both **BS 7671:2018** The IET Wiring Regulations, 18th Edition and the IET On-Site Guide provide tables for the maximum permissible current-carrying capacity for a range of conductor sizes. These tables do not, however, take into consideration all of the factors that affect the ability for a conductor to carry current. Calculations therefore need to be carried out to determine the suitable cable cross-sectional area (csa).

The following procedures use information from the IET On-Site Guide as it provides a simplistic approach for simple single circuit design. More complex circuit design procedures are covered in Book 2, Chapter 5.

Before looking at the procedure for selecting the correct conductor csa, we will consider the factors that affect the ability of a conductor to carry current.

Conductor insulation

It is the conductor insulation that governs the overall capacity of a conductor, as it is the weak point. Copper can carry large currents before it reaches melting point but the insulation cannot. General thermoplastic insulation has a maximum operating temperature of 70 °C so it is this temperature that sets the current limit. It is assumed that the initial temperature of any conductor, and therefore insulation, is 30 °C before any current is put through the conductor. As soon as current flows, the conductor temperature increases. The amount of current causing the temperature to reach 70 °C is the upper limit of the conductor.

Looking at Table F6 of the IET On-Site Guide, or the extract on the next page, it can be seen that a single-phase 4 mm² flat-profile cable installed in free air (method C, column 6) has a tabulated capacity of 37 A. This essentially means

▼ Current-carrying capacity (amperes) and voltage drop (per ampere per metre) — values taken from Table F6 in the IET On-Site Guide

Conductor cross-sectional area	Reference method 100 (above a plasterboard ceiling covered by thermal insulation not exceeding 100 mm in thickness)	Reference method 101 (above a plasterboard ceiling covered by thermal insulation exceeding 100 mm in thickness)	Reference method 102 (in a stud wall with thermal insulation with cable touching the inner wall surface)	Reference method 103 (in a stud wall with thermal insulation with cable not touching the inner wall surface)	Reference method C (clipped direct)	Reference method A (enclosed in conduit in an insulated wall)	Voltage drop
1	2	3	4	5	6	7	8
mm²	A	A	A	A	A	A	mV/A/m
1	13	10.5	13	8	16	11.5	44
1.5	16	13	16	10	20	14.5	29
2.5	21	17	21	13.5	27	20	18
4	27	22	27	17.5	37	26	11
6	34	27	35	23.5	47	32	7.3
10	45	36	47	32	64	44	4.4
16	57	46	63	42.5	85	57	2.8

that, given no further influences, 37A would cause the conductor's temperature to rise from 30 °C to 70 °C.

Ambient temperature

If cables are installed in ambient temperature other than 30 °C, the capacity of the cable's conductor is altered. If a cable is installed in an ambient temperature higher than 30 °C, the capacity of the conductor will reduce as the temperature of the cable, before any current flows, is higher. Rating factors (C_a) are used from Table F1 of the IET On-Site Guide. These rating factors are used to alter the capacity of the conductor depending on the ambient temperature of the cable and the type of insulation applied to the conductor. For any cable run which passes through a range of ambient temperatures, the highest shall be used to obtain a rating factor.

▼ Ambient temperatures — values taken from Table F1 in the IET On-Site Guide

Ambient temperature (°C)	Insulation			
			Mineral	
	70 °C thermoplastic	90 °C thermosetting	Thermoplastic covered or bare and exposed to touch 70 °C	Bare and not exposed to touch 105 °C
25	1.03	1.02	1.07	1.04
30	1.00	1.00	1.00	1.00
35	0.94	0.96	0.93	0.96
40	0.87	0.91	0.85	0.92

Thermal insulation

Where a cable is totally surrounded by thermal insulation, the cable's ability to dissipate heat is greatly reduced for any conductor that has a csa of 10 mm² or less. As a result, rating factors for thermal insulation (C_i) need to be applied as the heat build-up in the cable will have an effect on the temperature of the cable's insulation. Ideally, cables should be clipped to a joist, or other thermally conducting surfaces, or should be sleeved by conduit as they pass through the thermal

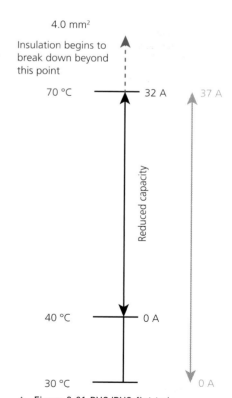

4.0 mm²

Insulation begins to break down beyond this point

70 °C — 32 A — 37 A

Reduced capacity

40 °C — 0 A

30 °C — 0 A

▲ Figure 3.61 PVC/PVC flat twin, 70 °C thermoplastic, 40 °C ambient insulation. In these situations, a thermal insulation rating factor need not be applied as the current capacity is allowed for within the selection tables depending on the method of insulation. As can be seen from Table F2 of the IET On-Site Guide, reproduced below, a rating factor of 0.5 is applied to any cable which passes through 500 mm or more thermal insulation. This effectively reduces a conductor's capacity by 50% so thermal insulation is best avoided if possible.

▼ Thermal insulation – values taken from Table F2 in the IET On-Site Guide

Length in insulation (mm)	Derating factor (C_i)
50	0.88
100	0.78
200	0.63
400	0.51
≥ 500	0.50

Grouping of circuits

Where circuits are grouped together, either in the same containment system such as conduit, or clipped close together on a wall, the heat from one circuit can have an effect on others. Rating factors for grouping (C_g) are applied to circuits using Table F3 of the IET On-Site Guide. As can be seen from the table, the more circuits that are grouped, the more the conductor's capacity is reduced. As an example, if a new circuit was installed into a cable basket with six other circuits, that would be seven in total. From the table, cables bunched in air (methods A to F) with a total number of circuits being seven has a rating factor of 0.54. This means that the conductor's capacity is reduced to 54% of the tabulated value.

Where the horizontal clearance between cables exceeds twice their diameter no factor need be applied. Cables are assumed to be equally loaded. Where cables are expected to carry no more than 30% of their grouped rating they may be ignored so far as grouping is concerned. The IET On-Site Guide contains more conditions regarding grouping.

▼ Grouping of circuits – values taken from Table F3 in the IET On-Site Guide

Arrangement (cables touching)	Number of circuits or multi-core cables										Applicable reference method for current-carrying capacities
	1	2	3	4	5	6	7	8	9	12	
Bunched in air, on a surface, embedded or enclosed	1.0	0.80	0.70	0.65	0.60	0.57	0.54	0.52	0.50	0.45	A to F
Single layer on wall or floor	1.0	0.85	0.79	0.75	0.73	0.72	0.72	0.71	0.70	0.70	C
Single layer multicore on a perforated horizontal or vertical cable tray system	1.0	0.88	0.82	0.77	0.75	0.73	0.73	0.72	0.72	0.72	E
Single layer multicore on a cable ladder system or cleats, etc.	1.0	0.87	0.82	0.80	0.80	0.79	0.79	0.78	0.78	0.78	E

Method of installation

Basic methods of installation are listed below but for a concise list, see Table 4A2 in Appendix 4 of **BS 7671:2018** The IET Wiring Regulations, 18th Edition. The method used to install cables can affect the ability for a cable to dissipate heat, meaning that cables could run too hot if the correct method or table selection isn't considered. When selecting a conductor csa from the relevant current capacity table, each column represents a different method of installation. The common methods include:

- **Method A:** enclosed in conduit in an insulated wall.
- **Method B:** enclosed in surface-mounted conduit.
- **Method C:** clipped directly to a surface in free air.
- **Method 100:** in contact with wooden joists or a plaster ceiling and covered by thermal insulation *not exceeding* 100 mm thickness.
- **Method 101:** in contact with wooden joists or a plaster ceiling and covered by thermal insulation *exceeding* 100 mm thickness.
- **Method 102:** in a stud wall containing thermal insulation with the cable touching the wall.
- **Method 103:** in a stud wall containing thermal insulation with the cable *not* in contact with the wall.

In order to determine a suitable conductor csa for a single circuit cable, the following procedure is used.

Step 1: Determine the design current (I_b). This is determined in several ways depending on the type of load and the allowance, if any, of diversity.

▼ How to determine design current for simple single-phase final circuits

Load type and nature of the final circuit	How to determine design current (I_b)	Notes
Standard resistive loads such as water heaters, electric space heaters and incandescent lighting	$I_b = \dfrac{\text{total watts}}{\text{volts}}$	For ceiling roses and similar in dwellings, each point is assumed as 100 W. See Table A1, IET On-Site Guide
Socket-outlet circuits such as radial or ring-final circuits	$I_b = I_n$	Where I_n is the rating of protective device
For discharge lighting such as circuits containing many fluorescent luminaires	$I_b = \dfrac{\text{total watts} \times 1.8}{230}$	The factor of 1.8 allows for increased current during starting, which all cables and switching devices must be capable of handling
For inductive loads where power factor is known, such as motors	$I_b = \dfrac{\text{total watts}}{\text{volts} \times \cos\theta}$	Where $\cos\theta$ is the power factor
Domestic cooker circuits	$I_b = \left(\dfrac{\text{total watts}}{\text{volts}} - 10\right) \times 0.3 + 10$	See table A1, IET On-Site Guide Add 5 A if the cooker control incorporates a socket outlet
Other	See Table A1, IET On-Site Guide	

Step 2: Select the nominal rating of protective device (I_n).

$$I_n \geq I_b$$

Step 3: Gather information such as:

- Circuit length – this is the full length of the circuit from the distribution board to the final point of use on the circuit in metres.
- Method of installation (see Section 7, IET On-Site Guide).
- Rating factor for ambient temperature C_a (Table F1, IET On-Site Guide).
- Rating factor for thermal insulation C_i (Table F2, IET On-Site Guide).
- Rating factor for grouping of circuits C_g (Table F3, IET On-Site Guide).
- Rating factor to be applied if the protective device is a BS 3036 semi-enclosed fuse (C_f) = 0.725 (see page 197).

Note: if a cable is not grouped or surrounded by thermal insulation, a factor of 1 is used. In reality, this can be ignored from any calculations as multiplying or dividing any value by 1 leaves the value unchanged.

Step 4: Apply the following calculation:

$$I_t \geqslant \frac{I_n}{C_a \times C_i \times C_g \times C_f}$$

where I_t is the tabulated current-carrying capacity specified (see step 5).

Note: any factor that is 1 may be ignored.

Step 5: Use the correct IET On-Site Guide cable selection table from:

- Table F4(i) for single-core cables having 70 °C thermoplastic insulation
- Table F5(i) for multi-core cables (non-armoured) having 70 °C thermoplastic insulation
- Table F6 for multi-core flat profile cables having 70 °C thermoplastic insulation.

Select, using the correct column number relating to the installation method, a value of tabulated current equal to or larger than that calculated in step 4.

This will now ensure that the cable selected is suitable for current-carrying capacity.

Step 6: Ensure that the selected csa is suitable for voltage drop. For the cable csa selection tables used in step 5, each has a corresponding voltage drop table or column. The values given in these tables or columns represent the value of voltage drop, in milli-volts per ampere per metre of circuit (mV/A/m). The circuit length determined in step 3 is required to determine voltage drop but remember that the circuit (cable) length is the length of the line conductor with the neutral, not added to the neutral. For example, if a twin cable has a length of 10 m, that is the length of the line with neutral. The line added to the neutral would be 20 m (there and back). The values of voltage drop per metre include the neutral resistance within the values.

Apply the following formula to determine the value of voltage drop:

$$\text{voltage drop} = \frac{I_b \times \text{length} \times \text{mV/A/m}}{1000} \text{(volts)}$$

Currently, **BS 7671:2018** The IET Wiring Regulations, 18th Edition states that values of voltage drop for installation supplied from a public network should not exceed:

- 3% of the supply voltage for lighting circuits (3% of 230 V = 6.9 V)
- 5% of the supply voltage for power circuits (5% of 230 V = 11.5 V).

As long as the voltage drop calculated for the selected cable csa is within the values above, the cable selected is suitable. If the calculated voltage drop is too high, the next largest csa is selected and calculated for voltage drop until a suitable csa is found.

Note: if a final circuit is supplied by a distribution circuit, the voltage drop for the distribution circuit will also need to be taken into account.

EXAMPLE

Determine a suitable csa for a conductor where a circuit is to be wired using flat-profile 70 °C thermoplastic insulated twin and cpc cable to supply a single-phase 6 kW water heater. The circuit is to be clipped directly to a plastered wall to a length of 21 m. The highest ambient temperature is 35 °C and protection for the circuit is by a BS EN 60898 circuit breaker. The cable passes through and is totally surrounded by 100 mm of thermal insulation. The circuit is not grouped with any other circuits.

Step 1

$$I_b = \frac{\text{total watts}}{\text{volts}}$$

So:

$$I_b = \frac{6000}{230} = 26A$$

Step 2

Select the nominal rating of protective device (I_n).

$$I_n \geq I_b$$

The next suitable nominal rating of protective device is 32 A.

Step 3

Gather information such as:

- Circuit length = 21 m.
- Method of installation = method C.
- Rating factor for ambient temperature C_a = 0.94.
- Rating factor for thermal insulation C_i = 0.78.
- Rating factor for grouping of circuits C_g = 1 (so can be ignored).
- Rating factor to be applied for the protective device = 1 (so can be ignored).

Step 4

Apply the following calculation:

$$I_t \geq \frac{I_n}{C_a \times C_i}$$

So:

$$I_t \geq \frac{32}{0.94 \times 0.78} = 43.7A$$

So the cable must be capable of carrying a current of 43.7 A as a minimum.

→

Step 5

Using Table F6 for multi-core flat profile cables having 70 °C thermoplastic insulation and looking at column 6 (method C), a suitable conductor csa is 6 mm^2 having a tabulated rating (I_t) of 47 A.

Step 6

To determine the value of voltage drop:

$$\text{voltage drop} = \frac{I_b \times \text{length} \times \text{mV/A/m}}{1000} \text{(volts)}$$

Column 8 of Table F6 gives a value of voltage drop in mV/A/m for a 6 mm^2 as 7.3 mV/A/m.

So:

$$\text{voltage drop} = \frac{26 \times 21 \times 7.3}{1000} = 3.99\,V$$

As this is within 11.5 V for a power circuit (5% of 230 V), this is suitable.

So the cable selected is a 6 mm^2 twin with cpc 70 °C thermoplastic insulated flat-profile cable.

ACTIVITY

Given a load current of 15 A, what would be the minimum value of I_n if the following fuse systems were used?

- BS 3036
- BS 88

INDUSTRY TIP

Some cartridge fuses have an indicator that shows when the fuse blows. This may be a bead that drops out to show that the fuse has blown.

Protective devices

Selecting protective devices is an important part of the design process. There is a wide selection within the many types available, including specialist devices with variable settings, which assist in meeting the precise requirements of **BS 7671:2018** The IET Wiring Regulations, 18th Edition.

Fuses

Fuses have been a tried and tested method of circuit protection for many years. A fuse is a very basic protection device that is destroyed and breaks the circuit should the current exceed the rating of the fuse. Once the fuse has 'blown' (i.e. the element of the fuse has melted or ruptured), the fuse needs to be replaced.

BS 3036 semi-enclosed rewireable fuses

In older equipment, the fuse may be just a length of appropriate fuse wire fixed between two terminals. These devices are now becoming uncommon as electrical installations are rewired or updated. One of the main problems associated with rewirable fuses is the overall lack of protection, including insufficient breaking capacity ratings. Another major problem is unreliability, which is often caused by using the wrong gauge of wire when changing the fuse. The overall reliability of such fuses cannot be guaranteed because a wide range of factors can cause them to fail. Typical factors include wire being labelled with an incorrect current rating and the number of times and the length of time that a fuse wire has been subjected to overload.

The lack of reliability of these fuses is a concern to designers and duty holders. Due to the lack of sensitivity, a special factor has been applied to Appendix 4 of BS 7671:2018 to account for **BS 3036** rewirable fuses. This rating factor (C_f) is 0.725.

BS 88 fuses

These modern fuses are generally incorporated within sealed ceramic cylindrical bodies (or cartridges). The whole cartridge needs to be replaced if the fuse ruptures. Although these devices have fixed time current curves, they can be configured to assist **discrimination**, which is often referred to as selectivity. The benefit of **BS 88** fuses is their simplicity and reliability, coupled with high short-circuit breaking capacity. These fuses are otherwise called High Rupturing Capacity (HRC) or High Breaking Capacity (HBC) fuses. They are available in two forms which are gG for general circuit protection and gM for motor-rated circuits.

▲ Figure 3.62 Fuse carrier without wire

Circuit breakers

Circuit breakers are thermomagnetic devices capable of making, carrying and interrupting currents under normal and abnormal conditions. They fall into two categories: miniature circuit breakers (MCBs), which are common in most installations for the protection of final circuits, and moulded case circuit breakers (MCCBs), which are normally used for larger distribution circuits.

Circuit breaker nominal ratings (I_n) relate to continuous service under specified installation conditions, although cables can carry higher currents for short periods without causing permanent damage.

▲ Figure 3.63 A range of BS 3036 rewirable fuses: 5, 15 and 20 A

KEY TERM

Discrimination: The arrangement of protective devices to ensure the device on the supply side of the fault operates before any other device.

There are three circuit breaker types: Type B, Type C and Type D. The current flow at which they trip depends upon the level of overcurrent and is usually determined by a magnetic device within the circuit breaker. The value of current at which each type of circuit breaker disconnects instantaneously is referred to as I_a.

Type B circuit breakers

Type B trips at between three and five times the rated current. These are normally used for domestic circuits and commercial applications where there is no inrush current to cause the circuit breaker to trip. As an example, the magnetic trip in a 32 A Type B circuit breaker could be 160 A. So $I_a = 5 \times I_n$.

Type C circuit breakers

Type C trips at between five and ten times the rated current. These are normally used for commercial applications where there are motors or fluorescent luminaires with some inrush current that would cause a Type B circuit breaker to trip. As an example, the magnetic trip in a 32 A Type C circuit breaker could be 320 A. So $I_a = 10 \times I_n$.

▲ Figure 3.64 Various lug arrangements on cartridge-type fuses

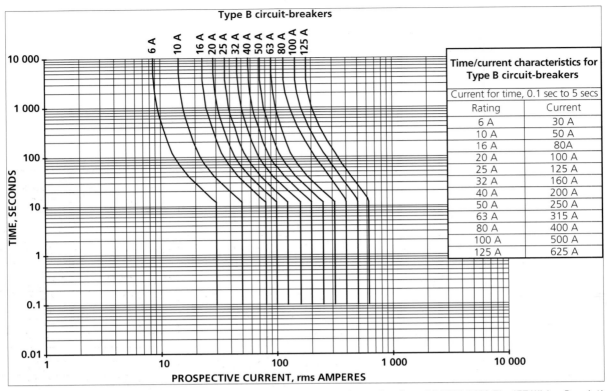

▲ Figure 3.65 Type B circuit breaker time current curve, derived from data taken from BS 7671:2018 The IET Wiring Regulations, 18th Edition (Appendix 3)

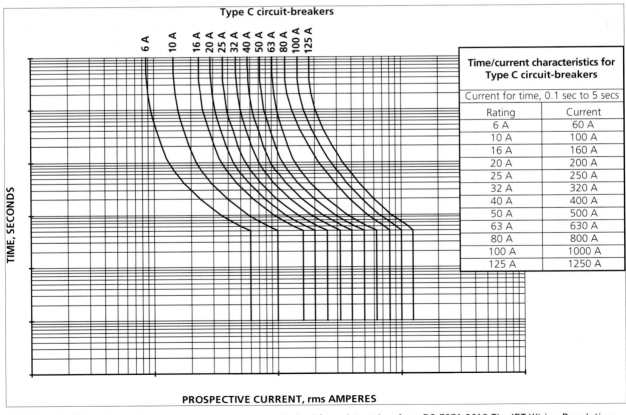

▲ Figure 3.66 Type C circuit breaker time current curve, derived from data taken from BS 7671:2018 The IET Wiring Regulations, 18th Edition (Appendix 3)

Type D circuit breakers

Type D trips at between ten and 20 times the rated current. These circuit breakers are industrial units and specified where there are large inrushes of current for industrial motors, x-ray units, and so on. As an example, the magnetic trip in a 32 A Type D circuit breaker could be 640 A. So $I_a = 20 \times I_n$.

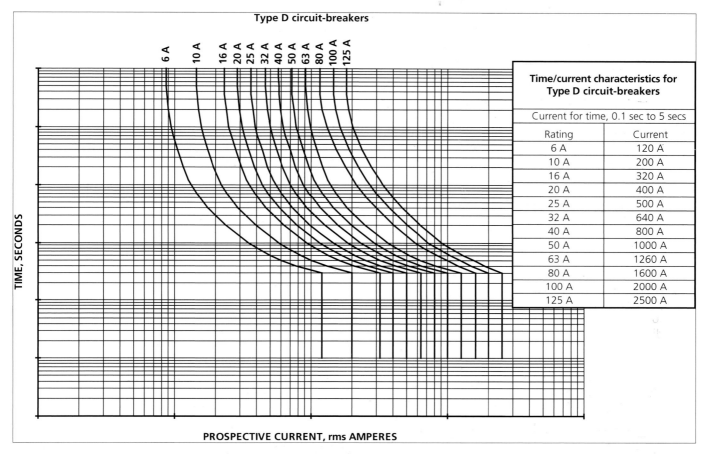

Type D circuit-breakers

Time/current characteristics for Type D circuit-breakers	
Current for time, 0.1 sec to 5 secs	
Rating	Current
6 A	120 A
10 A	200 A
16 A	320 A
20 A	400 A
25 A	500 A
32 A	640 A
40 A	800 A
50 A	1000 A
63 A	1260 A
80 A	1600 A
100 A	2000 A
125 A	2500 A

▲ Figure 3.67 Type D circuit breaker time current curve, derived from data taken from BS 7671:2018 The IET Wiring Regulations, 18th Edition (Appendix 3)

You can see from the graphs that a circuit breaker has a curve, then a straight line whereas the BS 88 device is fully curved. This demonstrates the two tripping mechanisms within a circuit breaker. The magnetic trip is represented by the straight line on the graph, indicating that a pre-determined value of fault current will disconnect the device rapidly. The curve represents the device's thermal mechanism. Like a fuse, the thermal mechanism reacts within a time specific to the overload current. The bigger the overload, the quicker the reaction.

The short-circuit rating or breaking capacity of the circuit breaker is generally no match for the high rupturing capacity (HRC) fuse. However, values from around 6–10 kA are normally available.

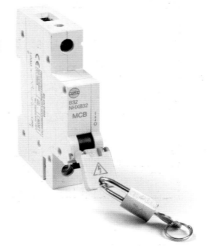

▲ Figure 3.68 Circuit breaker with lock

KEY TERMS

I_{cs}: the maximum short-circuit current the device can withstand and remain serviceable. If this value is exceeded, the device will require replacing.

I_{cn}: the maximum short-circuit capacity that the device will safely disconnect at. Any value of fault current above this could result in the device exploding or the contacts welding closed.

▼ Rated short-circuit capacities – values taken from the IET On-Site Guide

Device type	Device designation	Rated short-circuit capacity (kA)	
Semi-enclosed fuse to BS 3036 with category of duty	S1A S2A S4A	1 2 4	
Cartridge fuse to BS 1361 type I type II		16.5 33.0	
General purpose fuse to BS 88-2		50 at 415 V	
BS 88-3 type I type II		16 31.5	
General purpose fuse to BS 88-6		16.5 at 240 V 80 at 415 V	
Circuit-breakers to BS 3871 (replaced by BS EN 60898)	M1 M1.5 M3 M4.5 M6 M9	1 1.5 3 4.5 6 9	
Circuit-breakers to BS EN 60898* and RCBOs to BS EN 61009		I_{cn} 1.5 3.0 6 10 15 20 25	I_{cs} (1.5) (3.0) (6.0) (7.5) (7.5) (10.0) (12.5)

* Two short-circuit capacities are defined in BS EN 60898 and BS EN 61009:

BS EN 60947-2 is as specified by the manufacturer.

Residual current devices

The basic principle of residual current devices (RCDs) is a simple balance arrangement. When the load is connected to the supply through the RCD, the line and neutral conductors are connected through primary windings on a toroidal transformer. In this arrangement the secondary winding of the toroidal transformer is used as a sensing coil and is electrically connected to a sensitive

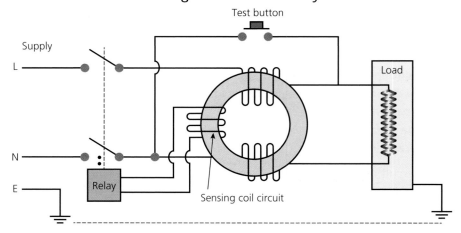

▲ Figure 3.69 Typical RCD arrangement

relay or solid state switching device, the operation of which triggers the tripping mechanism. When the line and neutral currents are balanced, as in a healthy circuit, they produce equal and opposite magnetic fluxes in the transformer core so no current is generated in the sensing coil. Where there is an imbalance of current in the line and neutral due to an amount of current leaking to earth, the magnetic flux in the transformer coil is detected by the sensing coil and the current generated triggers the tripping mechanism.

Tripping mechanisms in circuit breakers are triggered by large fault currents but an RCD tripping mechanism is triggered by very small currents. As these small currents have little energy, all RCDs have test buttons fitted which must be operated regularly. This frequent operation keeps the tripping mechanism free from sticking as any stuck mechanism may not react to such small trigger currents. It is essential that this button is pressed at regular periods to ensure that the RCD remains effective.

Residual current breakers with overcurrent

Residual current breakers with overcurrent (RCBOs) combine the residual current function of RCDs and the overcurrent protection function typical of circuit breakers in a single device. They are tripped by current leakage to earth, overloads and short circuits. They are self-protecting up to the maximum short-circuit current value indicated on the label.

Specialised equipment for installing wiring systems

There is a wide range of specialist tools and devices available to assist electricians. These include:

- conduit vice and bender – used to bend metallic conduits into shape and secure the conduit while cutting and threading
- stocks and die – used to thread metallic conduit
- tray bender – used to form internal and external bends to traywork
- MICC stripper – used to cut back and remove the outer copper sheath
- ringing tool – used to score the outer copper sheath of an MICC cable to provide a clean finish when the cable sheath is stripped
- potting tool – used to screw the pot onto the prepared end of an MICC cable
- crimping tool – used to compress and crimp the seal onto an MICC pot
- MICC straighter – used to straighten MICC cable prior to clipping to a surface.

Chapter 4 will look at some of these tools in greater detail.

ACTIVITY
How regularly should an RCD be operated by pressing the test button?

HEALTH AND SAFETY
An RCD exposed to an overload or short circuit will not trip, as both line and neutral currents are equal. An RCBO will trip as it senses the overload, and short circuit. As a result, RCDs should never be a sole means of circuit protection.

▲ Figure 3.70 Conduit bender and vice with stocks, dies and guides

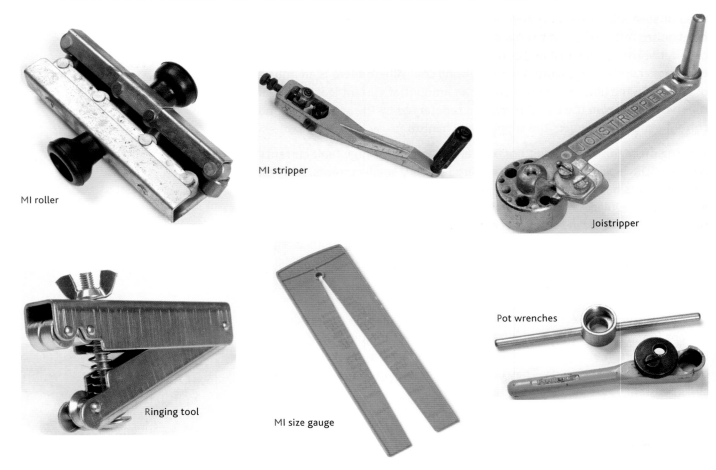

MI roller

MI stripper

Joistripper

Ringing tool

MI size gauge

Pot wrenches

▲ Figure 3.71 MICC installation and termination tools

KEY TERMS

Spacing factors: The amount of cable contained. So a 45% spacing factor means 45% cable, 55% air.

Bend: British Standard 90° bend. One double set is equivalent to one bend.

Spacing factors of wiring enclosures

Spacing factors are the free air allowances made around each cable installed in a conduit or trunking system. The method described below can be used to determine the size of conduit or trunking needed to accommodate multiple cables of the same or different sizes in accordance with **BS 7671:2018** The IET Wiring Regulations, 18th Edition. It employs a unit system, with a factor allocated to each cable size. Tables in Appendix E of the IET On-Site Guide can be used to satisfy this. The sum of the factors for all the cables needed in the same enclosure is compared against the factors given for conduit, ducting or trunking.

Any **bends** in a conduit run have a bearing on the calculation, so a choice must be made between three different categories:

- For case 1 (conduit straight runs not exceeding 3 m in length), each cable and conduit size is represented by one single factor.
- For case 2 (conduit straight runs exceeding 3 m, runs of any length with bends or runs that are offset (sets) within the containment system), each conduit size has a variable factor, depending on the length of run and the number of bends or sets.
- For case 3 (trunking), each cable size and trunking size is allocated a factor.

A number of other factors also have a bearing on the number of cables that can be installed in conduit and trunking. These include:
- level of care during installation
- use of the space available
- tolerance in cable sizes
- tolerance in conduit and trunking.

The tables on the following pages can only give guidance on the maximum number of cables that should be drawn into an enclosure but should ensure an easy pull with low risk of damage to the cables. However, this method does not assess the electrical effects of grouping. It may also sometimes be more economical to divide the circuits concerned between a number of enclosures rather than have one very large enclosure.

As the grouping factor (C_g) is used to determine the cable cross-sectional area, the more cables installed, the larger the cross-sectional area required. This has a negative impact, so it is ultimately better to install fewer cables in the same enclosure.

▼ Cable factors for use in conduit in short straight runs – values taken from Table E1 in the IET On-Site Guide

Type of conductor	Conductor cross-sectional area (mm²)	Cable factor
Solid	1	22
	1.5	27
	2.5	39
Stranded	1.5	31
	2.5	43
	4	58
	6	88
	10	146
	16	202
	25	385

▼ Conduit factors for use in short straight runs – values taken from Table E2 in the IET On-Site Guide

Conduit diameter (mm)	Conduit factor
16	290
20	460
25	800
32	1400
38	1900
50	3500
63	5600

▼ Cable factors for use in conduit in long straight runs over 3 m, or runs of any length incorporating bends – values taken from Table E3 in the IET On-Site Guide

Type of conductor	Conductor cross-sectional area (mm²)	Cable factor
Solid or Stranded	1	16
	1.5	22
	2.5	30
	4	43
	6	58
	10	105
	16	145
	25	217

EXAMPLES

With reference to Table E1 of the IET On-Site Guide, a conduit run of 3 m with no bends and a total of ten 4 mm^2 thermoplastic stranded cables has a cable factor of 10 × 58 = 580.

Reference to Table E2 indicates that the minimum size for the conduit would be 25 mm diameter, which has a conduit factor of 800.

With reference to Table E3 of the IET On-Site Guide, a conduit run of 10 m with two bends and a total of ten 4 mm^2 thermoplastic stranded cables has a cable factor of 10 × 43 = 430.

Reference to Table E4 indicates that the minimum size for the conduit would be 32 mm diameter, which has a conduit factor of 643.

▼ Conduit factors for runs incorporating bends and long straight runs – values taken from Table E4 in the IET On-Site Guide

Length of run (m)	Conduit diameter (mm)																			
	16	20	25	32	16	20	25	32	16	20	25	32	16	20	25	32	16	20	25	32
	Straight				One Bend				Two Bends				Three Bends				Four Bends			
1					188	303	543	947	177	286	514	900	158	256	463	818	130	213	388	692
1.5	Covered by				182	294	528	923	167	270	487	857	143	233	422	750	111	182	333	600
2	other tables				177	286	514	900	158	256	463	818	130	213	388	692	97	159	292	529
2.5					171	278	500	878	150	244	442	783	120	196	358	643	86	141	260	474
3					167	270	487	857	143	233	422	750	111	182	333	600				
3.5	179	290	521	911	162	263	475	837	136	222	404	720	103	169	311	563				
4	177	286	514	900	158	256	463	818	130	213	388	692	97	159	292	529				
4.5	174	282	507	889	154	250	452	800	125	204	373	667	91	149	275	500				
5	171	278	500	878	150	244	442	783	120	196	358	643	86	141	260	474				
6	167	270	487	857	143	233	422	750	111	182	333	600								
7	162	263	475	837	136	222	404	720	103	169	311	563								
8	158	256	463	818	130	213	388	692	97	159	292	529								
9	154	250	452	800	125	204	373	667	91	149	275	500								
10	150	244	442	783	120	196	358	643	86	141	260	474								

Additional factors:
▶ For 38 mm diameter use 1.4 x (32 mm factor)
▶ For 50 mm diameter use 2.6 x (32 mm factor)
▶ For 63 mm diameter use 4.2 x (32 mm factor)

INDUSTRY TIP

Always check the Notes under any table in BS 7671:2018 or the IET On-Site Guide

▼ Cable factors for trunking – values taken from Table E5 in the IET On-Site Guide

Type of conductor	Conductor cross-sectional area (mm²)	PVC BS 6004 Cable factor	Thermosetting BS 7211 Cable factor
Solid	1.5	8.0	8.6
	2.5	11.9	11.9
Stranded	1.5	8.6	9.6
	2.5	12.6	13.9
	4	16.6	18.1
	6	21.2	22.9
	10	35.3	36.3
	16	47.8	50.3
	25	73.9	75.4

EXAMPLES

With reference to Table E5 of the IET On-Site Guide, a trunking run of 10 m with two bends and a total of 20 × 4 mm² and 10 × 16 mm² thermoplastic stranded cables has a cable factor of (20 × 16.6) + (10 × 35.3) = 685.

Reference to Table E6 indicates that the minimum size for the trunking would be 50 × 38 mm, which has a trunking factor of 767.

Note that where other types of cable or trunking than specified in Tables E5 and E6 are used, an allowance of 45% space is needed.

▼ Factors for trunking – values taken from Table E6 in the IET On-Site Guide

Dimensions of trunking (mm x mm)	Factor	Dimensions of trunking (mm x mm)	Factor
50 x 38	767	200 x 100	8572
50 x 50	1037	200 x 150	13001
75 x 25	738	200 x 200	17429
75 x 38	1146	225 x 38	3474
75 x 50	1555	225 x 50	4671
75 x 75	2371	225 x 75	7167
100 x 25	993	225 x 100	9662
100 x 38	1542	225 x 150	14652
100 x 50	2091	225 x 200	19643
100 x 75	3189	225 x 225	22138
100 x 100	4252	300 x 38	4648
150 x 38	2999	300 x 50	6251
150 x 50	3091	300 x 75	9590
150 x 75	4743	300 x 100	12929
150 x 100	6394	300 x 150	19607
150 x 150	9697	300 x 200	26285
200 x 38	3082	300 x 225	29624
200 x 50	4145	300 x 300	39428
200 x 75	6359		

INDUSTRY TIP

All three of these calculations can be worked in reverse to check how many cables could be installed in conduit or trunking that has already been installed.

ACTIVITY

If it is safe to do so, look at existing trunkings and capacities, using cable capacity information to see how lightly/heavily loaded the systems are. You could do this around your training centre with the help of your tutor.

Earthing, bonding and protection

When protecting against electric shock and faults, there are a number **protection methods** outlined in **BS 7671:2018** The IET Wiring Regulations, 18th Edition. The most common method of protection used is automatic disconnection of supply (ADS).

ADS works by quickly disconnecting a circuit, should a fault to earth occur. For this to happen, a system of earthing and bonding is needed, as well as protective devices.

KEY TERM

Protection methods: methods given in BS 7671:2018 for protection against faults.

Extraneous conductive parts

Extraneous conductive parts are parts that are not part of electrical equipment or systems but may still introduce a potential path to earth. However, it is often difficult to establish what is or is not an extraneous conductive part. Some conductive parts may be partially insulated from earth – for example, where an insulated board that is part of the building construction creates a high resistance.

One method of checking if an extraneous part has a path to earth is to use an insulation resistance test instrument. Connect the instrument between a known point of earth, such as the earth pin on a socket outlet, and the part that needs to be defined. If the reading is low (e.g. below 0.05 MΩ or 50 000 Ω) when tested at 250 V DC, the part has a low enough resistance to earth to cause potential harm and may require some form of protective bonding.

Examples of extraneous conductive parts include:

- structural steelwork
- reinforcement bars
- pipes
- other metal components that are not part of an electrical system but introduce a potential path to earth.

Exposed conductive parts

Exposed conductive parts are parts that form part of the electrical system and should therefore be connected to earth. The only metallic parts of an electrical system that should not be connected to earth are metal casings of **Class II equipment**.

Examples of exposed conductive parts include:

- steel conduit
- steel trunking
- metal-cased distribution boards
- metal light switches
- metal casings of luminaires.

Automatic disconnection of supply (ADS)

This is the most common method of fault protection in electrical installation in the UK. In order for ADS to be effective, the following needs to be applied.

- **Basic protection** – by barriers, enclosures and basic insulation.
- **Main protective bonding conductors** – installed connecting extraneous conductive parts to the main earthing terminal (MET).
- **Circuit protection providing automatic disconnection** – by suitable protective device(s) providing earth fault, short circuit and overload protection.

KEY TERM

Class II equipment: equipment having basic insulation and a further reinforced layer of insulation around live parts meaning the risk of faults is minimal. Class II equipment is identified by the symbol:

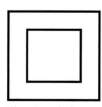

Class II symbol

- **Additional protection** – by an RCD not exceeding 30 mA which disconnects within 40 ms at a current $5 \times I_{\Delta n}$.
- **Circuit protective conductors** – connecting all exposed conductive parts to the MET.
- **Suitable means of earthing** – for the installation (as described earlier in this chapter, on pages 172–175).
- **Earth fault loop impedance (Z_s)** – which directly affects earth fault current and therefore disconnection time.

Basic protection

Basic protection is a means of stopping persons from receiving an electric shock by touching live parts directly. All electrical accessories such as socket outlets, switches, distribution boards and luminaires have live parts enclosed within a barrier or enclosure. These barriers must not have any holes in them allowing persons to put in a finger (IP2X), or, if there is a hole, the live parts must be distanced enough so the finger cannot touch them (IPXXB). There must also be no opening on their top surfaces allowing anything to drop into the live parts (IP4X and IPXXD).

All conductors or live parts not enclosed by a barrier or enclosure must have adequate insulation around them. The insulation must be suitable for the voltage of the circuit.

Main protective bonding conductors

Main protective bonding (MPB) conductors, sized in accordance with Section 4 of the IET On-Site Guide or Chapter 54 of **BS 7671:2018** The IET Wiring Regulations, 18th Edition must be installed connecting the MET and:

- metallic installation service pipes such as gas, water, oil
- metallic exposed structural steelwork of a building rising from the ground
- lightning protection systems (in certain situations)
- other extraneous parts as detailed in Chapter 41 of BS 7671:2018.

Where the MPB conductor is connected to the incoming service pipework, the connection must be made within 600 mm of any stop valve or meter, or point of entry to the building but before any branch pipe is installed.

Connection is made to the pipework using a specific bonding clamp which complies with **BS 951** Electrical earthing, Clamps for earthing and bonding. Different clamps have colour coding on them for installing in different conditions. They may be:

- red for dry conditions
- blue for damp locations.

In some situations, supplementary bonding may need to be installed but this will be covered in Book 2, Chapter 3, Electrical Design and Planning.

> **INDUSTRY TIP**
>
> Although various sizes of MBP conductor are required depending on the supply and installation, most domestic dwellings will be protected using 10 mm² conductors.

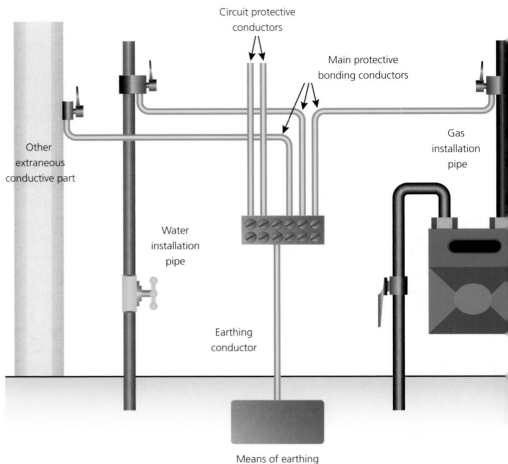

▲ Figure 3.72 Protective conductor arrangement for earthing and bonding

KEY TERM

Equal potential: the voltage between any two parts is within safe touch voltage levels, usually 50 V AC, depending on the location.

The purpose of MPB is to ensure **equal potential**. Should a fault occur in an electrical installation, the cpc for the circuit will become live to 230 V, as will the MET and any services or parts connected by MPB. If equal potential is present between the electrical installation and extraneous parts, current cannot flow between, especially through a person, meaning the risk of electrocution is reduced.

Circuit protection providing automatic disconnection

All final circuits must be protected by devices providing overload, short circuit and earth fault protection. In AC installations having a voltage to earth (U_0) of 230 V, protective devices selected must disconnect under earth fault conditions in the times given in the table.

Depending on the type and rating of protective device, different values of fault current are needed to cause disconnection in the required time (I_a). This is covered in more detail later in this chapter. The value of current likely to flow under earth fault conditions is limited by the total earth fault loop impedance (Z_s).

▼ Disconnection times of protection devices

System	Circuit		Rating (A)	Disconnection time as BS 7671 (s)	RCD rating	RCD functional tests (if fitted)	Times (ms)	Rating of RDC (mA)	Test	Time (ms)
					Automatic disconnection of supply (ADS)			**Additional protection**		
TN	Distribution		Any	5	≤30 mA for lighting (in dwellings), socket outlets or mobile outdoor equipment or suitable for $I_{\Delta n} \le \dfrac{50V}{Z_S \text{ or } R_A}$ where RCDs required by BS 7671	$1 \times I_{\Delta n}$ $0.5 \times I_{\Delta n}$	300 No trip	≤30	$5 \times I_{\Delta n}$	40
	Final: fixed equipment	≥32		5						
		≤32		0.4						
	Final: socket outlets	≥63		5						
		≤63		0.4						
TT	Distribution		Any	1		$1 \times I_{\Delta n}$ $0.5 \times I_{\Delta n}$	300 No trip			
	Final: fixed equipment	≥32		1						
		≤32		0.2						
	Final: socket outlets	≥63		1						
		≤63		0.2						

In some situations where the earth fault loop impedance is too high to generate a high enough current to cause disconnection of a protective device, an RCD is installed (or RCBO) as these disconnect at much lower current values. If an RCD is installed for this purpose alone and not additional protection, the RCD residual current setting ($I\Delta_n$) does not need to be 30 mA or less as long as the fault current causes disconnection within the required time for disconnection, or by the product standard.

Additional protection

Additional protection is required in situations where the risk of electric shock is increased, because any failure in basic protection presents a greater risk. For example, flexible cable might become damaged, cases to appliances might become broken, or risk of shock to the body might be increased due to the location, such as a bathroom full of steam.

Additional protection is provided on final circuits by an RCD or RCBO which has a residual current setting ($I\Delta_n$) not exceeding 30 mA. Additional protection is required in most electrical installations:

- where socket outlets rated at 32 A or less are used by ordinary persons
- where mobile equipment rated at 32 A or less is used outdoors
- in circuits supplying luminaires in dwellings
- in certain special locations such as bathrooms, swimming pools, saunas and construction sites
- where cables are buried in a wall or partition to a depth less than 50 mm and not protected by an earthed armouring or covering (see Chapter 52 **BS 7671:2018** The IET Wiring Regulations, 18th Edition for greater detail).

Circuit protective conductors (cpc)

These conductors are essential for ADS to work. As ADS requires disconnection within a specific time, a large enough fault current must flow to earth. This can only be achieved if the cpc has a low enough resistance, as Ohm's law states that resistance, with a set voltage, dictates current flow. If the current isn't high enough, disconnection will not occur in the required time.

When we consider the total earth fault loop impedance on the following pages, the cpc is referred to as R_2.

The cpc also performs a role of providing equal potential between all exposed conductive parts.

Suitable means of earthing

The type of protective device selected for ADS is completely dependent on the means of earthing. Together, the installation and supply make up the means of earthing. TN systems are regarded as having a reliable path to earth, so generating high earth fault currents. However, TT installations rely on the general mass of earth as an earth return path. As this resistance is likely to be high, when compared to TN systems, alternative protective devices may need selecting, such as RDCs, as they react more quickly to lower earth fault currents.

One part of the earthing system that also needs to be considered is the earthing conductor. This is the conductor that links the MET to the means of earthing. This may be the sheath of the supplier's cable (TN-S) or an earth electrode (TT) or the supplier's PEN conductor (TN-C-S).

Although the earthing conductor may be sized by calculation, the simple method is to ensure that it is:

- the same csa as the supply tails, where the tails are up to and equal to 16 mm^2
- at least 16 mm^2 when the supply tails are between 16 mm^2 and 35 mm^2
- at least half the csa of the supply tails if the tails are greater than 35 mm^2.

This can be seen on Table 54.7 of BS 7671:2018.

Earth fault loop impedance

The total earth fault loop impedance (Z_s) directly affects the amount of current under earth fault conditions. The component parts of the earth fault path are described below.

If the earth fault loop impedance is too high, then disconnection will take too long as the earth fault current will be too low. If earth fault loop impedance is low, the disconnection will occur quickly as the current is high.

Summary of ADS

If an earth fault occurs, a low resistance to earth generates a high fault current. The higher the fault current, the quicker the disconnection time. In certain situations, equal potential will reduce the risk until disconnection does occur.

Components of an earth fault loop impedance path

KEY FACTS

In TN systems, the earth fault loop impedance is a limiting factor to the operation of the protective device in an appropriate time. This value can be expressed as:

$$Z_s = Z_e + (R_1 + R_2)\ \Omega$$

where:

Z_s = impedance in ohms (Ω) of the fault loop, including the source impedance plus the installation impedances

Z_e = impedance in ohms (Ω) of the fault path external to the installation

R_1 = impedance in ohms (Ω) of the phase conductor

R_2 = impedance in ohms (Ω) of the circuit protective conductor.

Ohm's law tells us that the total earth fault loop impedance will directly affect the value of earth fault current, which in turn affects the disconnection time of the device as:

$$I_a \leq \frac{U_0}{Z_s}$$

where:

I_a = current in amperes (A) causing the automatic operation of the disconnecting device within the time specified

U_o = nominal AC line voltage to earth in volts (V), usually 230 V.

The earth fault loop impedance path for TN-C-S systems can be best described using the diagram in Figure 3.73.

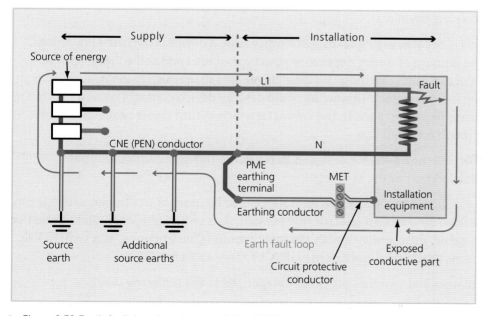

▲ Figure 3.73 Earth fault loop impedance path in a TN-C-S system

Isolation and switching

A typical electrical installation will contain many different switches or isolators. It is very important to know what the purpose of the device is in order to select the right type of device. **BS 7671:2018** The IET Wiring Regulations, 18th Edition requires isolation and switching devices for the following reasons.

- **Isolation** – a means of cutting off electricity to all or part of an installation to allow safe electrical work to be carried out.
- **Switching for mechanical maintenance** – a means of safely stopping a machine, equipment or appliance from operating in any way to allow safe non-electrical work that does not involve exposure to potentially live parts.
- **Emergency switching** – a means of quickly stopping a machine, equipment or appliance from operating in the event of an emergency.
- **Functional switching** – a means of controlling a machine, equipment or appliance such as turning it on or off.
- **Undervoltage protection** – a means of stopping a machine, equipment or appliance from automatically restarting following a loss or drop in voltage where sudden restarting could cause danger.

Selecting a switch or isolator

Before we look at the reasons for using a **switch** or **isolator** we need to understand the difference between them.

Isolators are rated for the current that can pass through them under normal load conditions but not the ability to switch that load. For example, a 63 A isolator could safely carry 63 A through it continually but it may not be capable of switching 63 A. If an isolator is to be used, equipment should always be powered down first.

This is because arcing takes place every time a device is operated under load. The amount of arcing depends on the type of load and some loads, such as inductive loads like motors, cause bigger arcs than resistive loads, such as immersion heaters. This arcing could destroy device contacts or, as a minimum, create high resistance to the contacts, which in turn creates excessive heat through constant use.

Switches, however, are designed to break the full load current without deterioration to the contacts.

If a device that is classed as an isolator is to be used, or positioned where it can be used by unskilled (electrically) persons, as a switch (on-load), then it must be de-rated, in accordance with the manufacturers' instructions, as a switch. This means that a 63 A isolator may only be rated to switch 20 A.

Isolators and switches are often categorised in the following way.

- **SP-N (single pole and neutral)** – a device that opens the line conductor only but the neutral can be isolated by a link.
- **Double-pole** – a device that opens the line and neutral conductors.
- **Triple-pole** – a device that opens the three-line conductors of a three-phase system and not the neutral.

- **TP-N (triple pole and neutral)** – a device that opens the three-line conductors and the neutral can be isolated by a link.
- **Four-pole** – a device that opens all three-line conductors and the neutral.

Isolation

Isolation must be provided for all installations in the form of a main switch. The type of installation will affect the type of isolator used.

▼ Suitable devices for isolating installations

Installation type	Supply and earthing arrangement	Type of device as a minimum
Dwelling and light commercial	Single-phase: TT, TN-S, TN-C-S	Double-pole switch capable of full load switching Must be rated to switch full load current
Commercial and industrial	Three-phase: TN-S, TN-C-S	TP-N device remote from CCU Triple-pole within CCU
Commercial and industrial	Three-phase: TT	Four-pole

Isolation may also be achieved for individual circuits by the protective device for the circuit.

In all situations, the device used for isolation must be capable of being secured in the open position. For main switches and circuit breakers, this is normally achieved with a padlock and, if necessary, a special locking device (Figure 3.70). Where circuits are protected by fuses, the fuse must be capable of being fully removed and it should be kept by the person who undertakes isolation. Devices are available which block off the fuseway meaning other fuses cannot be inserted while the circuit is securely isolated.

Whenever an electrically skilled person works on all or part of a circuit, the circuit should always be isolated at the start (origin) of the circuit.

Switching for mechanical maintenance

Mechanical maintenance covers many tasks carried out on a variety of different items of equipment. Provided that the work does not involve exposure to electrical terminals, it can be done by unskilled (electrically) persons. This work could include, among many other tasks:

- changing or adjusting drive belts on a machine
- installing or altering the pipework to a gas cooker
- cleaning air filters inside an air handling unit (AHU).

For these situations, a device capable of switching full load current must be located 'local' to the equipment. Ideally, the device must be in a position where the person carrying out the mechanical maintenance can keep the switch under their effective supervision so nobody else can switch it on. Common devices for this purpose include:

- fused spur connection units
- double-pole switches
- plugs and socket outlets.

If the device cannot be easily supervised then the device should be lockable – for example, a lockable rotary switch – removing the risk of someone else switching it back on.

> **INDUSTRY TIP**
>
> Local isolation is technically a myth as any device that is located within an installation, and is open to use by unskilled (electrically) persons, should be a switch, not an isolator.

Emergency switching

Emergency switching devices must be capable of cutting off full load current. If a resident within a dwelling had concern about the safety of an electrical system, such as might occur following a major water leakage, they are able to cut off the supply easily.

In other installations where items such as rotating machines can cause a danger, emergency switching is normally provided by stop buttons. When installing stop buttons to a machine or workshop, consider the following.

- Stop buttons must be near to the risk where someone using the machine can easily reach it if they are in danger.
- Stop buttons must not be capable of being reset from another place unless the reset is by a key switch.
- If this is not the case, the stop button should latch in the off position. This ensures that the person resetting the circuit has to be where the danger happened so they can see that the danger has been removed before resetting the circuit.
- Where the stop button could be reset by untrained persons, such as in a school workshop, it should be key operated.

Functional switching

Functional switching is probably the most common type of switch found in an installation. Functional switching allows equipment to be switched on or off as needed. It should be in a convenient place where manual operation is required. An example would be a light switch for a room, located by the entrance door to that room.

Where a room has more than one entrance, two-way and/or intermediate switching would normally be provided.

Functional switches may also be in the form of:

- timeclocks
- **passive infra-red (PIR)** sensors
- plugs and socket outlets (rated at or below 32 A)
- switches on socket outlets
- contactors
- standard switches
- push switches
- switched fused connection units.

Devices that should never be used as a functional switch include:

- fuses
- luminaire connections such as plug-in ceiling roses
- unswitched fused connection units
- socket outlets rated above 32 A.

Where a functional switch is controlling discharge luminaires, like fluorescent fittings, the switch must be rated high enough to manage the surge current that happens when the load is first switched on. For example, a 6 A lighting circuit with fluorescent lights will typically have 10 A switches for this reason. Care must also be taken when selecting dimmer switches as some do not perform well when underloaded as well as overloaded.

Test your knowledge

You may use the IET On-Site Guide with these tests.

1 Which non-statutory publication gives guidance on the methods of inspecting and testing new installations?

A EWR.

B BS 7671.

C HSE GS 38.

D On-Site Guide.

2 What type of diagram provides detailed information showing how components are connected together, including which terminal connects to another?

A Schematic diagram.

B Site plan.

C Circuit diagram.

D Sketch plan.

3 An internal wall measures 20 cm on a drawing and 10 m in actual size. What drawing scale is being used?

A 1:20

B 1:50

C 1:100

D 1:500

4 What is a transmission voltage?

A 400 V

B 11 000 V

C 25 000 V

D 400 000 V

5 Which is a renewable energy source?

A Gas.

B Oil.

C Solar.

D Coal.

6 Which is an extraneous conductive part?

A Metal conduit.

B Steel trunking.

C Plastic pipework.

D Copper service pipes.

7 What is the **maximum** permitted disconnection time for a 6 A circuit supplying luminaires in a commercial installation forming part of a TN-C-S earthing arrangement?

A 0.4 seconds.

B 0.8 seconds.

C 1 second.

D 5 seconds.

8 What is the maximum rating of RCD permitted where additional protection is required for 32 A socket-outlet circuits?

A 30 mA

B 150 mA

C 300 mA

D 500 mA

9 What symbol is used for total earth fault loop impedance?

A Z_e

B R_1

C R_2

D Z_s

10 What device is used to provide isolation by an electrician for an entire electrical installation?

A Supplier's meter.

B Circuit RCBOs.

C Main switch.

D DNO fuse.

11 State the wording that must be displayed on a label for a clamp that is connected to a metallic gas installation pipe connected to a main protective bonding conductor.

12 Draw an earth fault path for a circuit supplying a water heater fed from a consumer unit supplied by a TN-C-S earthing arrangement.

13 A ring-final circuit contains six socket-outlets which are wired directly in the ring. State for this circuit the:

a Spurs.

b Number of sockets permitted on a fused spur.

c Minimum permitted cable size for an unfused spur when using flat profile thermoplastic twin and cpc cable.

14 Draw a symbol for each of the following electrical accessories:

a One-way switch.

b Two-way switch.

c Double-pole switch.

d Two-gang socket-outlet.

15 Draw a circuit diagram for a light being controlled by two-way and intermediate switches.

Practical task

The external dimensions of a building measure 42 m × 28 m and it has three rooms on the ground floor.

Draw a plan of the building using an appropriate scale for each of the following paper sizes in order to maximise the paper.

a A3

b A4

c A1

ELECTRICAL INSTALLATION

INTRODUCTION

When you are undertaking the installation of wiring systems and electrical equipment, it is important to make sure that you show good workmanship and adopt the correct standards right from the start. When working on site, you may see incorrect and unsafe practices.

In this chapter, you will learn:

- the tools that you are expected to know about and their functions
- how to use these tools
- some of the basic installation practices.

This chapter will focus not just on domestic wiring techniques but also on some of those practices found in commercial and industrial types of installation.

This chapter also introduces you to the fundamentals of proving that your installation has been installed correctly by performing necessary tests on the installation before a live supply is connected. This is essential as it helps to identify errors and omissions before putting people at risk.

Learning objectives

This table shows how the topics in this chapter meet the outcomes of the different qualifications.

Topic	Electrotechnical Qualification (Installation) or (Maintenance) 5357	Level 2 Diploma in Electrical Installations (Buildings and Structures) 2365 Unit 204	Level 2 Technical Certificate in Electrical Installation 8202 Unit 203
1 Tools used to install wiring systems	(104/004) 3.1; 3.2; 3.3; 3.4; 3.5	1.1; 1.2; 1.3; 2.1; 2.2; 2.3	1.1
2 Installation of wiring systems	(104/004) 2.1; 2.2; 2.3; 2.4; 2.5	3.1; 3.2; 3.3; 3.4; 3.5; 4.1; 4.2; 4.3	2.1; 2.2; 2.3; 2.4; 3.1; 3.2; 3.3; 3.4
3 Terminating cables	(107) 1.2; 2.1; 2.2; 2.3; 2.4; 2.5		4.1; 4.2; 4.3
4 Inspection and testing (basic verification)		3.6; 3.7; 4.4; 6.1; 6.2; 6.3; 6.4; 6.5; 6.6	

1 TOOLS USED TO INSTALL WIRING SYSTEMS

Hand tools

The majority of electrical installation work involves the use of hand tools. Hand tools can be split into two main groups:

- those used in the *preparation* for installation tasks
- those used in the *installation* of wiring systems and electrical equipment.

Preparation hand tools

At the start of any installation work, it is important to identify where all the components will be positioned and where cables will run. This often involves a process of **marking out**.

The types of hand tool used for marking out include:

- *Rule* – this is used to measure distances and can be of either a rigid or flexible construction (Figure 4.1). The rigid type is often incorrectly referred to as a ruler, and the flexible design is often referred to as a tape measure.

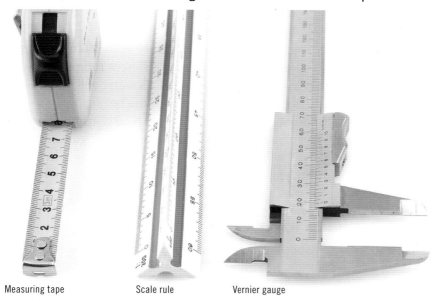

Measuring tape   Scale rule   Vernier gauge

▲ Figure 4.1 Types of rule

- *Level* – when we install components, we only have a short time to make sure that they are installed correctly and look presentable. The client has more time to study the quality of our work! Therefore, it is important to make sure that all items are correctly installed. A level will enable you to see whether you are mounting items straight, both vertically and horizontally. However, it is worth noting that some buildings are not built straight and the use of a level can emphasise this. On some occasions, positioning by eye can give a better result.

▲ Figure 4.2 Level

KEY TERMS

Datum line: a reference point or line from which multiple measurements are made.

Containment system: conduit or trunking installed to provide a level of mechanical protection to cables.

▲ Figure 4.4 Plumb line

- *Laser level* – this is a variation of the traditional level. It projects a line that can enable multiple items to be mounted to a common **datum line**. Not all construction sites permit the use of laser levels as they can cause issues with people's vision. Suitable eye protection must be used when using them.

▲ Figure 4.3 A laser level

- *Water level* – this is used to ensure two points are at the same level, when they are perhaps on different walls or even in different rooms. Two water gauges are connected by a hose of any length. As water will maintain the same level, this ensures two points are at the same height.
- *Plumb line* – this is normally used to see if something is installed on a perfect vertical line, but it can also be used to create a horizontal guide line. The cord can be coated in a chalk-like powder and then held firmly in place at either end. If it is stretched and allowed to snap back, it leaves a faint chalk line, which can be used as a reference for measurements.

Installation hand tools

After the initial marking out has been performed, installation work can start. Sometimes, the installation process can be split into two tasks:

- fabrication and installation of **containment systems**
- wiring.

Fabrication and assembly of the containment system uses tools that are designed for cutting metals and plastics.

Hand saws

There are various different types of hand saw used in our industry, and each has a different purpose. It is important to choose the correct saw for the job, as an incorrect tool can result in a poor installation.

- *Cross cut saw* – this is a general-purpose saw used for cutting timber.

▲ Figure 4.5 Cross cut saw

- *Floorboard saw* – this has a specially shaped blade with teeth on both the straight and curved edges to enable access points to be cut in floorboards.

▲ Figure 4.6 Floorboard saw
(photo courtesy of Axminster Tool Centre Ltd.)

- *Hacksaw* – this is used for cutting metal, and comes in both junior and regular sizes.

▲ Figure 4.7 Hacksaws

- *Coping saw* – this type of saw has a very small blade and can be used for performing very precise cuts. It is generally used for cutting mouldings.

▲ Figure 4.8 Coping saw

- *Keyhole saw* (also called a pad saw) – this saw is used to cut holes that need to be larger than a drill bit, and is often used to cut holes in plasterboard for recessed **pattresses**.

▲ Figure 4.9 Keyhole saw

KEY TERM

Pattress: the recessed container behind an electrical fitting, such as a socket, sometimes referred to as a back box.

Like any other tool, a saw must be kept sharp and in good condition. Missing teeth can result in a bad cut or a weakened blade.

The number and shape of teeth change according to the purpose of each type of saw. The number of teeth on a blade is referred to as the teeth per inch, or TPI; the softer the material that is being cut, the lower the TPI.

When cutting through mild steel (such as conduit and trunking), the TPI must be about 24 to 32 – this is the type of blade fitted to most hacksaws. Missing teeth or a bent blade can result in the blade becoming jammed and then breaking, so it is important to inspect your blade prior to use.

Files

When working with any metal, you should remove all sharp edges and any rough bits, which are generally referred to as burrs. As with hand saws, there are different types of file for different purposes.

- *Hand file* – normally a rectangular shape with one smooth edge. This smooth edge enables the file to be used in corners, without cutting both edges at the same time.

▲ Figure 4.10 Hand file

- *Half round file* – particularly useful for filing internal curves.

▲ Figure 4.11 Half round file

- *Three square file* (also known as a triangular file) – this file is in the shape of a triangle and is especially good for filing into corners.

▲ Figure 4.12 Three square file
(photo courtesy of Axminster Tool Centre Ltd.)

- *Knife file* – these are very thin and can fit into small gaps.

▲ Figure 4.13 Knife file

- *Square file* – normally used to file slots or key ways.

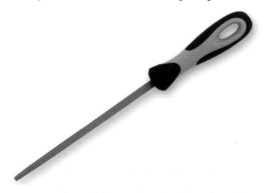

▲ Figure 4.14 Square file
(photo courtesy of Axminster Tool Centre Ltd.)

Files are not just described by their shape or purpose. They are also identified by coarseness – some files remove a lot of metal but leave a rough finish and others remove less but leave a very smooth finish.

The coarseness is identified as a grade of cut and the levels include:

- very smooth (finishing cut)
- smooth (dressing cut)
- second cut
- bastard
- rough.

As with any hand tool, it is important to make sure your cutting tool is safe to use. So, every file should be fitted with a handle which is secure and the faces should be kept clean with a stiff wire brush. Damaged or blocked teeth on the file will result in an uneven cut, which may cause slippage that results in injury to the user.

Wire strippers

Wire strippers provide a safe and reliable method of removing the insulation from a wire or cable without damaging the conductor. Wire strippers come in various designs, but the common principle is that

▲ Figure 4.15 Wire strippers

the cutting jaws only cut into the insulating material and not into the conductor.

Wire strippers can be either manually set (by a screw or dial) or automatically set. Side cutters can also be used but these often damage the conductor.

Cable cutters

There are a variety of different tools that can be used to cut cables, the most common being side cutters. Side cutters are probably one of the most important tools that electricians have in their tool kit. These are used for cutting cables to length, cutting sleeving and cutting nylon tie-wraps, for example. They work on a compression-force basis and are shaped so that the cutting point is along one side.

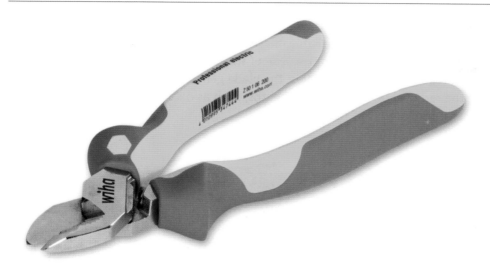

▲ Figure 4.16 Side cutters
(photo courtesy of Axminster Tool Centre Ltd.)

Depending on the size of the cutters, cable sizes up to 16 mm^2 can be cut easily. However, larger cables require larger types of cutters. These range from cable loppers, up to hydraulic, manual pump cutters for cables up to 300 mm^2.

Screwdrivers

An electrician will use a selection of different sized screwdrivers. They will all have one thing in common in that they will all be of approved standards.

The most common screwdrivers used include:

- terminal (3–3.5 mm)
- large flat (4–5 mm)
- pozi-drive (PZ2)
- and, more recently, a consumer unit screwdriver.

▲ Figure 4.17 Electrician's screwdrivers

Electrician's knife

The electrician's knife is available in various arrangements but usually has a folding blade. The most common use of the electrician's knife is for stripping the outer sheaths of *some* cables; the blade generally has a shaped section to aid the removal of certain cable sheaths, such as on armoured or some types of mineral-insulated (MI) cables.

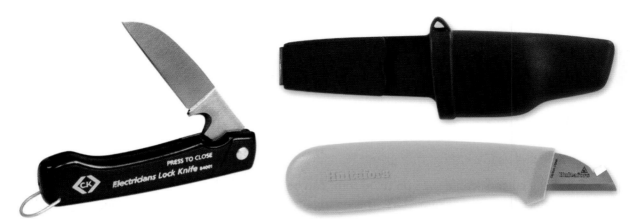

▲ Figure 4.18 Electrician's knives (photo courtesy of Axminster Tool Centre Ltd.)

An electrician's knife should never be used for stripping the insulation from cables or for cutting cables to length. It should always be used pointing away from the body.

Specialist tools

Certain activities require tools that are specifically designed for the task and are only used for this task. For example, when working with steel conduit, it is necessary to use stocks and dies, along with reamers and engineer's squares. These and other tools are explained later in the chapter.

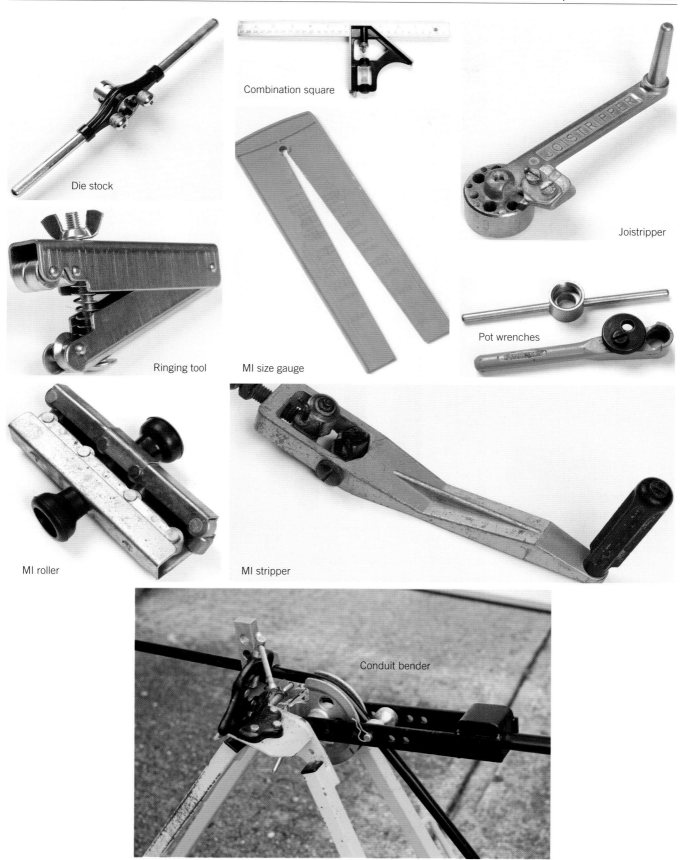

Combination square

Die stock

Joistripper

Ringing tool

MI size gauge

Pot wrenches

MI roller

MI stripper

Conduit bender

▲ Figure 4.19 A selection of electrical tools

Wrenches

A wrench is used to tighten nuts and/or bolts. They are often used in place of spanners, as one tool can be used on a variety of different sizes of nuts and bolts.

Some typical wrenches used in our industry include:

- *Pipe wrenches* – the jaws on these wrenches are designed to fit around a selection of different nuts and bolts, and so they are ideal for tightening lock nuts and cable glands.

Pipe wrench or Stillsons   Water pump pliers   Adjustable spanner   Adjustable spanner   Combination spanner

▲ Figure 4.20 A selection of wrenches

- *Footprints* – some larger cables require glands that are in excess of 60 mm in diameter and these are often too big for a pipe wrench. Footprints can be set to larger sizes and are well suited for tightening these larger glands and nuts.
- *Bush spanners* – when installing metal conduit, it is important to ensure that all connections are tight. Some of the assembly components are called bushes. These are often small and in positions that are not accessible to pipe wrenches and footprints. A bush spanner is designed to fit into these small areas, enabling the bush to be correctly tightened.

It is important to select the correct wrench for the job and to set it to the correct size. Incorrect use can damage the nut or bolt being tightened.

▲ Figure 4.21 Bush spanner

Hammers

There are several different types of hammer that can be used in our industry and these include:

- *Claw hammer* – this is the electrician's general-purpose hammer. It is used for driving and removing nails and tacks. It typically weighs between 16 and 28 ounces.
- *Ball peen (or pein) hammer* – this hammer is typically used when bending and forming metal. It weighs between 4 and 32 ounces.
- *Engineer's hammer* – sometimes referred to as a baby sledgehammer, this can be used during light demolition work. It typically weighs from one to five pounds.

Claw hammer

Ball peen hammer

Engineer's hammer

▲ Figure 4.22 A selection of hammers

The most common mistake that people make when using a hammer is that they do not grip it at the correct place. The handle is designed to be gripped at a point furthest away from the head of the hammer, enabling the full force to be applied.

If a hammer has a split or damaged handle, or shows signs of damage to the head, it should not be used – it should be replaced immediately. Never hit the heads of two hammers together as this may result in parts breaking off.

Basic safety

When using any hand tool it is important to ensure that you take basic safety requirements into account. Ask yourself these questions:

- Is the tool in good condition?
- Is the tool sharp?
- Is the tool the correct one for the job?
- Do I know how to use the tool?
- Is the tool safe to be used?

The key point when using any tool is to check to see if it is safe to be used. A cold chisel that has started to bend over where it is hit with the hammer is displaying 'mushrooming'. These bent-over pieces can break off, cutting your hand or, even worse, ending up in your eye.

Blunt tools require more force to perform cutting functions – which can lead to dangerous slips. If a saw blade is blunt, you may be tempted to apply more pressure, which in the case of a hacksaw could shatter the blade. The blade may also become hot, with the potential to burn you.

Using a screwdriver as a chisel or lever is not a correct use of the tool. The shaft of the screwdriver may snap or slip. You could also damage the handle, so that next time you use the screwdriver it ends up sticking into your hand.

Buying cheap tools can be a false economy; cheap tools tend to break easily, whereas good-quality tools often last a lifetime if looked after and used properly.

Power tools

Though most tasks can be performed using hand tools, the work is often performed more quickly with **power tools**.

Power tools can be hand held or can be larger machines that are mounted on a workbench or bolted to the ground.

Safety checks for power tools

When using power tools it is important to make additional safety checks to those already mentioned in connection with hand tools.
The use of electrical energy adds extra risk. Additional checks should include these questions:

- Is the item of equipment correct for the job?
- Is the casing of the tool without damage?
- Is the flex of the tool (if fitted) undamaged?
- Is the plug (if fitted) undamaged?
- Is the start/stop button in good condition?
- Are all guards in place (where fitted)?
- Is there a means of stopping the tool in case of emergency?

If you answer 'no' to any of these questions, the tool should not be used and the person responsible, such as your supervisor, should be notified immediately.

Drills

There are several different types of drill that can be used in our industry for different tasks. It is important to recognise the best drill for the job, making sure it is safe to use.

All drills have some common parts, such as a **chuck** and a **bit** (with some requiring a special tool called a chuck key to tighten and loosen the jaws of the chuck). In the past, the chuck key was often lost or misplaced, so some modern drills have a chuck that can be tightened by hand without the use of a special tool or key. It is important to make sure that, when securing a bit in the chuck, it is positioned straight and in the centre so that it does not wobble or come loose. For more about drill bits see pages 234–236.

Electric hand drill

This is the most common type of drill and can be mains powered from either 230 V or 110 V AC, as well as being available in battery form. This drill is probably the most versatile as it is a general-purpose tool.

The hand drill, as its name suggests, is portable and can be used in most positions. Consideration should be given to dust and **swarf** that may be generated during the drilling process – make sure suitable precautions are taken to minimise risk of injury.

KEY TERM

Swarf: chips and curly turnings of excess metal and wood produced when drilling.

▲ Figure 4.23 Electric hand drills

INDUSTRY TIP

Only use drill bits that are sharp. Make sure the drill is turning in the right direction.

Right-angled electric drill

The right-angled electric drill is a variation of the electric hand drill. This tool is often used when making alterations to an existing property and new holes need to be cut through the joists. The spacing of the joists often prevents access by a regular hand drill and so a right-angled drill is required.

The chuck is positioned at right angles to the rest of the body of the drill, so the space required for access is only that of the chuck and drill bit. Generally, this is less than the spacing of the joists.

▲ Figure 4.24 Using a right-angled drill

Hammer drill

For drilling into masonry walls or floors and ceilings, cutting is more efficient when accompanied by a controlled impact. This is achieved by the use of a hammer-action drill. Some electric hand drills have this option built in and it can be easily selected.

When in hammer mode, the chuck is subjected to a set level of vibration which is transmitted to the drill bit. This helps the drill knock the bit into the masonry. This function must not be used when drilling through timber or steel as it will damage the drill bit as well as the material. Masonry drill bits (for use in hammer mode) are designed and built to withstand these stresses.

▲ Figure 4.25 Hammer drill

Pillar drill

Where drilling needs to be very accurate and consistent, a pillar drill is often used. The pillar drill has an adjustable table which can be positioned at various heights. The table is used to clamp the material to be drilled so that it does not move, so a hole can be drilled in a precise location.

Drill bits

A drill bit is the item that is positioned in the chuck to perform the cutting process. Bits come in various shapes and sizes according to the material to be cut. When drill bits are used for cutting the wrong material they can become blunt, very hot and damaged.

When a drill bit becomes blunt, it is tempting to press down harder on the drill to make the bit cut. This increases the risk of injury – the bit could shatter. In addition, a blunt drill bit can become so hot that it melts, further increasing the risk of injury.

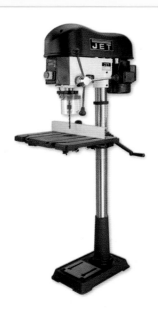

▲ Figure 4.26 Pillar drill

(photo courtesy of Axminster Tool Centre Ltd.)

All drill bits are designed to cut in a clockwise direction. If a reversible drill is used, care should be taken to ensure correct rotation. A drill bit should never be run in reverse as this will damage it.

Brad point bit

Brad point drill bits are specifically designed for drilling into wood and have a cutting face that includes a centre spike. This spike cuts a pilot hole for the drill to follow, making sure the drill bit stays on track. Without this centre spike, the drill bit could go off track due to the grain in the wood.

The flutes (which allow the cut wood to be fed out of the hole) are also of a wide design – the debris must be removed quickly from the hot tip or it may catch fire.

▲ Figure 4.27 Note the special design of the brad point drill bit that feeds cut wood away from the drill bit point

Hole saw

As the name suggests this is a saw bit for cutting holes. There are two types: one for wood and the other for steel. Both types fit onto an arbor which holds the saw in place. The arbor is fitted with a pilot drill to make sure the saw acts around a central point.

A hole saw is ideal for cutting holes in sheet material or housings where the material is relatively thin and the hole size is larger than that of a conventional drill bit. Some of the more common sizes are 16 mm, 20 mm, 25 mm, 32 mm and 60 mm diameter saws.

▲ Figure 4.28 Hole saws for cutting holes of different sizes

KEY TERM

HSS: high-speed steel.

Metal drill bit

Metal drill bits are sometimes referred to as **HSS** (high-speed steel) bits. These are intended to be strong enough to cut through steel and often have extra coatings, such as metal carbide, to make them resistant to wear.

The cutting face on an HSS bit is straight and there are two parts. A sharp HSS bit should look like a figure 8, when viewed from the cutting end. The cutting faces are set at an angle from the horizontal and the back of the face should be lower than the front, to ensure only the cutting edge is in contact with the material being cut. If the back edge of the cutting face is too high, it will rub against the material, causing excess heat and stopping the cutting by the drill bit.

HSS bits can be used to cut alloys such as aluminium, but they need to be run at a different speed.

▲ Figure 4.29 Note the design of the HSS drill bit, which minimises friction when cutting

Masonry drill bit

Use of an HSS bit on masonry will quickly blunt the bit, depending on the make of bit and the masonry being drilled. The flutes on an HSS bit are designed to remove swarf not masonry dust particles.

A masonry drill bit is designed with a larger cutting face diameter to that of the drill bit shaft. This enables the masonry dust to be pushed away from the cutting face and along the flutes.

The cutting face of a masonry bit is made from a hardened material that is designed to cut into hard materials and to withstand high temperatures.

▲ Figure 4.30 Note the special design of the masonry drill bit, which pushes masonry dust away from the drill bit

235

Core bit

To produce a large hole in a masonry wall, a core cutter is used. This is similar in design to a hole saw but is specifically designed to cut large holes in masonry. The cutting edge is either fitted with tungsten carbide teeth or is diamond coated.

▲ Figure 4.31 Core drill bit

Electric screwdriver

There is increasing pressure in electrical installation to work faster. Electric screwdrivers help to achieve this. They come in all shapes and sizes, but all have the following common characteristics:

- bi-directional for tightening and loosening
- adjustable torque setting to make sure screws are not over tightened
- variable speed, generally controlled by the start button.

Some electric screwdrivers can also be used as electric hand drills, if they have a drill setting. However, not all electric drills can be used as electric screwdrivers. If there is any doubt, they should not be used as screwdrivers.

▲ Figure 4.33 Drill used as a screwdriver

INDUSTRY TIP

Most battery-operated drills have a screwdriver function. Be careful when using them not to over-tighten the screw as you will either snap the screw head off or split the box you are fitting.

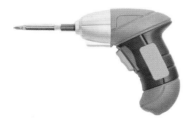

▲ Figure 4.32 Electric screwdriver

Personal protective equipment (PPE)

The use of power tools always presents a safety risk to the person using the tool and those around them. Consideration should always be given to the use of additional PPE.

When drilling, you should think about the damage that you may be doing to your ears, eyes, hands and lungs – always consider the use of ear protection, gloves and face masks. As a *minimum*, eye protection should be worn when using power tools. See pages 31–37 for more details about choosing appropriate PPE.

Before any drilling is undertaken, checks should always be made to ensure that you are not going to drill into something unexpected behind the surface being worked on. Always be alert when drilling as hidden dangers often lurk, such as buried cables, pipes or even asbestos.

The safe use of power tools

The aim of the Provision and Use of Work Equipment Regulations (PUWER) 1998 is to ensure that the equipment used to perform work is without risk to health and safety. In general terms, PUWER require that equipment which is provided for work activities is:

- suitable for the intended use
- safe for use and maintained in a safe condition
- used only by people that have received training on how to use the equipment correctly and safely
- accompanied by suitable safety measures, such as guards, marking and warning signs.

It is the responsibility of *both* the employer providing the tool and the employee using the tool to check the equipment is safe, appropriate for the job and used correctly.

Most plug-in power tools on a construction site are supplied from a 110 V AC supply and use a yellow **BS EN 60309** plug-and-socket arrangement for connection. The reason for the colour coding on the plug-and-socket arrangement is to ensure that the operator knows what voltage is required and available. The colours of the plugs and sockets used under BS EN 60309 are:

- purple – this is used on voltages below 50 V AC
- yellow – this is used on 110 V AC centre-tapped sources
- blue – this is used on 230 V AC standard single-phase supplies
- red – this is used on 400 V AC three-phase supplies.

As well as the colour coding, plugs and sockets have notches and bumps called keys which limit their connection. So, it is not possible to plug a piece of 110 V equipment into either a 230 V or 400 V supply. If a BS EN 60309 plug has been cut off and replaced with a standard **BS 1363** 13 A three-pin plug, you should not use the equipment and should bring it to the attention of your supervisor.

> **HEALTH AND SAFETY**
>
> When at work and at your training centre you must wear appropriate PPE. Failure to do so may have you removed from site or your assessment may be cancelled. Not only must you wear it, but it must be worn correctly. A safety helmet worn back to front is no good at all.

Section 704 in **BS 7671:2018** The IET Wiring Regulations, 18th Edition provides further information regarding the installation requirements for temporary power supplies and installations to be used by site workers during the construction stage.

Hazards associated with the use of electric power tools are shown in the table.

Hazard	Source
Impalement	If the bit of a hand drill is of the wrong type or blunt, the operator may be tempted to push too hard. Should the drill bit snap it may injure the operator.
Particles in the eye	When cutting wood, metal or other material with a power tool, debris such as wood splinters and swarf can be thrown up. If the operator is not wearing suitable eye protection they may get particles in the eye. Brick dust can cause severe damage, even blindness.
Electric shock	Power tools that are plugged into the mains are more likely to cause electric shocks than battery-operated ones. Some people do not realise the purpose of the protective device in a plug and may try to by-pass it if it keeps tripping or 'blowing'. They do not appreciate that the device is there to keep them safe in the event of an electrical fault.

Damaged leads, plugs and equipment casings can lead to electric shock. If a tool is not suitable for the particular work environment, this increases the risk of an incident happening. Only power tools designed for construction sites should be used on construction sites. |

Working at height

Working at height is obviously hazardous in itself and can compound hazards associated with the activity being performed.

The Work at Height Regulations 2005 were outlined, along with clarification of what working at height entails, on pages 45–48. The regulations apply when a person is working at any level, where if they fell from that level, they could be injured. Working on ladders or steps is a common feature of working at height, but this could also include working at ground level near an excavation.

Ladders

The HSE provides guidance for working on steps and ladders. Essentially, ladders should only be used as a means of access or where the work activity lasts for a short period of time typically not longer than 30 minutes and only for light loads not exceeding 10 kg mass.

Ladders come in various forms including:

- wooden
- aluminium
- fibre glass
- single extension
- double extension
- triple extension.

Inspection of ladders

Each ladder should carry a unique identification mark which is recorded on the company ladder register. The ladder register is used to record each ladder, along with the details of inspections, defects and actions taken. In addition, the record includes an entry every time someone takes a ladder and returns it.

Ladders should be inspected every three months as part of a routine inspection. They should also be checked annually by performing a thorough examination. This means looking carefully at the physical condition and noting any defects, such as splits, wear and corrosion. In addition, the user should always check the condition of the ladder before each use.

When using a ladder it is good practice to ensure that:

- it is placed on a firm and level surface
- the top and bottom of the ladder are secured or footed
- a stepladder is spread to its fullest extent
- there is a handrail or a hold at least 1 m above the top platform of a stepladder
- the ladder does not overreach
- tools are not left on the platform of steps.

Footing is when another person secures the bottom of a ladder by placing their foot on the bottom rung to make sure it stays in place and does not move.

When climbing a stepladder, both hands should be used and the user should face the steps, making sure that three points of contact are maintained whenever possible. Nobody should pass below the ladder when work is going on.

Blocks or boards should not be used to extend the height of a ladder. If a ladder cannot reach the desired height, a longer ladder should be used instead. When using a ladder to access a working platform such as scaffolding, the ladder must be secured at both the top and bottom, and must extend beyond the working platform to provide a handrail. This helps with safe mounting and dismounting. Figure 4.34 shows a typical installation of an access ladder on scaffolding. You will notice that the knee barrier has not been shown on the scaffolding. This would be at a height of 0.5 m above the working platform.

INDUSTRY TIP

Even if a ladder is destroyed, this should be recorded within the ladder register.

INDUSTRY TIP

If a ladder fails an inspection at any time it must be labelled and removed from service immediately; this includes failure of a user inspection.

ACTIVITY

Make a list of the checks you would make before using a wooden stepladder.

HEALTH AND SAFETY

The safety of others is a major responsibility that must be taken very seriously. Someone else's life will depend on you regularly on a construction site. At the same time, your life is in the hands of others and the way they behave. Always act responsibly!

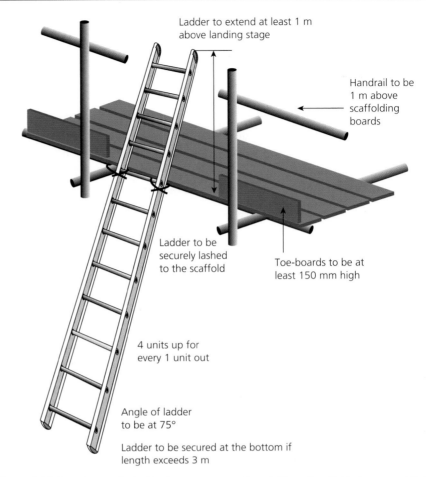

Ladder to extend at least 1 m above landing stage

Handrail to be 1 m above scaffolding boards

Ladder to be securely lashed to the scaffold

Toe-boards to be at least 150 mm high

4 units up for every 1 unit out

Angle of ladder to be at 75°

Ladder to be secured at the bottom if length exceeds 3 m

▲ Figure 4.34 Correct use of a ladder to access a working platform (scaffold clamps and fixings have been omitted for clarity)

Other working platforms

Installation often requires work at a height that would be accessible with a stepladder, but the duration of the work can be longer than 30 minutes. In such a case, an alternative form of access would be needed, such as a low-level working platform, a set of podium steps, a hop-up or trestle. The most common form of low-level access aid used in electrical installation work is the hop-up.

A hop-up platform has a typical construction of wood, aluminium or fibreglass and is designed to provide a working platform at a slightly raised level. Hop-ups do not have guards or rails and so will not prevent a fall; it is for this reason that they are not permitted on many construction sites.

▲ Figure 4.35 Hop-up

The risk assessment

Unless the various hazards are controlled and managed, working in the electrical installation business can be very dangerous. If people are not aware of the hazards around them, the risks increase. Everyone should be aware of the hazards and risks involved in the workplace. A risk assessment enables risks and hazards to be identified. Action plans are then put in place to prevent the risk from becoming a reality.

Most companies have standard risk assessments for the more routine work activities, such as use of hand tools and power tools, but it is important to understand that hazards and risks change depending on the circumstances of the individual job. Sadly, some companies perceive performing risk assessments as a necessary health and safety chore, rather than a business tool to help them plan and improve working processes. A lot of risk assessments are written and then stored away in a file on a shelf somewhere within the company. This is not the correct use of a risk assessment – it is intended to be a 'live' document which is constantly updated.

The HSE defines a risk assessment as a careful examination of what, in terms of work, could cause harm to people, so that precautions can be put in place to prevent that harm. Workers (and others) have a right to be protected from harm caused by a failure to take reasonable control measures.

There are five steps to a risk assessment:

1 identify the hazard
2 assess who might be harmed and how
3 evaluate the risk and decide how to prevent it
4 document the assessment and solution, and implement
5 monitor, evaluate and update as required.

Risk assessment is a *continual process*, because circumstances are constantly changing. Therefore, although the five steps are shown as a list above, the

ACTIVITY
When assessing the potential for harm, different locations may change the risk. A simple task, such as changing a lamp, needs to be carried out. Consider how this task differs where the lamp is in the following situations/locations:

- in a bus shelter
- in a 4 m high street light on a busy road
- above a furnace in a factory.

HEALTH AND SAFETY
Remember that a risk assessment is a continuous operation. Re-assess the risks as the work progresses.

process tends to be more of a cycle in practice (Figure 4.36). Pages 26–28 in Chapter 1 give full details of how to carry out a risk assessment procedure.

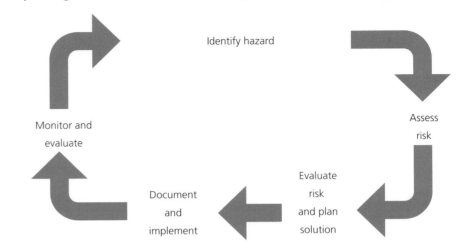

▲ Figure 4.36 Five steps of risk assessment

2 INSTALLATION OF WIRING SYSTEMS

Materials

When starting a task, it is important to be able to identify the **materials** that will be required to complete the work. As well as using your previous experience, you should be able to identify the components that are required by looking at the drawings and specifications. The project specification contains information about the type of finish that is required for the installation work and may include details of preferred manufacturers.

The drawings often include plans showing the positioning of components and accessories, as well as the routes to be followed when the wiring is being installed. This is important as none of the services being run through the installation must cause problems for other systems and services.

Materials take-off sheet

As well as showing where components and accessories are to be positioned, the drawings can also be used to create a **materials take-off sheet**.

The primary function of a materials take-off sheet is to enable components and accessories to be identified at the beginning of a project, to ensure that the correct materials will be available. This, in turn, helps in the preparation of accurate costings, so that a realistic tender or quotation can be produced. It also provides a 'shopping list' so that materials can be ordered prior to starting work, ensuring that all will be available on time and in the correct quantity.

The layout of the list varies from company to company but for each item all lists include:

- an identification number, generally an index number
- a description of the item
- the unit of measure for the item
- a quantity.

Some companies' lists also include the details of the supplier and a catalogue number for each component. The unit of measure for most components is EA (meaning each) and this means that the quantity is per item and not per packet. When items such as cable are supplied, the unit of measure is m for metres.

When producing the materials take-off sheet it is important to understand that not all materials required are identified on the drawings. Items such as patresses, conduit, trunking, cables, glands and protective devices are not shown on any of the drawings, but they still have to be listed. This is where prior knowledge and experience of installation work helps a lot. For the remainder of the components, it is important to understand how the drawings and specification identify the items to be installed.

Reading the drawings

In order to identify the materials to be used in an installation, it is important to be able to read drawings correctly. Standardised symbols are used for drawings in the electrotechnical industry. These symbols can be used to identify components and their locations.

The layout drawing is generally a plan view with a legend or key to clarify the types of components to be installed and their locations. Most designers use **BS EN 60617** symbols, but as the full range of symbols for this type of drawing are not covered by this standard, the designer may use their own variations. You must clarify with the designer any symbols that are not easily recognisable.

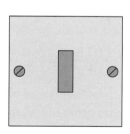

▲ Figure 4.37 One-gang one-way light switch, circuit diagram symbol and plan symbol

Some of the common symbols are shown on the inside cover of the Institution of Engineering and Technology (IET) On-Site Guide, but they are not explained. However, the IET On-Site Guide is a good memory jogger when looking at the drawings for an electrical installation.

Most of the symbols used in BS EN 60617 are made up of parts that have different meanings. Designers can add parts together to make a symbol for a component, should there be no symbol for what they are designing. Consider the light switch symbols shown In Figures 4.37 to 4.41.

When you consider a switch, you have to consider what the switch does. This includes:

INDUSTRY TIP

Make sure you remember items such as screws and wall plugs when drawing up a materials list. It is difficult to install without them.

- *ways* – is the switch one-way, two-way or intermediate?
- *gangs* – this is the number of actual switches mounted on the plate; a plate containing two switches is two-gang
- *poles* – this is the number of channels that a single switch operates in one switching action. As an example, if a switch operates both a line and neutral, it is double pole.

The circle in the plan symbol indicates that the device is a switch. The arm that sticks out of the circle shows how many ways this switch can operate. The flag at the end of the arm indicates the number of poles that the switch operates. The number of gangs is indicated by a number or letter (e.g. 2 for two-gang or M for multi-gang).

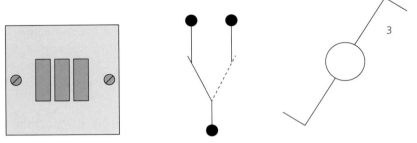

▲ Figure 4.38 Three-gang one-way switch, circuit diagram symbol (for one-gang) and plan symbol

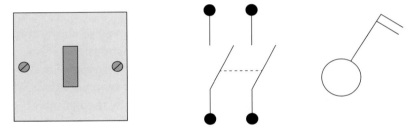

▲ Figure 4.39 One-gang one-way double-pole switch, circuit diagram symbol and plan symbol

Some multi-gang switches, often used in industrial and commercial installations, can have as many as 12 switches on the same faceplate! This type of multi-gang switch is normally called a grid switch (Figures 4.40 and 4.41) and is of a modular construction and would need to be assembled on site; generally all the parts are ordered separately.

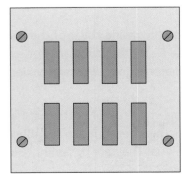

▲ Figure 4.40 Multi-gang switch

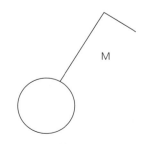

▲ Figure 4.41 Multi-gang plan symbol

Now you have looked at light switches and their symbols, you will find it easier to interpret other symbols – when a symbol has an arm with a flag on it (similar to that seen on the light switch examples) you know it is showing an accessory that can be switched. For example, Figures 4.42 and 4.43 show a non-switched socket outlet and a switched socket outlet.

For many pieces of electrical equipment, there is no dedicated symbol. Therefore, a lot of generic symbols are used. For example, a square identifies a piece of electrical equipment. When the square contains, for example, a figure 8 or infinity sign, the symbol indicates an electric fan. Always look at the key or legend of the plan to double check the meaning of the symbols used by the designer.

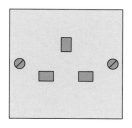

▲ Figure 4.42 Non-switched single (one-gang) socket outlet and symbol

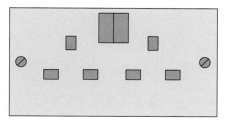
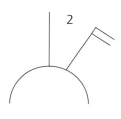

▲ Figrue 4.43 Switched twin (two-gang) socket outlet (switches are double pole) and symbol

INDUSTRY TIP

Socket outlets are not covered by BS EN 60617 so there is no standard symbol. Sometimes the symbol shown is used; sometimes the switch stick extends into the dome.

Drawing scales

The drawings may show the proposed cable routes that the installer is expected to follow. Using this information, it is possible to approximate the length of cable required.

All drawings and plans are produced to scale, to enable accurate measurements to be made. A scale is simply a ratio that is used to represent life-size structures at a smaller size. The ratio is generally expressed as two numbers and typical scales are:

- 1:20
- 1:50
- 1:100
- 1:500.

So, in the case of the third scale ratio above (1:100), one unit on the plan represents 100 life-size units. To apply the scale, make a measurement on the drawing and then multiply this by the second number in the ratio; this will give you the measurement in reality. An alternative method is to use a scaled rule with the same scale ratio as the plan. This will show the correct measurement. For example, on a scaled rule with a scale of 1:100, a measurement of 1 cm is shown as 1 m (100 cm).

Drawing dimensions

The correct positioning of materials is important. When a component must be fitted in a certain location, the layout drawing will give a measurement indicating where the component is to be installed. These measurements are generally referred to as dimensions and can be shown in one of two ways:

- datum form
- linear form.

Datum dimensioning is used where several accessories need to be installed at different heights and with varied spacing between them. A common start point, the datum, is identified and the dimensions are all measured from there. The common point may be the corner of a room or another fixed point, such as a piece of switch-gear.

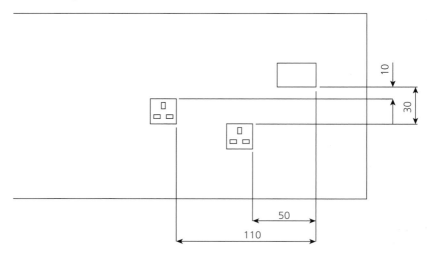

▲ Figure 4.44 Dimensioning to a datum point

KEY TERM

CRS: this means 'centres', indicating that the dimension given is between the centres of equipment.

Linear dimensioning is used when accessories need to be installed at the same height and at set distances apart. A common method of referencing the accessories is to use the term '**CRS**', which means a given dimension is measured from the centre of each component.

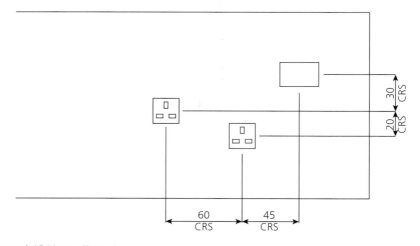

▲ Figure 4.45 Linear dimensions

Understanding the different dimensioning styles is important, as it affects the actual positioning of the components.

Marking out

When dimensions are given, it is important that these are observed when marking out and performing the first fix. Make sure that you understand all dimensions and have applied them correctly. Clarify any confusions or unclear information with either the site supervisor or the installation designer before proceeding.

Dimensions are not just used to show where components are to be installed; they are also used to plan where walls need to be cut for cables to be installed. These cuts in walls are called chases. If a wall is chased out in the wrong location, it can damage the wall and result in a lot of extra work and expense.

Marking out chases

When cables are placed in walls, they are generally installed within prescribed zones as identified in **BS 7671:2018** The IET Wiring Regulations, 18th Edition Regulation group 522.6. The IET On-Site Guide (Chapter 7) refers to these as 'safe zones'. The chasing out of walls is a very dusty and dirty job and it is normally performed for electricians and electrical installers by a builder. However, the builder will only chase out the wall once the information as been drawn on – so it is important to transfer the dimension information correctly onto the wall. To transfer dimensions onto a wall it is common practice to use:

- a pencil
- a straight edge
- a long or short level
- a tape measure
- and a plumb line.

Marking out on a finished wall should be kept to a minimum as unsightly marks have to be removed after the work is completed. Faint pencil marks (that are clear enough to be seen for installation purposes only) should be used. Faint marks are easier to remove if necessary. A useful tool at this stage is a laser level; this projects a level line over a long distance. However, not all construction sites permit the use of laser levels due to health and safety concerns.

On an unfinished wall that needs chasing, it is important to make the marking out clear; often the part of the wall to be removed is indicated by hatching the area. This is when diagonal lines are drawn across the part of the wall to be cut away.

> **INDUSTRY TIP**
>
> Remember the old adage of 'measure twice, cut once'! Errors in component siting will have to be corrected without additional cost to the client.

> **INDUSTRY TIP**
>
> A chalk line is a common tool for marking out horizontal and vertical lines. Held taut at both ends and then pinged, a chalked string will leave a faint line across the surface.

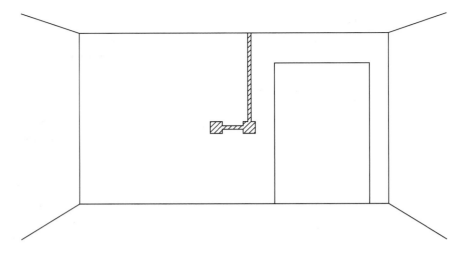

▲ Figure 4.46 Illustration showing a wall section marked out for chasing

Figure 4.46 shows how a wall may be marked out for a builder to chase out for electrical wiring. Additional information, such as the depth of the chase may also be advised, but builders generally work to the maximum permissible depth as identified in Part A of the Building Regulations.

Electricians and electrical installers must also observe parts of the Building Regulations with regards to the positioning of certain components. If the installation may be used by a person with disabilities, in order to comply with certain aspects of the Equality Act 2010, Part M of the Building Regulations must be applied.

Part M applies if:

- a non-domestic building or dwelling is newly erected
- an existing non-domestic building is extended or significantly altered
- an existing building or part of a building is altered due to a change of purpose or use.

This means that new houses must comply, but the regulation does not necessarily apply to changes within an existing home if the purpose of the room is not being changed.

If Part M applies, electrical accessories must be installed on walls in a zone which is from 450 mm to 1200 mm, measured from the finished floor level (as detailed in Section 8 of the approved document M of the Building Regulations).

So, if you are installing within a new dwelling, these heights must be observed. However, if the building is an existing dwelling and the new switch or socket outlet is an addition or alteration to an existing circuit, there is no need to comply with part M if the building did not already comply with it. Another exclusion is a socket outlet for a dedicated piece of equipment, such as a wall-mounted television, which you would not expect to unplug and plug in continually. If this is the case, the height restrictions do not apply as they focus on the general-purpose socket outlets.

Think about the purpose of a ground-floor conversion or extension when wiring. If the extension or conversion is for a person with disabilities, consideration should be given to the application of the minimum and maximum heights.

Installing wiring systems

There are many types of wiring systems used. A wiring system could be a cable on its own or a means of containing or managing the cable. Some of the terms you may hear include the following.

- **Wiring system** – this is the term used to describe the type of cable used and the method of supporting it. For example, single-core cable in conduit is a wiring system but, equally, so is a twin and cpc cable clipped to a wall.
- **Support system** – this is the method of supporting a cable and could include systems such as cable tray, basket or simply clips or cleats.
- **Cable management system** or **cable containment system** – this is a method of supporting and protecting cables, usually by enclosing the cables in conduit, trunking or ducting, for example.

Conduit systems

In some environments, the installation may be subjected to external influences which may cause damage to the wiring. Damage could result, for example, from impact or corrosion from oils and other solutions. A common method of providing an additional level of mechanical protection from such damage is the use of conduit.

Conduit is available in both steel and PVC and comes in many different forms including:

- solid steel extruded
- solid steel rolled
- rigid PVC
- flexible steel
- flexible PVC.

Both steel and PVC conduits come in both light gauge and heavy gauge, providing a degree of flexibility with regards to application and cost. Conduits come in a variety of finishes: PVC is available in various colours and steel can be galvanised or brushed, or finished in black enamel.

Solid steel extruded conduit is seamless (there is no join). This makes it an expensive option and it is only used for special gas-tight, explosion-proof or flame-proof installations. Conduit is typically supplied in 3 m or 3.75 m lengths and, unlike pipework for water and gas services, is identified by the overall *external* diameter.

Typical conduit external diameters are 16 mm, 20 mm, 25 mm and 32 mm, though it can be supplied in sizes up to and including 63 mm.

> **INDUSTRY TIP**
>
> First fixing refers to the initial installing of boxes and cables and so on. Second fixing, the connection of accessories, occurs when the plastering and other work has been completed.

Flexible conduit is predominately used in short runs as the level of mechanical protection is greatly reduced by the need for movement. A typical application for flexible conduit is in the final run of containment from wall-mounted conduit to a machine which vibrates, such as a motor.

Advantages of conduit

One of the main advantages of using conduit is the increased mechanical protection that is afforded by the containment. Flexible conduit offers the least level of mechanical protection, followed by PVC conduit, with steel conduit providing the highest level of mechanical protection.

Other advantages of using conduit include the ability to perform rewiring relatively easily, providing all cables are being replaced at the same time – cables should all be drawn into the conduit at the same time and not pulled past one another. **BS 7671:2018** The IET Wiring Regulations, 18th Edition Regulation 522.8 states that cables should not be subjected to stress, strain or damage during installation, maintenance or use.

Using steel conduit in an installation can minimise the risk of spread of fire, providing all joints, fittings and accessories are constructed to a minimum of IP33. Another benefit of steel conduit is the fact that steel conducts electrical energy. The conduit can be used as the circuit protective conductor (cpc), provided it is of a suitable size. BS 7671:2018 Regulation 543.1.1 explains how to establish whether the conduit is big enough to act as the circuit protective conductor.

Although steel conduit is rarely used as a sole protective conductor, steel conduit may act as a protective conductor when installed buried in a wall to satisfy the requirements of Regulation group 522.6 of BS 7671:2018, or when acting as a high integrity protective conductor on circuits having high protective conductor currents.

Disadvantages of conduit

The first disadvantage of using conduit is cost. The time and materials required to install a conduit system are often far greater than that needed to install the cables directly (clipped direct) in the fabric of the building or onto the surface of the building. It is important, therefore, to establish the need for mechanical protection prior to installing.

PVC conduit is directly affected by UV rays and can become discoloured and brittle in direct sunlight. In addition, PVC expands a lot if it gets hot. It becomes very brittle if subjected to temperatures below 4 °C and can become very soft if subjected to temperatures above 60 °C.

Both PVC and steel conduit can also be affected by acids, alkalis and corrosive fumes, with steel conduit being susceptible to rust from moisture which can form on the inside of the conduit.

Steel conduit has one big disadvantage that PVC conduit does not; eddy currents can form in it.

When a current passes along a conductor, a moving magnetic field forms around it. If a single current-carrying conductor is placed within a steel conduit, the moving magnetic field can induce an electromotive force, e.m.f., in the steel conduit. This induced e.m.f. can create an eddy current which, in turn, can produce heat as well as the potential for electric shock. Therefore, BS 7671:2018 Regulation 521.5.1 states that, when a circuit is carrying AC current, all circuit conductors must pass through the same conduit. This means that no run of steel conduit has just one line or neutral conductor running through it.

Accessories

Steel and PVC conduits use some common accessories in their installations, but they also have their own specific accessories.

Conduit boxes

There are various forms of box that are used in the construction of a conduit system.

- *Junction box* – this is sometimes called a termination box as it is used at the end of a conduit run. This box has one spout coming out of the side of the box. This type of box can be used when you come to the end of a run or when you need to get tight into a corner.

> **INDUSTRY TIP**
>
> Conduit boxes are often referred to as besa (pronounced 'beezer') boxes.

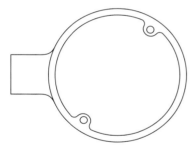

▲ Figure 4.47 Junction box

- *Through box* – this box has two spouts coming out of opposite sides of the box. This type is often used as a through link and can be used over longer runs to make drawing cables easier.

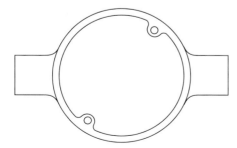

▲ Figure 4.48 Through box

- *Angle box* – this box has two spouts at 90° to each other. This is useful for going round corners or changing direction in situations away from corners.

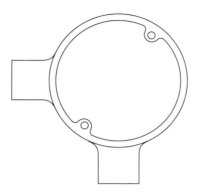

▲ Figure 4.49 Angle box

- *Tee box* – as its name suggests, this box has three spouts forming the shape of the letter 'T'. It has the benefits of both an angle and a through box.

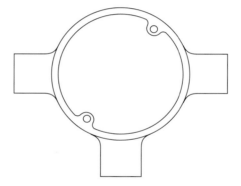

▲ Figure 4.50 Tee box

- *Four-way box* – this is sometimes called a cross box and it has four spouts at 90° to one another.

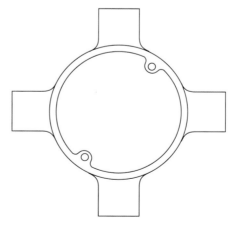

▲ Figure 4.51 Four-way box

- *Tangent-through box* – similar to the standard through box, this box has the spouts at one edge. This allows the box to be placed closer to a wall joint or ceiling.

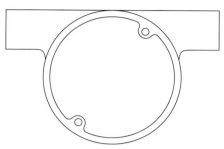

▲ Figure 4.52 Tangent-through box

- *Tangent-angle box* – again this is the same as the angle box, but with the spouts positioned towards one side of the box.

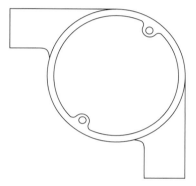

▲ Figure 4.53 Tangent-angle box

- *Tangent-tee box* – this box has the same shape as the tee box but, due to the opposing spouts being on one edge, the conduit can be positioned closer to a ceiling or other edge.

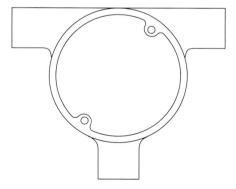

▲ Figure 4.54 Tangent-tee box

INDUSTRY TIP

There is a selection of lids available to fit standard conduit boxes. These range from plain pressed steel covers to hook plates and dome lids.

- *H-box* – this box has four spouts forming a letter 'H', with two on either side. It is similar to two tangent-through boxes joined together.

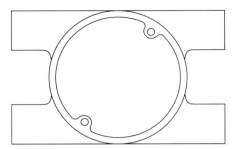

▲ Figure 4.55 H-box

- *U-box* – this box allows for a 180° change of direction, by having two spouts on the same side of the box.

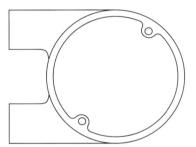

▲ Figure 4.56 U-box

- *Y-box* – similar to the U-box, the Y-box allows for a 180° change of direction and has the benefits of a through box. The through spout is centrally located on the opposite side from the other two spouts.

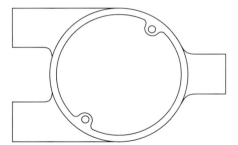

▲ Figure 4.57 Y-box

- *Back-entry box* – this type of box is similar to a junction box with the spout or spouts located in the back of the box section. It can house up to four spouts.

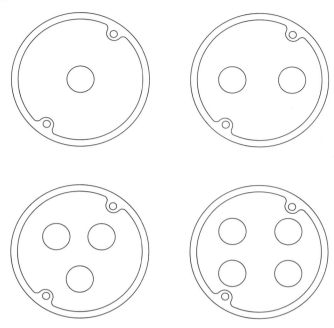

▲ Figure 4.58 Back-entry boxes

Bends and elbows

You will learn to fabricate your own bends during assembly. However, pre-manufactured bends and elbows are also available and they tend to be one of two types:

- inspection joints
- non-inspection joints.

Inspection joints are usually preferred as these allow the cables to be drawn in easily during installation. However, if the bend is on a short run and within 0.5 m of the accessory, a non-inspection bend can be used.

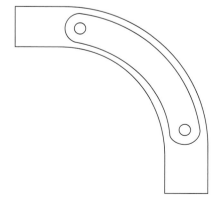

▲ Figure 4.59 Inspection elbow bend

Whatever type of bend you use, bear in mind compliance with **BS 7671:2018** The IET Wiring Regulations, 18th Edition Regulation 522.8.6 (you need to ensure that the cables are not subjected to excessive strain or stress during installation by providing adequate means of access to perform the drawing in of cables). So, if inspection joints are not used, there must be adequate alternative access points along the conduit run. These can be provided by accessory boxes, such as through boxes on long runs.

Joining components

Conduit is available in standard lengths of 3 m and 3.75 m; often runs need to be joined. The basic method of joining two runs of conduit together is by use of a coupler.

The coupler that is used on PVC conduit is a piece of tube that has a larger diameter than that of the conduit. The conduit is simply slotted into either side of the coupler and secured in place with adhesive. Note that, under COSHH Regulations, the adhesive that is used in the joining of PVC conduit must be recorded on the National Poison Centre records; this should be done by the supplier or wholesaler selling the adhesive.

In a long run, PVC must be able to expand and contract. When the run of PVC conduit is 5 m or longer, an expansion coupler must be used.

This is similar to a normal PVC conduit coupler but is twice the length and adhesive is only used on one side, allowing the other side to move in and out of the coupler. It is important to ensure the level of penetration is correct, allowing for contraction without leaving the coupler and expansion without buckling.

Steel conduit also uses couplers but the use of adhesive is not permitted. All joints in steel conduit must be tight and electrically sound, so that if one part did become live, the installation would still be connected to earth and the protective devices would operate.

To achieve this, the coupler is slightly larger than the conduit and there is a thread all the way through the inside of the coupler. This means that, once a thread is cut on both pieces of the conduit to be joined, the components can be screwed together.

In some joints it is necessary to have the thread on the outside and this is called a nipple. A nipple has the same size external diameter as the conduit, and the thread is cut along its length. It is normally about the same length as a coupler and can be connected with a coupler.

Erection components

There are various different ways of securing conduit to walls. If the conduit is to be covered by plaster, it is common for a crampet to be used to secure the conduit temporarily in place. This looks like a bent-over floorboard nail.

When the conduit is to be installed on the surface, it is common to use a type of saddle to hold it in place. All of the saddles shown here are available for PVC and steel conduits.

INDUSTRY TIP

You can see that it is essential that metal conduit joints are tight. This not only ensures good continuity, but prevents the conduit becoming loose.

▲ Figure 4.60 Crampet

- *Strap saddle* – this is used when the conduit is to be secured in place directly on the surface, with no gap at the back. It is also referred to as a stamp saddle.
- *Spacer bar saddle* – this is the most common form of saddle and is available for both steel and PVC conduit. It allows the conduit to be positioned slightly away from the surface of the wall. The holes on the saddle part or stamp are designed so that, when the fitting is mounted horizontally, the stamp will drop into place to make installation easier.
- *Distance saddle* – this is similar to the spacer bar saddle, but with a thicker back-plate. It can be used to install conduit on surfaces that undulate.

▲ Figure 4.61 Strap saddle

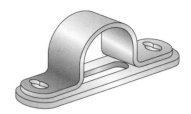

▲ Figure 4.62 Spacer bar saddle

▲ Figure 4.63 Distance saddle

- *Hospital saddle* – this is used where it is important to be able to clean behind the conduit. The hospital saddle has no horizontal edges for dirt and dust to accumulate and also moves the conduit away from the wall, just far enough to make cleaning possible. As its name suggests, these are used in some areas of hospitals, but they are also suitable for use in commercial kitchens, along with stainless steel conduit.

The saddles used on PVC conduit can be different to those for steel conduit. Some do not need screws to fix the saddle part to the back-plate.

Bushes

For steel conduit there are two types of bush, both of which are made of brass. These are:

- *Male bush* – this has the thread on the outside and it is used to thread into a coupler or accessory.
- *Female bush* – this has the thread on the inside and is used where the thread on the conduit protrudes into an accessory and the cable needs to be protected from the rough edges.

For PVC conduit, the bushes are combined with the adaptors for securing the conduit. There are two types of bush and adaptor combination:

- *Male adaptor* – this uses a part similar to the brass male bush used on steel conduit assemblies and screws into the adaptor.
- *Female adaptor* – this uses a part similar to a lock ring and screws onto a thread already on the adaptor.

> **INDUSTRY TIP**
>
> The spacer bar stamp normally has a keyhole slot to allow one-handed fitting. The maximum distance between fixings for conduit is set out in Table D3 in the IET On-Site Guide.

▲ Figure 4.64 Hospital saddle

Sizing of conduit

When you are installing conduit, consider the fact that the cables will get *hot* and need to be able to cool, to avoid overheating. There must be sufficient air space around the cables.

The maximum fill capacity for conduit allows for air flow around the conductors.

Appendix E of the IET On-Site Guide has standard tables giving the number of conductors that can be installed in a piece of conduit, taking into account all the relevant factors. The tables take into account the size and type of conductors, the length of the conduit run and how many bends there are in the run (see Chapter 3, pages 204–207).

To use Appendix E, follow this simple and logical approach:

1 Identify all the different sizes of wiring conductors used.
2 Identify the quantities of each different size.
3 Using the tables in Appendix E, select the correct cable factor for each size of cable.
4 Multiply the factor by the number of conductors of that size.
5 Add the totals together.
6 Select the conduit with the next factor up from the total value.

EXAMPLE

An easy way to follow these steps is to use a simple table, filling in the data as you go. The example below is a conduit sizing calculation for three lighting circuits and one ring final circuit, wired using solid copper thermoplastic cables in a run of 2.5 m. The steel conduit is to be used as the protective conductor for each circuit.

Conductor size in mm²	Number of conductors	Cable factor	Total cable factor
1.5	$3 \times 2 = 6$	27	$6 \times 27 = 162$
2.5	$2 \times 2 = 4$	39	$4 \times 39 = 156$
		Total	$162 + 156 = 318$

Example of a conduit sizing calculation

The calculation gives an overall conduit factor of 318. Again, using the appropriate table in Appendix E of the IET On-Site Guide, it can be seen that a piece of 20 mm conduit can be used as it has a spacing factor in excess of the 318 required.

However, if, after checking BS 7671:2018 Regulation 543.1.1, you discovered that 20 mm conduit did not have sufficient area to act as the protective conductor, you would have to repeat the calculation allowing additional circuit protective conductors for each circuit.

Preparation for conduit installation

When installing any form of conduit system, it is important to plan the route effectively and accurately. This often involves positioning accessories in their appropriate locations and using chalk to project straight lines for the conduit to follow.

With the chalk lines in place, the saddles can be positioned along the run to aid with the erection and measurement process. The IET On-Site Guide provides guidance, in Appendix D, on the maximum distance between saddles – this is dependent on the size and type of conduit.

Conduit systems are rarely an exact copy of their dimensioned drawings, as buildings are not often built exactly to the planned dimensions. Marking out the proposed route helps to establish the exact measurements of the conduit system. Once all the accessories and saddles are in place, it is possible to commence building the conduit system.

Cutting of conduit

Cutting PVC conduit is easier than cutting steel conduit; conduit cutters for PVC conduit are readily available and come in several forms. These include the snip style of conduit cutter and the ringing tool style. Both of these tools will cut the conduit, leaving a clean edge on the end with no burrs and minimal deformation of the conduit shape.

Steel conduit is not so easy to cut and typically a hacksaw is used. Ideally, a hacksaw blade should be between 24 and 32 TPI for cutting steel conduit. Other tools that you will require include:

- graphite pencil for marking out
- tape measure
- engineer's square
- hand file (generally second cut grade)
- pipe vice
- conduit reamer
- appropriate signs and obstacles for making sure the work area is cordoned off.

The pipe vice is generally mounted on the tool that is used for bending the steel conduit, called a conduit bender. The pipe vice jaws must be well maintained and clean, or they will damage the conduit. If the vice is mounted on the conduit bender, the conduit may be at an angle, not necessarily horizontal.

Having established how long the conduit should be, measure and mark the conduit using the tape measure and pencil, remembering to allow for 15 mm of thread at each end.

Position the conduit in the pipe vice, allowing sufficient space to perform the cut but ensuring the cut will be close to the vice, so that the conduit does not vibrate excessively while being cut. If the conduit does vibrate too much during cutting, the hacksaw blade can become jammed, damaged or broken.

Stand square to the job and make sure your movement is unobstructed. Position the blade next to the mark (on the piece of conduit not required) and grip the hacksaw lightly, applying light pressure on the forward cutting stroke. Use the full length of the blade to cut, as this helps to ensure that the blade does not overheat and become jammed in the cut.

Once the cut has been performed, there will be rough and sharp edges on the outside and on the inside of the conduit. You should never insert your finger into the conduit to check for rough edges. To remove these rough edges, position the required conduit in the pipe vice. Stand square to the conduit. Use a hand file and, with one hand on the file handle and the other hand supporting the file at the other end, draw the file across the cut end of the conduit.

Repeat this until the face of the conduit is smooth and square. To check to see if the face is square to the length of conduit, use an engineer's square. Once the end of the conduit is square, use a hand file to remove the rough burrs on the outside edge.

This time the file must be held at an angle of around 40° to the face. Draw the file across the sharp edge, slowly working around the end of the conduit, to create a slight chamfer. This chamfer will aid with cutting the thread later.

Once the outer sharp edges and burrs have been removed, the inner burrs must also be removed (these could damage the cables as they are installed). You will sometimes see people doing this with a round file but the correct tool to use is a reamer. The reamer is like an oversized drill bit, but it is designed to be inserted into the conduit and rotated by hand until all the sharp edges have been removed.

Simply rotate the reamer through 120° and back again several times to remove all the burrs. Check that the burrs have been removed with a quick visual inspection. Don't touch the end and risk cutting your fingers.

Threading conduit

Once a piece of steel conduit has been sawn to length, it is necessary to cut a thread for the accessory to be mounted on the conduit. The pipe vice is used, and to cut the thread you will also need:

- a set of stocks and dies
- some cutting compound
- a soft wire brush.

Before cutting the thread, you need to check the die and make sure it is clean and clear of swarf. If the die is not clean it can damage the thread to be cut. The stock and die set should also be checked to make sure everything is tight and has not loosened since last use.

With the die clean and tight, apply a liberal coating of cutting compound to the end of the conduit, making sure it is coated all the way round. The cutting compound acts as both a lubricant and cooling aid during thread cutting and is generally applied using a brush.

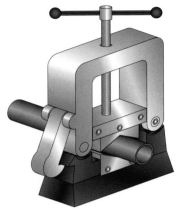

▲ Figure 4.65 Conduit in pipe vice about to be cut (top) and the edge of the cut conduit being dressed to remove sharp edges (bottom)

HEALTH AND SAFETY
Use gloves and/or barrier cream to protect your skin from the cutting compound when cutting threads on steel conduit.

With the end of the conduit covered in cutting compound, slide the stock and die set over the end, making sure that the guide slides on first. Stand square to the end of conduit and grip the stocks and, using the heals of both hands, press the die onto the end of the conduit. Rotate your hands in a slow clockwise direction, allowing the die to bite into the conduit.

Once the die has started to cut, change your hand movement to a push and pull rotation, and steadily turn the handles through 180°. Repeat this once or twice and then turn the handles in an anticlockwise direction for just half a turn. You will feel a bit of resistance and then the die will come loose. This helps clear the swarf so the die does not jam or overheat.

Repeat this process slowly and steadily. If you rush you will cause the die to overheat and the thread will be damaged. Remember, a superior thread can be cut by using a little care and patience.

Repeat the process of two to three turns forward and then half a turn back until two whole threads protrude from the back of the stock and die. Once you have reached this point, slowly unscrew the stock and die from the conduit, again taking care not to damage the thread. It is tempting to try to spin the stocks and die off the conduit, but this can cause the die to wobble and damage part of the thread.

Once the stock and die have been removed from the conduit, make sure the thread is clean and clear of all swarf. Use a soft wire brush (and safety glasses in case the residue flicks up towards your eyes). With all the swarf removed, use a rag to clean away any excess cutting compound.

The stock and die set should be cleaned in preparation for their next use. Some people do this by banging the tool against the leg of the conduit bender. However, this can result in damage to both the bender and the stock and die set.

In the absence of the tool's correct cleaning brush, a stiff bristle bottle brush makes a suitable alternative. Again making sure your eyes (and the eyes of those around you) are protected, simply push the brush into the die and then withdraw. Repeat this process twice, leaving the die clean and ready for future use.

Before removing the conduit from the vice, check the thread is even and undamaged throughout. You can screw on a coupler to protect the thread if you are not ready to screw on the accessory.

When you have to cut a longer thread, it is important not to let the stocks and die overheat. To help, you can reduce the number of forward turns to one or two half turns forwards for every half turn back.

Using a running coupler

As mentioned earlier, once you have cut your thread, you normally thread on the accessory. To do this, you have to be able to turn one piece, if not both pieces, of conduit and the accessory. In some circumstances this is not possible – for example, if alterations are being made to a conduit system with bends that can not be dismantled sufficiently to turn it.

ACTIVITY

What are the common sizes of dies used for threading conduit?

In this instance, the conduit to be added is built and then joined to the existing assembly using a running coupler. A running coupler is a type of joint that uses:

- a coupler
- a lock nut or ring
- (and sometimes) a nipple.

A thread must be cut on the conduit to be added that is long enough to thread on both the lock nut or ring and then the coupler. When connecting into an accessory, a nipple must be threaded into the accessory so that half the thread is sticking out. Then the new part of the conduit assembly is offered up to the nipple and the coupler is unscrewed from the conduit. For joining to a second piece of conduit rather than to an accessory, the nipple is not required; the conduit already has another thread to accept the coupler.

As the coupler is unscrewed, it threads onto the second thread so that it can be threaded tightly onto either the accessory or conduit while still being threaded onto the new piece of conduit. To stop this from coming loose, a lock nut or ring is tightened up behind the coupler to hold it in place.

Once in place, the exposed thread must be painted to stop it from corroding. The paint used must complement the conduit finish, for example black enamel.

Bends

Though there is a wide selection of pre-manufactured bends and accessories, you also need to be able to bend conduit to meet specific circumstances.

Bending conduit is more about technique than brute force. It is vital that you know how to bend the conduit in the correct place. Sometimes people bend the conduit and then cut it to size; this not only creates excess waste, but also takes longer as it creates more work.

Before we look at how to bend conduit, we will study the names of different bends. These include:

- *The set* – sometimes referred to as a 'dog-leg', this consists of two bends, each of about 45°, on top of one another with one bend going one way and the other in the reverse direction. It is used to bring the conduit out further from the surface that it has been running on and is common at distribution panels and other switchgear.

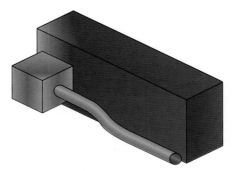

▲ Figure 4.66 Set bend

- *The kick* – this is similar to the set, but only has one bend at an angle up to 45°. This is normally used when a conduit runs up a wall and along the roof structure, such as along a rafter.
- *The bubble set* – sometimes referred to as a 'camel' set, this is normally used when there is no other option but to run a conduit over another conduit or pipe. The two conduits (or conduit and pipe) should not come into direct contact, and the centre of the bubble should be directly above the centreline of the underlying structure. There should be a minimum of a finger-gap clearance behind the top conduit, but no more clearance than the diameter of the conduit itself.

▲ Figure 4.67 Kick bend

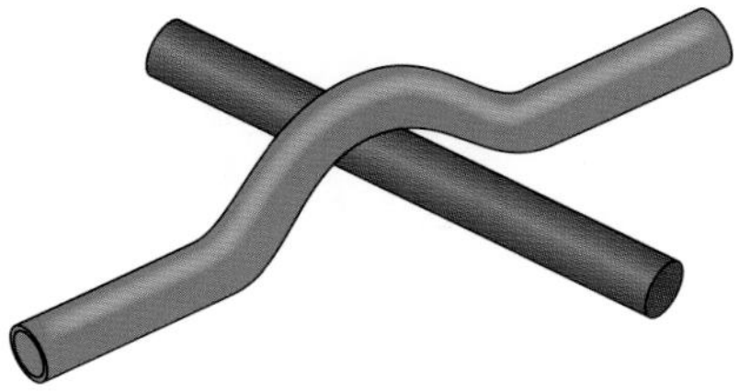

▲ Figure 4.68 Bubble set bend

- *The 90° bend* – this is probably the most common of the bends as conduit often goes around corners of *about* 90°. Note that very few buildings have exact 90° corners, so the bends must be sized to fit.

INDUSTRY TIP

A door frame is a useful test piece when bending conduit. The door frame should have a 90° internal angle.

▲ Figure 4.69 90° bend

- *The swan neck* – this bend is normally used where the conduit runs up a wall and over an overhang at the top. It is not a very common bend. Because of the shape of the bend, it can be difficult to remove it from the conduit bender.

▲ Figure 4.70 Swan-neck bend

The conduit bender

Though there are many different types of conduit benders, they all have similar components and work in a similar manner. Learning how to bend conduit using a bender can save time and money – bending conduit is quicker than cutting and applying a thread to use a manufactured bend.

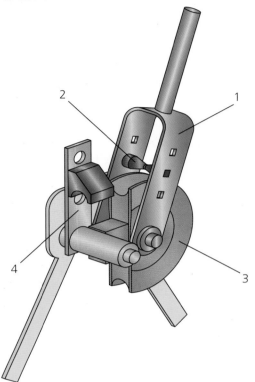

▲ Figure 4.71 View of a conduit bender (with part of the frame removed for clarity)

All steel conduit benders have some key components and in the illustration above, the following items can be seen:

1 the roll bar, where the pressure is applied to bend the conduit and which has several holes for the different size conduits
2 the roller, which sits within the roll bar and can be positioned in the relevant hole depending on the size of conduit to be bent
3 the former, which is sized to ensure that the minimum bend radius of 2.5 times the conduit diameter is achieved
4 the stop bar, which consists of a bar with several holes giving different positions for the different size conduits, and the stop used for holding the conduit in place while it is bent.

The conduit bender is designed so that the former can be changed and so that the position of the stop and roller can be altered to suit different sizes of conduit. It is important to make sure the former is in place and the correct positions for stop and roller are selected before attempting to bend any conduit.

All conduit benders are built to bend downwards, but when bends have runs of over 1 m on either side, the conduit can be fouled by the floor. Therefore, most conduit benders have the ability to bend conduit upwards with the stop bar and roll bar both being able to pivot through angles in excess of 180°. Before setting the equipment to bend upwards, make sure you have sufficient clearance above the conduit bender!

Making a 90° bend

Bending a piece of conduit requires good technique. To reduce waste, make sure you get it right first time! Some people bend the conduit and then cut it to size, but this can result in more work and a longer process as two threads must be cut.

In order to avoid this, use a method called 'measuring to the back of bend' (Figure 4.72). First, measure the distance from the conduit or accessory that will accept the bent piece, to the far side of the conduit saddle where the conduit will be after the bend, as shown opposite.

Use this dimension to mark a pencil line on the piece of conduit. The conduit is then inserted into the bender from the former side and rested under the stop bar.

Using an off-cut of the same size of conduit and an engineer's square, move the conduit until the pencil mark lines up with the back of the off-cut, making sure both parts remain square, as shown in the diagram.

With the conduit now in place, remove the engineer's square and the off-cut and use the roll bar to bend the conduit around the former. You can use one hand on the conduit to aid with the bend; this reduces the pressure on the roll bar and, therefore, reduces the potential for damage to the finished conduit. It is important to apply a steady constant pressure and not to swing on the roll bar. Bending your knees as you pull the roll bar can also assist with the bend.

When the roll bar is sticking out horizontally, the conduit bend is about 45°; when the roll bar is parallel with the front legs, the bend is about 90° (Figure 4.73). Remember that the conduit will spring back a little when the pressure is removed, so it is important to check the angle visually. This is done by standing side-on to the bend and squatting down so your eyes are level with the bend. It is tempting just to turn your head to see the angle of the bend, but if your eyes are not horizontal then it will be harder to see if the bend is correct.

It is possible to remove the conduit from the bender to check its shape against its intended location. To increase the bend, the conduit can be repositioned easily in the conduit bender without having to measure everything again.

Though it is possible to correct bends that have been bent too far, this should be avoided as it weakens the conduit and can cause damage. So, take your time and do it right first time. Should you have to correct a bend that has been bent too far, you should position the conduit in the bender in the opposite direction and correct using just your hands (not the roll bar) as this will give you a better idea of how far it has bent.

A set

There are two types of set: one with angles of 30° and 60°, and the other with angles of 45° each. The set with 30° and 60° angles is the most commonly used as it places less strain on the cables when they are being drawn in. It is this set that we will consider here.

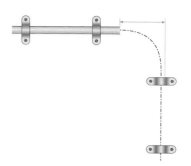

▲ Figure 4.72 How to measure for a back of bend

▲ Figure 4.73 How to position the conduit at the right point for bending

ACTIVITY

What is the minimum internal radius of bend for 16 mm, 25 mm and 32 mm conduit? (Answer in mm.)

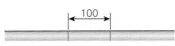

▲ Figure 4.74 Marking out for bends for a set

If we know what height is required for the set, we can identify the distance between the two bends: the height of the set multiplied by two bends. For example, if the set is required to be 50 mm, the distance between the two bends will be 100 mm (50 × 2 = 100).

Having worked this out, we can mark out the conduit to be bent. Make the first mark where the first bend should go. Then, measure the distance required (from the calculation) and make the second mark (Figure 4.74).

Having made the two marks, position the conduit in the conduit bender with the first mark aligned with the centre of the bender. This can be checked using an engineer's square placed on the conduit. This may not be directly at the top of the bender, as it will depend on whether the conduit sits horizontally within the bender.

▲ Figure 4.75 Top view showing positioning within bender ▲ Figure 4.76 Side view showing positioning within bender

Steadily, bend the conduit down to an angle of around 30°, making sure you check the angle as you go.

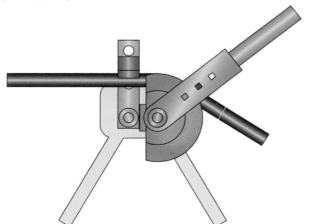

▲ Figure 4.77 Angle of first bend

INDUSTRY TIP

When bending conduit it is virtually impossible not to damage the paint or galvanised finish. The surface should be cleaned and repainted.

Once the first bend has been performed, the conduit must be rotated by 180° while still in the bender. Line up the second mark with the centre of the former and proceed as before.

It is important, at this point, to check that the conduit is correctly aligned in the conduit bender. Using a spirit level, ensure that the first bend is sitting perfectly vertical and not leaning to one side or the other. If it is not vertical, the second bend will not line up with the first and the two bends will be offset.

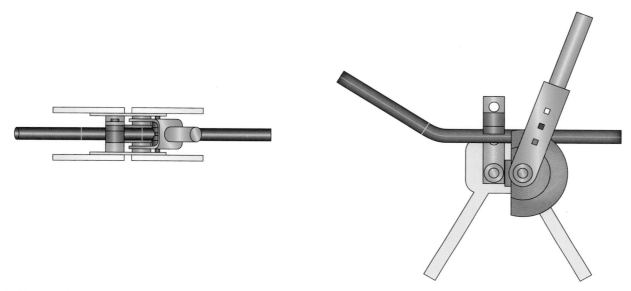

▲ Figure 4.78 Top view showing conduit positioned for second bend   ▲ Figure 4.79 Side view showing conduit positioned for second bend

With the conduit correctly aligned, the second bend can be performed until the start of the conduit runs parallel with the end of the conduit.

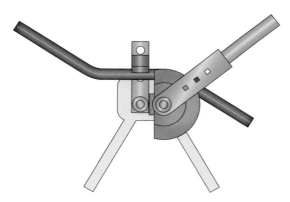

▲ Figure 4.80 Bend until the start and finish are parallel

A bubble set

This type of bend can be performed in one of two ways. The first is by performing two separate sets; this method can be used where lots of pipes (or other obstacles) are to be crossed in one go.

This second, more common, method requires three bends to be made within close proximity to one another. This takes a bit of practice to get right. Start off by placing a 45° bend in the conduit at the point which will be the centre of the bubble set. Then remove the conduit from the conduit bender and place it flat down on a surface to mark out the other two bends. Using a straight edge and tape measure, position the straight edge 30 mm below the inside edge of the first bend, making sure it is kept an even distance from the bend for both straight sections of conduit on either side of the bend. (This dimension will provide a very close bubble set for 20 mm conduit and can be increased if a larger clearance is required.) Mark on the conduit where the straight edge meets the conduit on both sides of the bend.

Using an engineer's square, mark all the way round the conduit with a pencil on each side of the first bend. Now position the conduit back in the conduit bender with the top of the first bend pointing downwards and at the stop end.

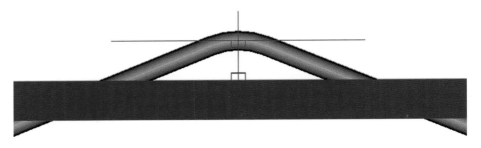

▲ Figure 4.81 Positioning of straight edge for marking the second and third bends

Using the same method for lining up the conduit in the conduit bender as the method for a set, ensure the second mark is positioned correctly on the conduit bender. The second bend of 45° can now be applied.

The third bend is a repeat of the process for the second bend but on the other side of the first bend. With the bends correct, the conduit run before and after the bubble set should line up. Fine adjustment can be performed by replacing the conduit in the conduit bender and adjusting by hand.

Bending PVC conduit

The bending of PVC conduit is very similar to bending steel conduit as you use the same bends and measure them out in the same way. The difference is that a conduit bender is not used. Instead, a bending spring is used to help the conduit maintain its shape.

The bending spring is used to support the inside of the conduit to stop it from becoming squashed during the bending process. Bending springs come in two types:

- white tipped – for light gauge PVC conduit
- green tipped – for heavy gauge PVC conduit.

Bending springs are designed so that they can be fed easily into the conduit prior to bending. Before use, it is important to attach a means of removing the spring after the bend. Simply tie a piece of string to the loop at one end or,

▲ Figure 4.82 Bending spring

preferably, use a cable. The string or cable should be fed through the centre of the bending spring. This will stop the spring being stretched and damaged during the removal.

Nowadays, some PVC conduits can be cold bent. However, it is sometimes necessary to heat the conduit prior to bending and this can be done in several ways. Some electricians use a heat blanket for PVC conduit; however, simply rubbing the conduit where it will be bent produces enough heat to help with the bend.

Having measured the bend, insert the spring into the conduit so that it is positioned where the bend is to be performed. One way to do this is to measure how much string or cable should be drawn into the conduit to hold the spring in the correct place.

With the spring in place, heat the conduit by simply rubbing your hand up and down the position for the bend. Don't rub too hard or you will get blisters – hands can be protected by using gloves or a cloth. The internal bend radius should be the same as for steel conduit, so an internal radius of 2.5 times the conduit diameter is needed. The easiest way to bend PVC conduit is over the top of the thigh. Remember to remove items such as money, phone and keys from your pockets before starting.

The bend should be performed in one go. So, once the conduit is heated, place your hands 300–400 mm apart on either side of section to be bent, and then bend the conduit over your thigh. The bend will need to go a little past the required angle as the conduit will tend to spring back once released.

If the wrong bending spring is used for a particular type of conduit, the bend can become damaged. Ribbing of the bend happens when the spring creates indentations on the outside of the bend. It can also result in kinks on the inside of the bend where the conduit tries to flatten during the bend.

When creating sets, it sometimes helps to mark a line so that the bends will always be square to one another. The line can be used to make sure that your hands and bending force are in the correct place before you bend. PVC conduit usually has writing along one side and this can be used as a guide. However, it is important to ensure that any lines are removed before installation and any printed writing is positioned to the back of the conduit, out of sight.

Wiring conduit systems

One of the key points to remember when installing a conduit system is that the wiring cannot be installed until the conduit is complete.

Bear in mind that **BS 7671:2018** The IET Wiring Regulations, 18th Edition states that we need to protect the insulation of the wiring from damage during installation and maintenance. This is important when guiding wiring into conduit sections. All the wires for a piece of conduit need to be installed, or 'drawn in', at the same time. Wires are fed into the conduit and pulled from the other end at the same time.

The wires are drawn in using a push-and-pull technique that requires a draw tape to be placed in the conduit from the finish end first. A draw tape is a nylon cable that has a metal cap attached to a spring at one end. This end is pushed into the conduit from the finishing end of the run. It is pushed right into the conduit until the end with the metal cap emerges at the beginning of the conduit run. If the conduit run is long, this may have to be done in stages (draw tapes tend to come in 5 m, 10 m and 20 m lengths).

Once the draw tape has been pushed through the conduit, and the end is sticking out of the conduit, the wiring can be attached to the draw tape. The most effective attachment method is to use PVC tape, taping the wires together with staggered starts and leaving one wire longer than the others. This single wire will be used to secure the wiring to the draw tape.

▲ Figure 4.83 Wires staggered for taping

With the wires grouped together, the long wire can now be secured to the draw tape by feeding it through the hole in the end of the cap and then bending it back on itself to form a tight loop. This loop is then secured in place with more PVC tape. It is important to ensure this attachment is as secure as possible. Depending on the cross-sectional area of the wire, several passes may be made through the hole in the cap on the end of the draw tape.

Once the wiring is secured to the draw tape, it is ready to be drawn into the conduit. This can be a two-person job: one person feeds the wiring into the conduit, guiding the wiring through the conduit, as the other person gently pulls on the draw tape. The draw tape should never be used to pull the wiring into the conduit (without a corresponding push) as this places a strain on the wires. This could generate heat which might damage the wires and, in the case of PVC conduit, the conduit as well.

The push-and-pull technique is repeated until the wiring reaches its destination and the draw tape exits the conduit. Shorter runs, no longer than a metre, with no bends, can be installed without the use of a draw tape. The wiring is still grouped as it would be when using draw tape, but in this case the long wire is simply bent over to form a loop. This presents a curved edge which helps the wires move through the conduit and prevents the wire getting stuck on any edges at accessories or joints.

Trunking systems

As with conduit systems, there are various forms of trunking system. Trunking is the generic name given to the enclosure of a wiring system that has a rectangular or box section fabrication and a removable lid. The purpose of trunking is to enable wiring to be installed easily and to provide the wiring with a level of mechanical protection.

Trunking systems come in two forms:

- PVC
- steel.

The different types of trunking that are available and their uses are identified on page 191. Here we will look at terminating steel trunking only. Termination of other trunking, such as lighting trunking and Powertrack, are outside the scope of this course and further training should be sought before attempting this type of work.

There are benefits that trunking can provide which are not available when using conduit. Trunking can be used to install cables and wires alike, and in the same run. Trunking can have wiring added in at a later point, without the need to remove existing wiring. Trunking can be used to contain wiring for different voltage bands and different types of circuit as the trunking can be split into sections. This is called segregation. It is a requirement to remove potential interference from sensitive circuits and circuits of different voltage bands.

In any metallic containment system, line and neutral, or line and switch line conductors must be run together in order to avoid induction of eddy currents in the metal containment system.

Steel trunking comes in a variety of different types of finishes including:

- hot-dipped galvanised coating
- grey enamel on zinc coating
- silver enamel on zinc coating
- stainless steel.

Other finishes are also available to order. Trunking is typically stocked and sold in 3 m lengths, with the exception of lighting trunking which is also stocked and sold in 5 m lengths.

Trunking comes in a variety of different sizes, with some of the more common ones being listed within Appendix E of the IET On-Site Guide. Most suppliers will provide less common types and sizes of trunking, but only on special request and subject to a minimum order quantity.

Sizing of trunking

Even in trunking the cables get hot. There must be sufficient air space around the cables so they do not become overheated.

The maximum fill capacity for trunking differs from that of conduit as trunking is designed to carry both wires and cables. There is a constant fill capacity of 45% which means that air must occupy at least 55% of the internal area of the trunking.

The IET On-Site Guide has standard tables which take into account all the factors that govern how many conductors can be installed in a piece of trunking. The tables take into account the size and type of conductors, but do not consider bends or length as this does not affect the size of trunking to be used. The IET On-Site Guide also does not include the factors for different cables and the trunking manufacturer's data sheets should be consulted in these cases.

> **INDUSTRY TIP**
>
> Mini-trunking is just a version of plastic trunking. It is sometimes supplied with a self-adhesive backing for quick fixing. The surface to be fixed to must be clean and free of dust otherwise it won't stick.

When using Appendix E, follow this simple and logical approach:

1 Identify the different sizes of the wiring conductors.
2 Identify the quantities of each different size.
3 Using the tables in Appendix E, select the correct cable factor.
4 Multiply the factor by the number of conductors.
5 Add the totals together.
6 Select the trunking with the next factor up from the total value.

EXAMPLE

One easy way to follow these steps is to use a simple table, filling in the data as you go. The example on the next page is a trunking sizing calculation for a steel trunking system carrying eight single-phase circuits. Four circuits have conductors with a size of 25 mm^2 and the other four are 16 mm^2. The circuits are wired in single-core cables and each circuit needs a circuit protective conductor of the same size as the live conductors.

Conductor size in mm^2	Number of conductors	Cable factor	Total cable factor
16	4 × 3 = 12	47.8	12 × 47.8 = 573.6
25	4 × 3 = 12	73.9	12 × 73.9 = 886.8
	Total		573.6 + 886.8 = 1460.4

Example of trunking calculation

The final result is an overall trunking factor of 1461. Using the appropriate table in Appendix E of the IET On-Site Guide, it can be seen that 75 mm × 50 mm or 100 mm × 38 mm trunking can be used – both have a spacing factor in excess of the 1461 required.

Unlike conduit, there are often several choices of trunking size to use. The final choice will have to take into account the installation and aesthetic qualities required.

Trunking accessories

Trunking comes in all shapes and sizes, but the most common is the box type. The lid of the trunking forms a slight overlap at either edge, giving strength to the overall construction.

The lid of the trunking can be held in place with different types of fastenings, the most common being the turn buckle. Turn buckles are designed so that a simple hole can be drilled in the lid and the body of the turn buckle inserted into the hole. The body has a head that fits either a large flat screwdriver or, preferably, a pozi-screwdriver, and this needs to be on the outside of the lid. On the inside, the turn buckle has a hole into which a rod is inserted that is slightly smaller than the width of the trunking. This rod grips under the return edge of the trunking when the turn buckle is turned, retaining the lid in place.

Unlike conduit, the main accessories for steel trunking are used to join trunking pieces together rather than to mount accessories. Where accessories are to be connected, a conduit coupler and male bushes are used. The male bush is often tightened using a bush spanner.

ACTIVITY

How many 16 mm^2 cables could be installed in a 75 mm × 50 mm trunking?

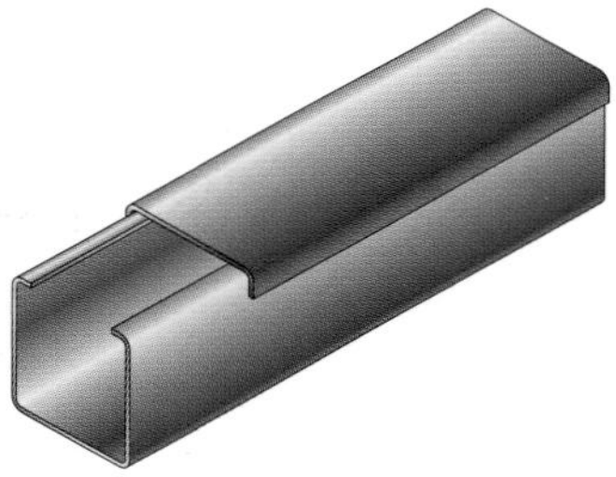

▲ Figure 4.84 Steel trunking

▲ Figure 4.85 Connection methods

- The *flat bend* is simply a 90° bend with the lid on the front of the trunking. This can be used when trunking is run up a wall and then turns to run along the wall.
- There are also variations of the flat bend in terms of the position of the lid. The *internal bend* has the lid positioned on the inside of the bend and the *external bend* has the lid on the outside.

▲ Figure 4.86 Flat bend

▲ Figure 4.87 External and internal bends

- As well as the flat bends, there are the *relieved bends* where, instead of trying to force the wiring around a tight 90° bend, the bend is spaced out as two 45° bends. Again, there is the choice of the lid being on the flat, internal or external faces of the bend.

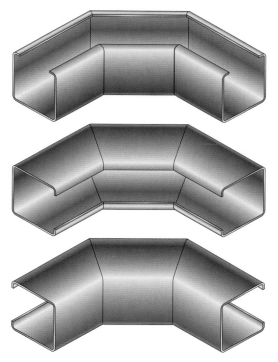

▲ Figure 4.88 Flat, internal and external relieved bends

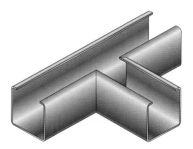

▲ Figure 4.89 Tee piece

- The *tee piece* is used to join a piece of trunking onto the side of another piece of trunking at a perpendicular angle. Unlike the bends, the tee piece only comes in one form, with the lid on the front.
- The *relieved tee* is similar to the relieved bends in that, instead of presenting a sharp 90° bend, it presents two spaced 45° bends to make the bend suit the wiring bend radius.

▲ Figure 4.90 Relieved tee

When the installation changes from one size of trunking to another, *reducers* can be used.

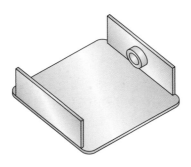

▲ Figure 4.91 End cap

When two pieces of trunking are to be joined, it is necessary to use two *couplings*. A coupling is simply a plate designed to fit inside the trunking and bridge the join. To enable the trunking to be secured, the coupling has two threaded bushes mounted on one side. This removes the need to position a nut inside the trunking and so a short length bolt can be used. The bushes on the inside are sometimes called 'captive nuts' or, more correctly, 'hank bushes'.

When a join is performed, it is important to ensure the electrical continuity of the steel. Trunking manufacturers generally recommend, that to ensure continuity, an *earth strap* is installed on the outside of the trunking across the gap.

At the end of any run of trunking, an open end is normally present and this needs to be closed off. To do this an *end cap* is often used, but in some cases it is possible to cut and bend over the lid of the trunking. The end cap has similar bushes to those used in the coupling – a short screw can be screwed into the end cap, through the trunking, once it is in place.

Just as with conduit bends, trunking bends can be fabricated or purchased. Sometimes fabrication is best when a custom angle or a slightly different bend is required. As mentioned earlier, most buildings are not exactly square, meaning that pre-manufactured bends do not always fit properly. It is essential, therefore, to know how to manufacture these bends.

Cutting trunking

Trunking is a hollow structure that must be complete to exhibit its full strength. This can be problematic when cutting.

It is important to cut the trunking and the lid separately as, often, the lid needs to be a different size or shape.

Just as when cutting steel conduit, there are certain tools that are required for cutting steel trunking. These include:

- hacksaw with a blade between 24 and 32 TPI
- graphite pencil for marking out
- tape measure
- engineer's square
- hand file (generally second cut grade)
- small round file (generally smooth cut grade)
- engineer's vice or clamps
- steel rule
- appropriate signs and obstacles for making sure the work area is cordoned off.

You will need plenty of appropriately sized wood to pack the trunking – this prevents it from distorting when it is being clamped for cutting. The wood should be positioned on the clamping side of the cut line, so as not to foul the hacksaw as it cuts.

When cutting trunking, make sure that all cuts are even and square; accurate marking out is essential. To perform a simple cut, first use the tape measure to measure the required length and mark with the pencil. Then, using an engineer's square, mark round the trunking making sure that the line is perpendicular to the trunking on all sides.

Once marked out, the trunking must be secured ready for cutting. This can be achieved by placing the trunking in an engineer's vice or by using clamps to secure the trunking onto a work bench. The position of the clamp should be close to the cut line, but not so close as to impede cutting. Trunking should never be held in one hand while cutting with the other.

▲ Figure 4.92 Using an engineer's square to mark out trunking

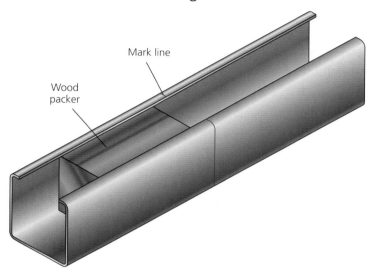

▲ Figure 4.93 A wood packer being used to support trunking

To cut steel trunking, a similar cutting style to that used when cutting steel conduit should be adopted, making sure the hacksaw does the work by using the entire blade length.

Once cut, the engineer's square should be used to check that the edges are square and even. If they are not, they need to be filed. When using the hand file, work evenly across the end of the trunking as this will minimise the amount of noise the filing makes.

Once square, all sharp edges and burrs should be removed from both the outside and inside of the trunking. Internal corners may require the use of the small round file to remove burrs.

Making a tee piece

To create a tee piece, the same tools as for cutting conduit will be required. Figure 4.94 shows how the trunking should be marked out for cutting. Note the size of the parts to be removed are based on the width of the trunking. The hole in part A should be the same width as the part of part B which is entering part A. The depth of the back and front of part B that needs to be removed is the width of part A minus 5 mm. This allows for the internal bend of the trunking.

Use a junior hacksaw to cut down the edge of part A to the side wall and then use the hand file to remove some of the metal from the back of the bend (where the return edge is to be removed) until it can be bent easily with pliers.

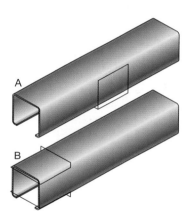

▲ Figure 4.94 Marking out for a tee piece

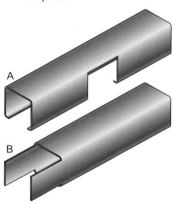

▲ Figure 4.95 Cut parts for a tee piece

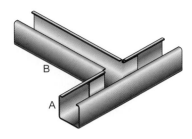

▲ Figure 4.96 Assembled tee piece

Bend this piece back and forth a few times until the edge comes away. The same technique can be used for removing the wall section, but you will need to cut at least one side of the back of part B.

Remember to remove the sharp edges and burrs from your cuts.

The two parts are joined together by slotting part B into part A and then bending the ends of part B over to create flaps. These can be bolted to the wall of the trunking or riveted.

Making an internal bend

An internal bend is arranged with the lid on the inside. Therefore, the trunking is cut on the inside and the top edges need to be removed.

Mark a line all the way around the trunking, using a pencil and engineer's square. Then, using the width of the trunking as a measurement, measure out and mark a line on both sides of the mark line. Measure 5 mm from either side of the centre line and then mark down as shown in the diagram below left.

▲ Figure 4.97 Marking out for internal bend (left), cutting for internal bend (middle), and assembled internal bend (right)

Using a junior hacksaw, cut the return edges at the outside marks and then file the edges until the return edges can be bent and removed with pliers. Next, cut the walls along the angled lines to form a vee shape.

Once you have cut the trunking, make sure all burrs and sharp edges have been removed and then simply bend one side up, slotting the walls of one side into the other. Bolt or rivet in place.

Making an external bend

An external bend is made so that the lid is placed on the outside; the marking and cutting is performed from the back.

Mark a line all the way around the trunking, using a pencil and engineer's square. Then, using the width of the trunking as a measurement, measure out and mark a line on the back on both sides of the mark line.

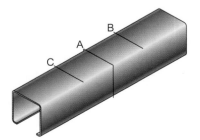
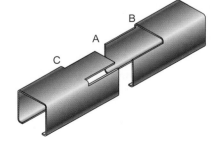

▲ Figure 4.98 Marking out for external bend (left) and cutting for external bend (right)

Cut right through the trunking at line A and cut through the back only at point B. File between A and B on one of the back corners to weaken it, then bend the flap of metal until it breaks off. File down the edges. File away the corners between A and C and make the resulting slots smooth.

Now slot part AB over part AC and bend the flap up. Bolt or rivet through both walls and the flap to secure in place.

Making a flat bend

A flat bend is made so that the lid of the trunking is fitted to the top of the trunking; the marking out and cutting is made to one side wall only.

As with all the bends, start off by marking a line, line A, all the way around the trunking using an engineer's square. Then measure and mark, using the width of the trunking as a measurement, a line on either side (lines C and B in the diagram). Also, mark a line 10 mm from line A on one side only. This is line D.

Using a hacksaw, cut down line A leaving one wall intact. Then cut along line B, cutting only one wall. File the back corner between A and B to weaken it. Bend the flap of metal until it breaks off. File down the edge where the metal breaks.

▲ Figure 4.99 Assembled external bend

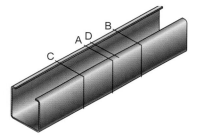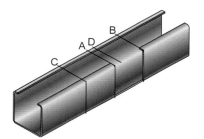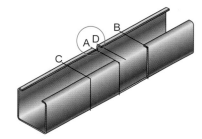

▲ Figure 4.100 Marking out for flat bend (left), removal of wall (middle) and removal of return lip (right)

Now, using a hacksaw, cut the return edge on line D, use a file to weaken the corner, then bend the flap of metal between A and D until it breaks off. File down the edge where the metal breaks.

Using a hand file, remove the lip between line C and line A. File away the lower corner, leaving a flap on the same side as the removed wall.

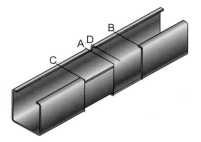

▲ Figure 4.101 Removal of return edge from flap (left) and assembled flat bend (right)

Now the trunking can be bent along the remaining edge of line A into a 90° bend with the flap bent to seal the bend. The back and flap can be bolted or riveted into place.

INDUSTRY TIP

The minimum IP rating for metal trunking is IP2X but if used in areas where there is a high risk of spread of fire this is increased to IP33.

▲ Figure 4.102 Conduit secured to ceiling using ceiling flange

Mounting trunking

There are various methods that can be used for mounting trunking. Most trunking is either:

- surface mounted

or

- suspended.

Surface mounting simply requires mounting holes to be drilled through the trunking so that it can be screwed to the surface. For suspended mounting, other methods and accessories can be used.

One method is a simple run of conduit which is secured in place to the ceiling and then joined to the trunking using a coupler and male bush arrangement. Another common method uses channelling, normally called uni-strut, and threaded shafts, called studding.

The uni-strut is secured in place and captive nuts are positioned within the uni-strut. The studding is threaded into the uni-strut and is cut to the required length. The trunking is then attached to the studding by use of a trunking suspension bracket. The benefit of this system is that the trunking is installed level and true, despite deviations in the height of the uni-strut mounts.

PVC trunking

The benefit of PVC trunking is that manufacturers can mould many different forms, finishes and colours. For example, skirting board trunking can either replace the skirting board of the room completely or can supplement the existing skirting board. The versatility of PVC trunking means that it is commonly favoured over steel trunking, especially in situations where it is visible.

Cutting PVC trunking

PVC trunking can be cut using a variety of different methods, including:

- trunking snips
- electrician's knife
- hacksaw.

Cutting PVC trunking with trunking snips can be quick, but in colder environments this can cause splitting and cracking. Cutting with snips can also cause the trunking to deform, but this can be overcome by cutting the lid and the trunking at the same time.

An electrician's knife can be used to cut through light gauge trunking, but this can also result in deformation of the trunking (and injury to fingers). However, when removing a part of the trunking wall, scoring the inside edge with an electrician's knife can make the wall weak enough to snap off, leaving a very clean break with few sharp edges to be removed.

Using a hacksaw minimises the risk of deforming the trunking (and of injuring the installer). A trunking mitre block is generally used and this can be secured in an engineer's vice. The mitre block typically has two 45° slots and one at 90°, meaning that it is not necessary to measure and mark out these cuts. The base of the mitre block has ridges at set intervals and of set sizes, so that the lips of the trunking and lid slot in. This prevents them moving while being cut.

Trunking is normally cut at 45° for corners and at 90° for joins. Lids tend to be cut at 90° as the joints and corners are typically covered with an accessory, forming a moulded section to ensure the relevant **IP code** rating can be maintained.

It is important to make sure that the trunking has a continuous form around all joins and bends. Lids, however, are cut to stop about 5 mm from each join or bend, allowing a cover to be fitted. This space allows the cover to clip onto the trunking.

To remove a wall of the trunking, simply cut down the sides using a junior hacksaw and then, using a sharp electrician's knife, score the inside wall at the base between the two hacksaw cut lines. Bend back the wall section and snap it off.

Segregated trunking

Irrespective of whether the trunking is steel or PVC, it can be split into different sections or compartments. The use of different compartments or sections can assist with **segregation** of specific circuits, for example keeping data cables segregated from low-voltage power circuits, and keeping band 1 and band 2 circuits separate.

Segregated trunking can even have separate lids for each of the sections, so that only the lid of the partition being worked on is opened. This is particularly useful in rooms where computers are positioned around the perimeter of the room. This type of trunking, when installed on the surface, is commonly referred to as dado trunking.

Cable tray

Sometimes the size or type of cable and the route or the required bend radius means that the use of trunking or conduit is not practicable. Instead, armoured cables are used in conjunction with cable tray.

Cable tray comes in several forms including:

- heavy gauge perforated
- light gauge perforated
- heavy gauge solid
- light gauge solid.

The choice of cable tray depends on the application, but heavy gauge is used for larger cables or where there is a greater mass of smaller cables to be supported. Heavy gauge cable tray tends to have walls with bent over edges, whereas the walls on light gauge are straight with no returned edge.

The difference between solid and perforated cable tray is simply that the perforated version has a series of slots and holes in the back and walls to enable cable fixings such as P-clips and cable ties to be used to secure the cables in place.

▲ Figure 4.103 PVC trunking mitre-cutting block

KEY TERM

IP code: international protection code used to identify accessories and equipment according to their resistance to penetration and water ingress.

KEY TERM

Segregation: keeping things apart in different compartments.

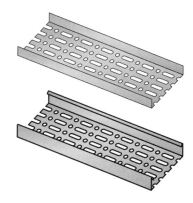

▲ Figure 4.104 Light gauge and heavy gauge cable tray

▲ Figure 4.105 Flat bend

▲ Figure 4.106 Relieved flat bend

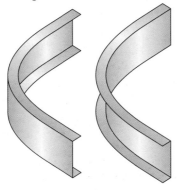

▲ Figure 4.107 Internal and external bends

Cable tray accessories

As with all other forms of support system, it takes time to fabricate parts during cable tray assembly. Some standard items are pre-manufactured. Some of these accessories are shown here. It is important to note that the end of each run of cable tray is shaped to fit into the next one, providing a standard overlap and creating a much stronger join.

The 90° bend comes in several forms, the first being the flat bend. This is where the cable tray can be connected to another piece that is in the same plane, but at an angle of 90°.

As with trunking, sometimes the flat 90° bend is too sharp for larger cables and, therefore, a relief bend is required to give a slower bend, permitting a larger bending radius.

Often cable tray has to curve upwards or downwards by 90°, for example when the tray has been run along a wall and across a ceiling. In this instance, the bend radius still needs to be observed; either an internal or external bend can be used.

Cable tray also uses the tee piece in both the flat and relieved form, shaped in a similar way to trunking.

Cutting cable tray

Cable tray is available in 3 m lengths and must be cut to length as required. The process for measuring the length required is the same as that for trunking. To cut the tray, it must be supported on a flat surface – use clamps to hold it in place or screw the tray to a wooden surface.

As with some of the earlier containment systems covered in this unit, the accessories for cable tray can also be manufactured if required. When making these joints, all sharp edges and burrs must be removed before cables are installed. The technique used for removing the walls of the cable tray is the same as that used for trunking: a hand file is used to weaken the edge, allowing the walls to be bent off by hand.

The tools needed to cut and form cable tray include:

- hacksaw with a blade between 24 and 32 TPI
- junior hacksaw
- graphite pencil for marking out
- tape measure
- engineer's square
- hand file (generally second cut grade)
- engineer's vice or clamps
- steel rule
- appropriate signs and obstacles for making sure the work area is cordoned off.

When marking out, most of the dimensions are made from the back of the tray as this provides a flat surface to work on. However, cutting the tray often involves just the walls, so the use of the engineer's square is essential to transfer the measurements to the side walls.

INDUSTRY TIP

Cables can be mounted on trays by using P clips or cable ties. One cable should not be tied to another as this puts undue strain on the cable and tie.

For joining cable tray, one end of each run of tray is reduced in size so that it can sit inside the other end. This creates an overlap which is typically the same as the width of the tray being used. Therefore, when any join or bend is being manufactured, this is the minimum amount of overlap that is required. Using a junior hacksaw, the walls can be cut and then a hand file used to break or weaken the back of the bend so that the wall can be bent down and removed by hand.

Mounting of cable tray

During the erection of cable tray, it is important to ensure it is run level. Pre-manufactured saddles or, preferably, uni-strut and studding can be used. As with trunking, this means that the cable tray can be kept level and adjustment can be made where required.

Cable basket

Cable basket is also referred to as 'basket tray' and is commonly used for the structured cabling of information technology systems. The benefit of this containment system is the ease with which wiring can be installed. It offers little in the form of mechanical protection to the cables. The main purpose of it is to provide support rather than protection.

▲ Figure 4.108 Cable basket

The cable basket is specifically designed so that cables are just laid within the basket. No further means of securing the cables is required. This enables the addition of further cables at a later point without creating any installation issues. Cable baskets can be run overhead in the space above suspended ceilings or even underneath raised floors in office areas. This is common in open-plan offices in large buildings, where there needs to be a certain amount of flexibility as to the location of the network points.

Ladder systems

Where large cables are to be installed, then the use of cable tray may not be suitable and a more robust containment system is required. The favoured choice

INDUSTRY TIP

BS 7671:2018 The IET Wiring Regulations, 18th Edition 521.10.202 states:

'Wiring systems shall be supported such that they will not be liable to premature collapse in the event of a fire.'

This is because wiring systems hanging across access or exit routes may hinder evacuation and firefighting activities in the event of an emergency. As a result, cables installed in or on steel cable containment systems are deemed to meet the requirements of this regulation. However, the use of non-metallic cable clips or cable ties as the only means of support is not suitable where cables are clipped direct to exposed surfaces or suspended under traywork. Suitably spaced steel or copper clips, saddles or ties are examples that will meet the requirements of this regulation.

for many is ladder racking. It is so-called as it resembles a ladder and the cables are run across what would be called the rungs of a ladder. Due to the nature of how it is manufactured it is not possible to bend this on site, unlike the other containment systems.

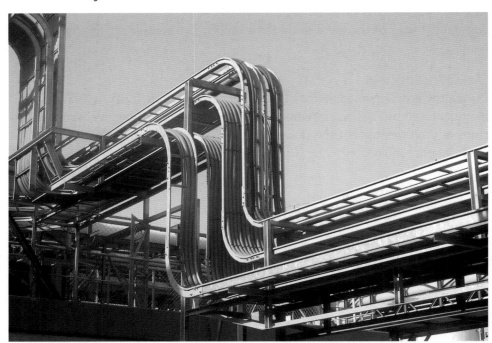

▲ Figure 4.109 Ladder racking carrying SWA cables

INDUSTRY TIP

The bends in the ladder rack must accommodate the required bending radius of the cables.

Bends in ladder racking must be designed and ordered accordingly, before the ladder racking is installed. Due to the strength of this type of containment it is used predominantly for large power cables.

Ladder racking is similar to cable basket, in that it does not offer much mechanical protection to the cables and so the use of armoured cables is common in such situations.

Busbar and Powertrack systems

In some installations there is a need to provide a level of flexibility regarding where the loads are to be connected to the supply. An example of this is in an open office area with a raised floor and or a suspended ceiling. For an open-plan office area, the system installed needs to be flexible enough to support any arrangement of office layout.

One method of achieving this is with the use of a system called Powertrack, which is made up of modular sections. Each section is a rectangular shape, rather similar to trunking, but inside the casing three conductors are arranged, each connected to connection sockets at regular intervals along their length. These connection points can be connected via a form of plug and socket arrangement to what are called floor points, in the raised floor. The raised floor is made up of a series of tiles mounted on pillars, which are secured to the floor. This arrangement results in sections of floor that can be moved from one position to another very easily. The system not only saves time in wiring, but also gives great flexibility in the layout of modern offices.

Powertrack also comes in an arrangement that can be installed above suspended ceilings and is used for connecting office lighting. This makes installation and maintenance of the lighting much safer and more efficient, as a luminaire can be disconnected and then brought down to ground level to be worked on.

Powertrack typically only comes in ratings up to and including 40 A and is single-phase only. Larger applications such as three-phase machinery need a different solution. Rather than running lots of cables the length of a factory, an alternative is the use of busbar trunking. This form of trunking does not house cables. Large copper bars are mounted within the trunking and these are used to carry the current.

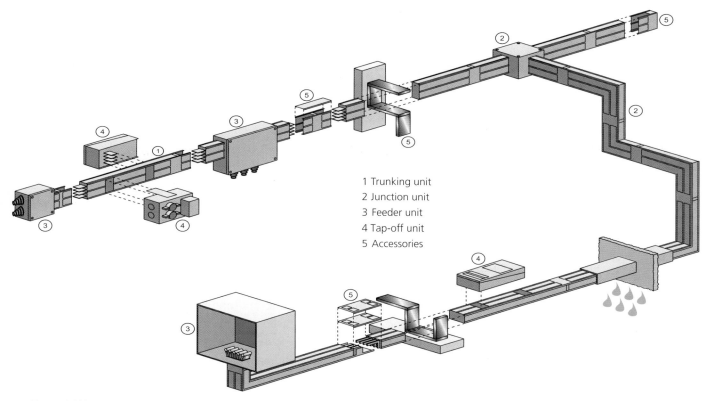

1 Trunking unit
2 Junction unit
3 Feeder unit
4 Tap-off unit
5 Accessories

▲ Figure 4.110 Typical busbar trunking arrangement

Busbar trunking can be designed to carry currents of 2000 A and beyond and so offers a simpler approach to providing high power than cable. In industrial applications, the use of busbar trunking to provide distribution of power to lots of pieces of machinery within the same room is quite common.

Each connection to a final piece of equipment can be made using a variety of different tap-off units, which can include the provision of local isolation and protective devices such as fuses. The final connection to the machinery can also be made using a variety of different cables or installation methods.

Thermoplastic (PVC) cables

Thermoplastic cables are commonly referred to as PVC (polyvinyl chloride) cables and they come in various shapes, sizes and forms, including:

- single-core cable
- twin and cpc flat-profile cable
- three-core and cpc flat-profile cable
- multi-core flexible cable.

In domestic installations, the most common cables are the twin and cpc, and the three-core and cpc flat-profile cables.

Cables have three main parts:

- *Sheath* – this is on the outside and holds the conductors in one cable, as well as providing minor mechanical protection to the inner conductors
- *Insulation* – this is on the live conductors only and is used to provide basic protection against electric shock, as well as being a means to identify the conductors
- *Conductor* – this is what carries the current and is commonly made from copper, but may be made out of other materials such as aluminium.

▲ Figure 4.111 Flat-profile cables

Conductors can be formed in several ways. Conductors with a cross-sectional area (csa) up to and including 2.5 mm^2 can be formed out of a single piece of copper. Above 2.5 mm^2 csa, conductors tend to be made out of multiple strands of copper. It is important to be able to bend the cable and a solid piece of copper would be hard to bend. However, for conductors of 300 mm^2 csa and over, the conductors tend to be solid again; bending is not normally required.

If a cable is to be bent a lot in use, for example in the flex connecting a vacuum cleaner to a plug, the conductor is normally made of many fine strands of copper, each no thicker than a hair. The reason for this is that as the conductors keep getting bent, they will eventually break. In a solid conductor, the circuit would be broken. In a conductor consisting of many strands, if one breaks, it will not significantly change the amount of current the cable can carry.

Stripping single-core thermoplastic cables

Removal of the thermoplastic sheath, or insulation, is relatively easy with thermoplastic cables.

Use of wire strippers is recommended for single-core conductors. Automatic wire strippers tend to rip the insulation from the conductor and can damage the

INDUSTRY TIP

The minimum size of aluminium conductor allowed by BS 7671:2018, is 16 mm^2.

insulation. Manual wire strippers are preferred and these come in various forms, but all work on a similar principle.

Manual wire strippers all have two blades that cross one another like scissors. The blades each have a notch so that they cut *around* the conductor in the middle. They also all have a means of adjustment so that different sizes of conductor can be stripped (Figure 4.112).

Before using wire strippers, they must be set to the correct size. This can be done using a scrap or off-cut piece of wire of the correct size. With the jaws of the wire strippers together, turn the adjustment screw until the hole in the jaws is just bigger than the size of the conductor to be stripped. Test the setting on the off-cut piece of wire by placing the wire strippers over the wire and squeezing the handles to close the jaws. Then slightly release the jaws and try to slide the insulation off using the wire strippers to assist.

If the wire strippers are correctly set, the insulation will come off easily and there will be no damage to the copper conductor. If the aperture is set too small, the insulation will slide off easily but the conductor may be damaged. If set too large, the insulation will not come off easily. A simple adjustment of the adjusting screw will correct these problems. Now, with the correct setting, the wires can be stripped safely.

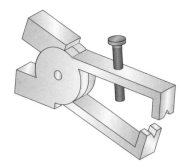

▲ Figure 4.112 Cable stripper jaws

Stripping flat-profile thermoplastic cable

Flat-profile cable is stripped in the same way as single-core cables, once the sheath has been removed. There are several ways to do this, but some methods can damage the cables.

First identify how much cable is required within the accessory. A conductor should terminate with ease at any termination point within the accessory. As a rule of thumb, measure the diagonal size of the accessory and then add 10% to allow for termination. However, if the accessory is a distribution panel, wires must be able to reach not just the required position, but also any point on the panel if the sequence of the board were to be changed.

Next, identify how much of the sheath should be removed from the cable. The purpose of the sheath is to provide some mechanical protection for the insulation on the conductors. However, too much sheath within the accessory takes up space and will put excess strain on the conductors. The sheath should be stripped back almost to where the cable enters the accessory, leaving only 10–15 mm, a thumb's width, of sheath within the accessory.

Having decided on the length of sheath required and with the cable in place, score round the sheath so that it can be removed. This can be done with a scriber or, preferably, with an electrician's knife. Care should be taken not to cut into the wires. Use a pair of side cutters to snip down the centre of the cable, splitting the line to one side and the neutral to the other.

Pull the live conductor and sheath apart, tearing the sheath as far as the score mark. Once at the score mark, bend the sheath back and snap it off, leaving a clean finish.

Now the inner wires can be stripped in the same manner as single-core wires, making sure that green and yellow sleeving is applied to the bare copper cpc.

An alternative – but less safe – method of removing the sheath is to score round the sheath with an electrician's knife, then to split the sheath by running a knife along the route of the protective conductor. The risk with this method is of damaging the cpc or even the insulation of the conductors.

Stripping thermoplastic flexible cables

Stripping the actual wires of flexible cables is done in the same way as stripping single-core wires. However, the sheath is stripped in a different way from the other cables already mentioned here.

The outer sheath can be removed with the use of a ringing tool which comes in various shapes and forms, the most basic of which is shown in Figure 4.113. This tool slides over the end of the cable, to the required position, and is then rotated around the cable, cutting it slightly. The ringing tool is removed, the cable is bent to finish the cut and the sheath is then pulled off.

There may be times when a ringing tool is not available, so it is important to know how to do this without a special tool. Thermoplastic has a tendency to split when it is damaged and stretched, and this feature can be used to help strip the sheath.

Bend the cable tightly at the point where the sheath is to be removed. Using a sharp knife gently score the top of the bend, noticing the thermoplastic split open like a little mouth. Gently work the split until the inner cables are visible and then unbend, rotate and re-bend the cable at about 90° from the first point. Repeat the score until again the wires become visible. Taking care not to cut into the wires, repeat the bend and cut until the sheath can be removed. Once the sheath has been removed, the wires can be stripped as mentioned previously.

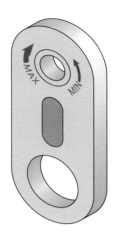

▲ Figure 4.113 Flexible cable sheath-stripping tool

> **INDUSTRY TIP**
>
> The requirement for connections is that they are electrically continuous and mechanically strong. Many purists do not like the use of strip connectors (chocolate blocks) but, like anything else, used correctly they are fine. Given the very large numbers in use, they must have something going for them.

③ TERMINATING CABLES

Terminating cables and BS 7671:2018

Section 526 of **BS 7671:2018** The IET Wiring Regulations, 18th Edition is the key section relating to the terminating and connection of conductors. Below is a regulation-by-regulation discussion of the requirements.

526.1

This requires that all connections shall provide 'durable electrical continuity and adequate mechanical strength and protection'. This is a general requirement and is in keeping with requirements of the EAWR. The subsequent regulations give further information on how this can be achieved.

526.2

This requires that 'the means of connection shall take account of, as appropriate:

i   the material of the conductor and its insulation

ii   the number and shape of the wires forming the conductor

iii the cross-sectional area of the conductor

iv   the number of conductors to be connected together

v *the temperature attained at the terminals in normal service, such that the effectiveness of the insulation of the conductors connected to them is not impaired*

vi *the provision of adequate locking arrangements in situations subject to vibration or thermal cycling.*

Where a soldered connection is used, the design shall take account of **creep**, *mechanical stress and temperature rise under fault conditions.'*

There are a large number of considerations covered by this regulation. Take each of these points in turn.

The material of the conductor and its insulation

Different metals or, more correctly, dissimilar metals may react with each other, resulting in corrosion of the termination. It is therefore important to make sure that compatible materials are used when terminating cables and conductors. A further discussion of corrosion takes place later in the book.

The number and shape of the wires forming the conductor

Conductors come in many formats, round or triangular, solid or stranded. It is important to select terminations that are compatible with the cable and/ or conductor. Failure to use compatible parts may result in the conductor becoming loose within the termination.

When crimping lugs onto cables, it is important that the correct size lugs and crimp dies are used, to ensure that a sound mechanical and electrical connection is made. Shaped lugs are available to use with triangular and half-round conductors used in some two-, three- and four-core cables. Where flex is being terminated, it is important that the ends are treated, for example, fitting ferrules to ensure that the individual strands are not spread and are all contained within the termination.

The cross-sectional area of the conductor

The cross-sectional area (csa) of the cable was carefully selected at the design stage, to ensure that the current-carrying capacity of the conductor was adequate for the circuit load. When terminating the cable, it is important to ensure that all strands of the conductor are contained within the terminal, to ensure that the current-carrying capacity is maintained. Terminals need to be large enough to house the conductor and to be suitably rated to carry the circuit load. Failure to meet either of these requirements could result in overheating at the termination that, in turn, may cause damage to the equipment and/or the insulation of the cable. This in turn may pose a fire risk.

The number of conductors to be connected together

The termination must be suitable for the number of conductors to be connected together. Attempting to fit more conductors into a terminal than it is designed to hold will invariably result in one or more of those conductors not being properly connected and becoming loose over time. In the case of accessories, where more than one conductor is intended to be connected, manufacturers will provide a number of linked terminals to house all of the conductors so that a sound mechanical and electrical connection is made.

KEY TERM

Solder creep: sometimes called 'cold flow', 'solder creep' can occur when the termination is under constant mechanical stress and the solder can literally move or 'creep'. Incidence of creep increases with temperature.

▲ Figure 4.114 Tube lug and bell mouth lug designed to capture all strands of conductor

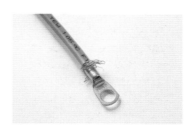

▲ Figure 4.115 Terminals that are not large enough to accommodate all the strands are problematic

▲ Figure 4.116 Ceiling rose showing linked terminals

The temperature attained at the terminals in normal service, such that the effectiveness of the insulation of the conductors connected to them is not impaired

The maximum operating temperature of conductors with thermoplastic insulation (usually PVC) is 70 °C, whilst the maximum operating temperature of thermosetting cables (XLPE) is 90 °C. When using thermosetting cables at 90 °C, it is important to make sure that the terminals are capable of withstanding this temperature, as the majority of electrical accessories are designed to operate at 70 °C. Other specialist cables may operate at higher temperatures, so it is important to check the maximum operating temperature of the terminals.

The provision of adequate locking arrangements in situations subject to vibration or thermal cycling

Where conductors are terminated into machinery, this can cause vibration and may adversely affect the terminations. In such cases, it is common practice to use, for the final connection, one of these options:

- a flexible cable – where added mechanical protection is required, braided flex such as SY flex can replace standard flex
- a flexible conduit – where unsheathed cables make the final connection or where additional mechanical protection is required, the cables can be housed within flexible conduit
- an anti-vibration loop in the cable – with cables such as MICC, or other cable types which are not flexible, a loop is included in the cable to allow the cable to absorb any vibration.

> **HEALTH AND SAFETY**
>
> The correctly sized terminal is critical as ill-fitted examples can cause a fire risk.

> **INDUSTRY TIP**
>
> The use of a flexible connection will also allow the motor to be moved when adjustment or alignment is required.

▲ Figure 4.117 Motor connected by means of SY flex (left), by means of flexible conduit (middle) and showing an anti-vibration loop (right)

> **KEY TERM**
>
> **Thermal cycling:** heating and cooling of metal (in this case) which causes expansion and contraction and, eventually, the loosening of terminals.

Thermal cycling can also harm the terminations. It is greatest when cables are run at or near their maximum operating temperature. The effects can be reduced by ensuring that terminations and connections are kept tight.

526.3

Poor and loose terminations cause many fires of electrical origin. Regulation 526.3 requires every connection or joint to be accessible for inspection, testing and maintenance, with the exception of:

- joints designed to buried in the ground
- a compound-filled joint
- an encapsulated joint
- a cold tail of a floor or ceiling heating system

- a joint made by welding or soldering
- a joint made with an appropriate compression tool
- spring-loaded terminals complying with BS 5733 and marked with the symbol (MF).

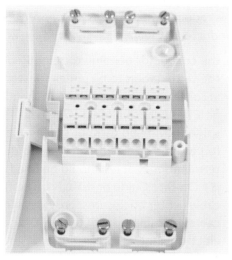

▲ Figure 4.118 Types of connection not required to be accessible

526.4L

This requires that the insulation of the cable must not be adversely affected by the temperature attained at a connection. An example of this may be where a cable with an insulation temperature rating of 70 °C is connected to a bus-bar which has been designed to run at a higher temperature. In this case the insulation on the cable would be removed from the cable to a suitable distance and replaced with insulation capable of withstanding the higher temperature.

526.5

All terminations and joints in **live conductors** must be enclosed within a suitable enclosure or accessory. There are no exceptions. This requirement applies to both low-voltage and extra-low voltage connections, but sadly it is not uncommon to see poor examples of connections, especially where down-lighters are fitted.

KEY TERM

Live conductor: a conductor intended to be energised in normal service, and therefore includes a neutral conductor.

▲ Figure 4.119 Example of an acceptable connection

In harmony with regulation 526.5 is regulation 421.1.6, which requires that all enclosures have suitable mechanical and fire-resistant properties.

526.6

This requires that there is *'no appreciable mechanical strain in the connections of conductors.'* Mechanical strain may come about due to:

- the conductor bending too tightly before entering the terminal, causing the termination to be under constant stress. Cables must be installed in accordance with the minimum bend radii, as given in the IET On-Site Guide or IET Guidance Note 1 (GN1).
- cables having no form of strain relief fitted. This can exert mechanical forces on the termination, due to the weight of the cable pulling on the termination. Cables should be fixed at the maximum distances given in the IET On-Site Guide or IET GN1. The fitting of suitable cable glands, where cables enter enclosures, provides strain relief on the terminations.
- cables terminated without any slack. Under faulty conditions large electromagnetic forces, due to high fault currents, are exerted on the cables, with the highest forces being exerted at the 'crutch point', where cables come together at the outer sheath. If there is little slack in the cables, the forces will be transferred to the terminal.

▲ Figure 4.120 Using a gland to provide strain relief

526.7

This requires that, where a joint in a conductor is made within an enclosure, the enclosure must provide adequate mechanical protection as well as protection against relevant external influences. The minimum requirements for an enclosure to meet the requirements for basic protection are that:

- the bottom sides and face meet at least IP2X or IPXXB
- for the top surface, the enclosure must meet IP4X or IPXXD.

However it may also be necessary to take into account external influences such as water ingress or dust ingress, depending on the location.

The IP Code

The IP Code is an international code specifically aimed at manufacturers of enclosures and equipment. It applies to degrees of protection provided by electrical equipment enclosures with rated voltages not exceeding 72.5 kV. The abbreviation 'IP' stands for International Protection, so the full title is International Protection Code (IP). The code is defined in IEC 60529 (BS EN 60529).

The three general categories of protection given in the standard are:

1 the ingress of solid foreign objects (first digit)
2 the ingress of water (second digit)
3 the access of persons to harmful electrical or mechanical parts.

When referring to the IP code in wiring regulations, 'X' is used in place of the first or second numeral, to indicate that:

1 the test is not applicable to that enclosure or
2 in the case of standards, the classification of protection is not applicable to this standard.

For example, IP2X means that protection against the ingress of solid objects must meet at least IP2 but the requirement for water ingress protection is not applicable in this case. Manufacturers will provide a full code, such as IP44, for the enclosure.

Code letters

International protection

First numeral 0–6 or letter X

Protection of persons by prevention or limiting ingress of parts of the body

Limitation of the ingress of solid objects

Second numeral 0–8 or letter X

Resistance to the ingress of water

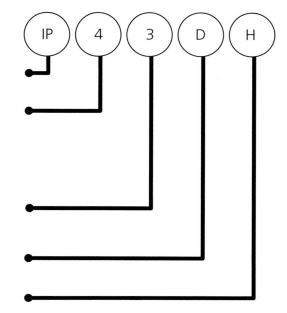

▲ Figure 4.121 How the IP Coding system works

▼ First numeral 0–6: Ingress of solid objects

IP	Requirement	Example
0	No protection	
1	Full penetration of **50.0 mm** diameter sphere not allowed and shall have adequate clearance from hazardous parts. Contact with hazardous parts not permitted.	Back of the hand
2	Full penetration of **12.5 mm** diameter sphere not allowed. The jointed test finger shall have adequate clearance from hazardous parts.	A finger
3	The access probe of **2.5 mm** diameter shall not penetrate.	A tool such as a screwdriver
4	The access probe of **1.0 mm** diameter shall not penetrate.	A wire
5	Limited ingress of dust permitted. No harmful deposit.	
6	Totally protected against ingress of dust.	Dust-tight

INDUSTRY TIP

Do not use your own finger to test equipment. Use a proper test finger.

▲ Figure 4.122 The first digit relates to the ingress of solids

▼ Second numeral 0–8: Ingress of water

IP	Requirement	Example
0	No protection	
1	Protected against vertically falling drops of water	
2	Protected against vertically falling drops of water with enclosure tilted 15° from the vertical	
3	Protected against sprays to 60° from the vertical	
4	Protected against water splashed from all directions	Outdoor electrical equipment
5	Protected against low-pressure jets of water from all directions	Where hoses are used for cleaning purposes
6	Protected against strong jets of water	Where waves are likely to be present
7	Protected against the effects of immersion between 15.0 cm and 1.0 m	Inside a bath tub
8	Protected against submersion or longer periods of immersion under pressure	Inside a swimming pool

▲ Figure 4.123 The second digit relates to the ingress of water

ACTIVITY

What is meant by the code IP44?

▼ Additional letter A–D: Enhanced protection of persons

IP	Requirement	Example
A	Penetration of 50.0 mm diameter sphere up to guard face must not contact hazardous parts	The back of the hand
B	Test finger penetration to a maximum of 80.0 mm must not contact hazardous parts	A finger
C	Wire of 2.5 mm diameter × 100.0 mm long must not contact hazardous parts when spherical stop face is partially entered	A screwdriver
D	Wire of 1.0 mm diameter × 100.0 mm long must not contact hazardous parts when spherical stop face is partially entered	A wire

INDUSTRY TIP

You will find that over the years, many creatures find their way into enclosures. This might seem impossible, but insects are common and mice not unknown.

ACTIVITY

When removing knockouts from a plastic box the slot should be as tight to the cable as possible. What tool should be used to cut the slot?

INDUSTRY TIP

When cables enter the top of an enclosure, IP4X must be maintained, meaning there should be no gap larger than 1 mm.

The IP codes that you are most likely to come across are:

1 For protection against the ingress of solid objects and protection to persons. The codes are IP2X and IP4X used in relation to barriers and enclosures.
2 For protection against the ingress of water the codes are IPX4, IPX5, IPX6, IPX7 and IPX8.
3 For enhanced personal protection IPXXB and IPXXD again used in relation to barriers and enclosures.

▲ Figure 4.124 Cable entry at electrical accessory not meeting the IP code

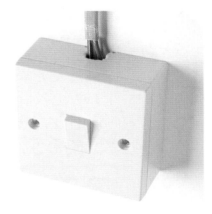

▲ Figure 4.125 Non-sheathed cables outside of an enclosure

526.8

Where the sheath of a cable has been removed the cores of the cable must be enclosed within an enclosure as detailed in 526.5. This also applies to non-sheathed cables, contained within trunking or conduit.

526.9

The group of regulations designated 526.9 relates to the connection of multi-wire, fine-wire and very-fine wire conductors.

526.9.1

To stop the ends of multi-wire, fine-wire and very-fine wire conductors from spreading or separating, this regulation requires that suitable terminals, such as plate terminals, or suitable treating of the ends be undertaken. One suitable method is to fit ferrules on the ends of the conductor. Manufactures will almost always fit some form of ferrule to the ends of flexes so that the conductor can be terminated in a screw terminal.

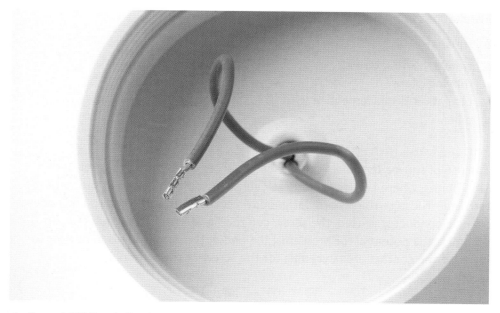

▲ Figure 4.126 Ferrule fitted to pendent flex

526.9.2

Soldering or tinning of the ends of multi-wire, fine-wire and very-fine wire conductors is not permitted if screw terminals are used.

526.9.3

The connection of soldered and non-soldered ends on multi-wire, fine-wire and very-fine wire conductors is not permitted where there is relative movement between the two conductors.

Connection methods

Allowable connection methods include:

- screw
- crimped
- soldered
- non-screw compression.

Each method has its advantages and disadvantages.

INDUSTRY TIP

Soldering flex forms a hard mass that, when subjected to vibration, may work loose. Most appliances come fitted with a 13A plug but occasionally, a flex with soldered ends is supplied. In this situation, advice from the manufacturer should be followed.

ACTIVITY

Why should solid 2.5 mm^2 not be twisted together when terminated at 13A sockets?

Screw terminals

When cables are terminated into electrical equipment, the type of terminal used must be taken into account. Most accessories use a grub screw, which is screwed down onto the conductor to ensure it is retained in place. Problems arise, however, when the terminal is designed to take more than one conductor, or a conductor of a larger csa than the one being installed.

The common types of screw terminal used in the accessories within electrical installations are:

- square
- circular
- moving-plate
- insulation-displacement
- pillar.

Square base terminal

The use of square terminals can be seen in accessories such as socket outlet face-plates, where two or even three cables can be terminated in the same terminal. These terminals are designed to accept up to three cables and, where a single conductor is used, the screw can miss or damage the conductor. To minimise the potential for problems, the end of the conductor is bent over to increase the contact area available for the screw of the terminal.

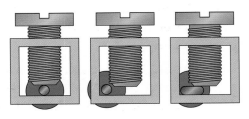

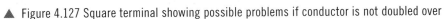

▲ Figure 4.127 Square terminal showing possible problems if conductor is not doubled over

▲ Figure 4.128 Bending over the end of a conductor: bend is too big, correct and too small

To bend over the end of the conductor, it is necessary to remove twice the normal amount of insulation from the cable. Then a pair of long-nosed pliers is used to fold the exposed conductor in half. It is important to make sure that both sides of the bend are the same length, as shown in the centre diagram of Figure 4.128. If the return edge is too long, as shown in the left diagram of Figure 4.128, the conductor may protrude from the terminal, causing a shock hazard. If it is too short, as shown on the right diagram of Figure 4.128, the bend will be redundant.

Circular terminal

Circular terminals can be seen in accessories such as light switches and ceiling roses, where single cables or small cables are terminated. They are also commonly found in consumer units and distribution boards on both the neutral and earth bars.

▲ Figure 4.129 Circular base terminal

Circular bottom terminals are designed to ensure that the conductor is positioned directly beneath the terminal screw, so there is no need to bend the end of the conductor over.

Moving-plate terminal

Moving plate terminals are often used on protective devices, such as circuit breakers and fuse holders that are mounted in consumer units and distribution boards.

The option of bending over the end of the conductor depends not only on the size of the conductor, but also on the size and type of the terminal. If the terminal is the type where the bottom moves up towards the top when the screw is tightened, it is not necessary to bend over the conductor, as the terminal tightens evenly. If, however, the terminal has a plate that moves towards the bottom of the terminal as the screw is tightened, small conductors should be doubled over. If there is any doubt, refer to the manufacturer's recommendations.

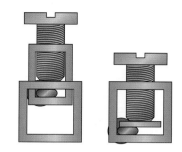

▲ Figure 4.130 Two types of moving-plate terminal

No matter what type of terminal is being used, different conductor types should never be mixed within the same terminal. If flexible cable is terminated within the same terminal as solid or stranded cable, the screw may fail to clamp on the flexible cable and only a few of the fine strands may be secured. This could result in a poor electrical connection and the wire might come loose.

If there is no option other than to mix flexible cable and solid or stranded conductors in the same terminal, the flexible conductor must be fitted with a ferrule. This is a small, metal tube that is crimped onto the end of a flexible conductor to hold the strands together.

Terminating copper and aluminium conductors within the same terminal should also be avoided due to the electrolytic reaction between the two different metals.

Whatever types of terminal and conductor are being used, always make sure that the screw tightens on the conductor and not the insulation. To ensure this, the insulation should stop at the opening of the terminal. Take care not to stop the insulation too early, leaving the conductor exposed, with the possibility of faults occurring.

INDUSTRY TIP

Common faults with terminations are exposed conductors or screwing down onto the insulation.

Advantages and disadvantages of screw terminals

The advantages of screw terminals are:

- they are cheap to produce
- they are reliable
- they are easily terminated, with basic tools
- the terminals are reusable.

The disadvantage of screw terminals are that:

- over-tightening could result in damage to the terminal or the conductor
- under-tightening of the terminals can result in overheating and arcing
- terminals can become loose, due to movement of the conductor in use or due to mechanical vibration
- terminations need to be accessible for inspection.

Crimps

Crimps and crimp lugs come in two basic forms, insulated and uninsulated.

Insulated crimps

When using a cable crimp lug, the wire's insulation must be stripped back about 5 mm. This enables the crimp to be installed with the correct amount of conductor within the crimp and with the insulating section being sealed down onto the insulation of the wire.

Crimp lugs come in three colours for the different sizes of wires:

- red – 1 mm^2 to 1.5 mm^2 wires
- blue – 1.5 mm^2 to 2.5 mm^2 wires
- yellow – 4 mm^2 to 6 mm^2 wires.

The jaws of the crimping tool are shaped to apply a different crimp style and pressure to the conductor and insulation sides of the connection. The crimping tool applies the correct amount of pressure through a ratchet that cannot be defeated unless the correct amount of pressure is used, or the release button is pressed.

Once a crimp has been installed, it must be checked to ensure that the conductor of the wire protrudes from the crimped part of the lug and that the insulation has been trapped on the other end, so that no exposed conductor is showing. Bearing in mind that crimps are used in applications where vibration occurs, they should never be used on solid conductors.

Uninsulated crimps

Uninsulated crimps are used on conductors with cross-sectional areas from 6 mm^2 upwards. It is important that the crimp is sized in accordance with the cross-sectional area of the conductors and is compatible with the conductor material being terminated.

Whilst a hand crimper may be suitable for smaller-sized conductors, on larger conductors a hydraulic crimper will be required. Battery-operated crimp tools are available that take all of the hard work out of crimping a lug onto a conductor.

Method

Select the correct size cable lug for the conductor. Cable lugs with different-sized holes are available, so be sure to always choose one that has the correct-sized hole for the connection bolt or screw.

▲ Figure 4.131 Insulated crimps

ACTIVITY

It is not unknown for pliers to be used for fitting crimps, instead of the correct crimping tool. List two possible faults that could occur.

▲ Figure 4.132 Uninsulated crimps

▲ Figure 4.133 Make sure the appropriate lug is selected

Strip enough insulation from the cable so that the copper conductor meets the end of the cable lug, while the sheath of the cable fits tight to the base of the lug.

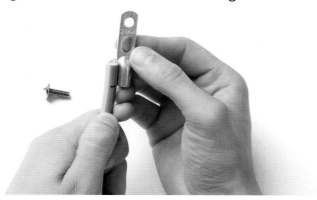

▲ Figure 4.134 Measure how much to strip

Ensure that the cable reaches right to the end of the lug tube.

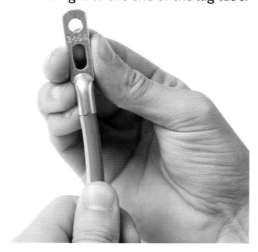

▲ Figure 4.135 Fitting the lug to the conductor

The cable lug is then crimped to the cable, using a proprietary cable crimper that suits the size of the lug.

▲ Figure 4.136 Lug being crimped to the conductor

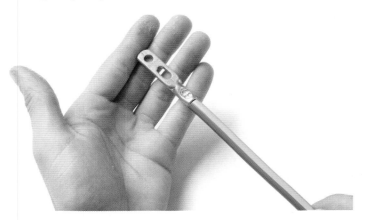

▲ Figure 4.137 Conductor ready to be connected

Advantages and disadvantages of crimped connections

The advantages of crimped connections are that they:

- are quick and convenient to install
- provide a secure termination
- do not need to be accessible for inspection.

The disadvantages of crimped connections are that:

- special tools are required
- tools for larger sizes are expensive to purchase
- crimps cannot be reused.

Soldered terminations

In the past, lugs were soldered to cables, but this has mainly been replaced by the use of crimp lugs.

Soldering is mainly used in the assembly of electronic components and equipment.

▲ Figure 4.138 Soldering is mainly used with electronic equipment

Advantages and disadvantages of soldered terminations
The advantages of soldered terminations are that:

- they provide a good electrical connection
- they offer good mechanical strength
- large numbers of connections can be made within a small area as the joint area is very small.

The disadvantages of soldered terminations are that:

- a heat source is required
- there are hazards associated with molten metals
- there may be damage to conductor insulation
- there may be damage to components.

Non-screw compression
Non-screw compression connectors, including push-fit connectors, have been used in many associated industries such as lighting manufacturing for a number of years and have proved to be robust and reliable, both electrically and mechanically. In recent years these have been used more and more in electrical installations and, depending on choice of connector, can be used for joining:

- solid conductors to solid conductors
- flexible conductors to solid conductors
- flexible conductors to flexible conductors.

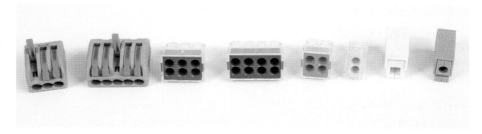

▲ Figure 4.139 A wide range of push fit connectors is available, to cope with various cable types

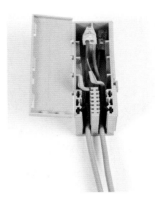

▲ Figure 4.140 Termination of PVC cables into purpose-designed box

Manufacturers make a range of accessories to go alongside these connectors that ensure the termination of cables can be both speedy and reliable.

Advantages and disadvantages of non-screw compression terminations

The advantages of non-screw compression terminations are that:

- they are quick and convenient to install
- they provide a secure termination
- they are not affected by vibration
- no special tools are required
- they are reusable
- they are maintenance free.

The disadvantage of non-screw terminations are that:

- they are generally not available for cable sizes exceeding 6 mm^2.

Proving that terminations and connections are electrically and mechanically sound

It is important that when terminations are complete, they are verified to be both electrically and mechanically sound. The procedures to follow will include both inspection and testing.

Inspection will include checks such as:

- making sure that all terminations are tight. This includes both live and protective conductors and can be accomplished by both careful scrutiny and by 'tugging' the conductor to ensure it is securely fastened in the termination
- checking visually that the electrical connection is made to the conductor rather than the insulation
- checking visually that conductive parts are not accessible to touch
- checking that the correct termination methods have been used and that they are suitable for:
  - the type of conductors being terminated
  - the environment in which the termination is to be used.

Testing will need to be carried out to ensure that:

- conductors are continuous
- there are no shorts in the conductors
- conductors are connected to the correct points.

The appropriate tests would be:

- continuity, including that of the cpc and ring final circuit conductors
- insulation resistance
- polarity and phase rotation.

It is important that correct inspection and testing procedures are followed.

> **INDUSTRY TIP**
>
> Careful and thorough inspection will identify the majority of faults with terminations.

The consequences of terminations not being electrically and mechanically sound

If the termination of cables and conductors is not electrically or mechanically sound, the consequences can be disastrous. The cause of terminations not being electrically and mechanically sound is usually high-resistance joints or corrosion.

The effects of high-resistance joints

The most common cause of high-resistance joints is a loose connection. When current is passed across such a joint, it heats up. This is likely to cause damage to the cable insulation and/or the connected electrical equipment. In the worst case, this may result in the overheating of adjacent material, resulting in a possible fire. It is important to make sure that cables are seated properly in the terminals and that the terminals are correctly tightened. Manufacturer's instructions need to be consulted to check whether a torque setting is given for connections, which must then be complied with.

It should be remembered that, even with a sound connection, when current is flowing, the conductors and terminations will heat up, resulting in expansion of the metal, which can lead to loosening of the terminal. This is why terminals, apart from those exempted by regulation 526.3, must be accessible for maintenance and inspection.

Another cause of loose terminations is vibration from such things as machinery. It is important that initial terminations are correctly made and tightened and that regular maintenance is carried out to ensure that loose connections cannot occur.

Corroded terminals will also result in high-resistance joints.

The effects of corrosion

The most common form of **corrosion** is rusting, which occurs when iron combines with oxygen and water. Most metals, with the exception of precious metals such as gold and platinum, do not exist in metallic form in nature, but rather exist as ore.

For corrosion to occur four conditions must exist.

1 There must be an anode (corroding) and a cathode (protected) component.
2 There must be an electrical potential between the anode and the cathode.
3 The anode and cathode must be connected by a metallic path of low resistance.
4 The anode and cathode must be immersed in an electrolyte, which is an electrically conductive fluid.

When two dissimilar metals, such as aluminium and brass, are in contact with one another points 1, 2 and 3 are met. In the presence of a fluid such as moisture, corrosion will occur. The tendency to corrode can be reduced by plating terminals or using a corrosion inhibitor on the jointed parts. Bimetallic crimps are available for joining aluminium to copper. These are engineered to

▲ Figure 4.141 Lug bolted to casing of equipment

KEY TERM

Corrosion: the breaking down or destruction of a material, especially a metal, through chemical reactions.

obtain the best possible transition between the metals and corrosion is reduced by coating the inside of the crimp with grease.

Techniques and methods for the safe and effective termination and connection of cables

The following section will describe and illustrate the preparation of cables for connection. Various methods will be illustrated. The health and safety requirements for each method and the tools required to terminate will be discussed for each type of cable. The cables covered in this section are:

- thermosetting insulated cables, including flexes
- single and multicore thermoplastic (PVC) and thermosetting insulated cables
- PVC/PVC flat profile cable
- MICC (with and without PVC sheath)
- SWA cables (PILC, XLPE, PVC)
- armoured/braided flexible cables and cords
- data cables
- fibre optic cable
- fire-resistant cable.

Terminating some of these cables requires the use of glands and shrouds as described below.

Cable glands

Cable glands are available in a range of sizes and formats and with a bewildering array of designatory letters and numbers: BW, CW, CX, CXT to name but a few. So what do all these letters mean? These tables provide the answers.

▼ First letter

Code	Definition
A1	For unarmoured cable with an elastomeric or plastic outer sheath, with sealing function between the cable sheath and the sealing ring of the cable gland.
A2	As type A1, but with seal protection degree IP66 – means 30 bar pressure
B	No seal
C	Single outer seal
E	Double (inner & outer) seal

▼ Second letter

Code	Designation of cable armouring
W	Single wire armour
Y	Strip armour used
X	Braid
T	Pliable wire armour

From the tables, these meanings can be gathered.

BW   A gland without seals suitable for single wire armour cable. SWA for indoor use.

CW   A gland with a single outer seal for single wire armour cable. SWA for outdoor use.

CX   A gland with a single outer seal for braided cables such as SY flex for outdoor use.

Additional letters may be used to signify a method of termination, such as in the designation 'CXT'. In this case the third letter, T, signifies that the braid is formed into a tail (T) and the gland is designed to terminate by this means.

The construction and fitting of each type of gland will be discussed separately, alongside each cable termination type.

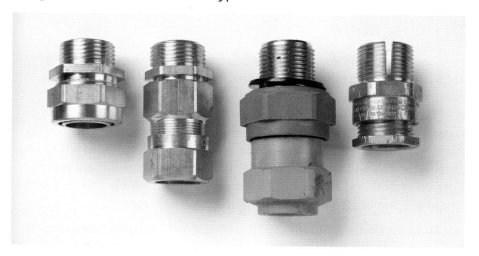

▲ Figure 4.142 From left to right BW, CW, CX and CXT gland

When cables are terminated for use in potentially explosive atmospheres, glands must be suitable for the environment and comply with ATEX or IECEx standards. Operatives must be trained and competent to **CompEx Scheme** standards.

What is the purpose of a cable gland?

A gland can be used to:

- maintain the IP rating of an enclosure
- provide continuity of earth
- provide strain relief to terminations.

A gland is an integral part of the termination of a cable and so must be fitted correctly. Incorrect fitting could result in:

- water being allowed to enter an enclosure
- the connection to earth not being adequate and posing a shock risk in the event of an earth fault
- strain being placed on cables and the cables pulling out of terminals, creating either a short-circuit fault or an earth fault.

KEY TERM

CompEx Scheme: the global solution for validating core competency of employees and contract staff of major users in the gas, oil and chemical sectors. This covers both offshore and onshore activities.

ACTIVITY

Identify the tool shown in Figure 4.143.

▲ Figure 4.143

Shrouds

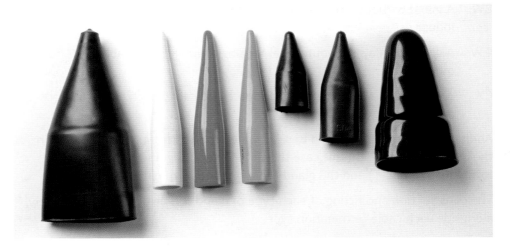

▲ Figure 4.144 A range of shrouds

▲ Figure 4.145 A badly fitted shroud could trap moisture

What does a cable shroud do? A cable shroud can aid the process of keeping the surface of the gland clean and free from the build-up of dirt. It does not, however, necessarily improve the ingress protection (IP) rating of the cable gland. In fact, the gland will invariably have been tested and rated without the installation of a cable shroud. The shroud may provide corrosion protection to cable armour or the sheath, but it must be installed in such a way as not to trap moisture under itelf and thus increase the corrosion potential. If the shroud is too loose on the cable sheath, moisture may enter the assembly and, as the fit with the gland is going to be tight, the moisture will be trapped.

Shrouds generally come in PVC or LSF (low smoke and fume) varieties, in a range of sizes to match the gland, and in a range of colours to match the sheath colour of the cable, with black being the most common.

How to fit a shroud to ensure a tight fit to the cable sheath
The same method of fitting a shroud can be used with all cables.

STEP 1 – Push the shroud lightly on to the cable so that a small bulge appears where the cable end is. Do not push too hard, as this will stretch the shroud and you will end up cutting in the wrong place.

STEP 2 – Cut the shroud at the bulge, with a pair of side cutters or, better still, a pair of cable croppers.

STEP 3 – Push the shroud onto the cable. The top of the shroud should now be a snug fit to the outer sheath of the cable. Remember, when assembling the gland and shroud combination, the shroud goes on before the gland.

▲ Figure 4.146

The next section describes the common methods of terminating cables.

Cable entry to an enclosure

When a cable enters an enclosure, the integrity of the enclosure should not be compromised. The entry may have to meet one or more of these criteria.

- The point of entry must not cause damage to the cable. Rough edges should be removed and, as a minimum, rubber grommets should be used on all cable entries. Cable glands are a better alternative.

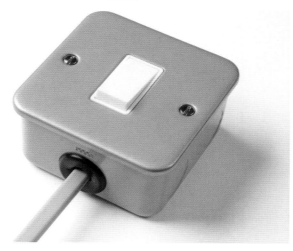

▲ Figure 4.147 Rubber grommet to enclosure. Grommets protect cables from rough edges

- The entry of the cable should not compromise the IP rating of the enclosure. For basic protection this is:
  - top surface IP4X – a 1mm diameter wire will not enter
  - front, sides and bottom, a 12.5 mm diameter object will not enter.

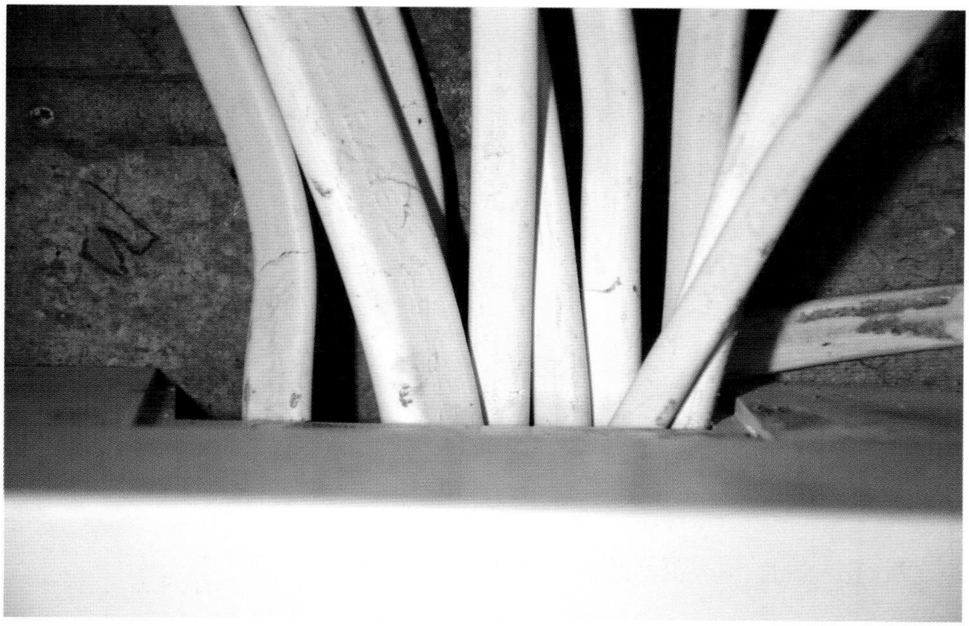

▲ Figure 4.148 Non-compliance on cable entry

INDUSTRY TIP

Grommets are available as open or blind. An open grommet has a hole in the middle. A blind grommet has a thin membrane, through which the cable is pushed, and provides a better seal. Grommet strip is also available for irregular-shaped holes.

- There may be IP ratings that are applicable for the ingress of water:
  - for an enclosure outside a building, it is likely to be IPX4 splash proof
  - for an enclosure where water jets are used, the rating should be IPX5.

There may be requirements for fire protection. Where there is a fire risk due to powders or dust being present in locations such as a carpenter's workshop, the minimum IP rating is IP5X.

▲ Figure 4.149 Splash proof socket

Terminating flexes

Tools required

- ringing tool (Method 1)
- stripping knife (Method 2).

▲ Figure 4.150 Ringing tool (Method 1) Stripping knife (Method 2)

Safety considerations

- Cuts to hands from use of knife

The use of gloves and eye protection is recommended.

Method

Before the flex can be terminated, the outer sheath must be removed.

Two methods of removing the outer sheath are outlined here.

Method 1 – The outer sheath can be removed with the use of a ringing tool. These come in various shapes and forms, the most basic of which is shown in the diagram. This tool slides over the end of the cable to the required stripping position, and is then rotated around the cable, cutting it slightly.

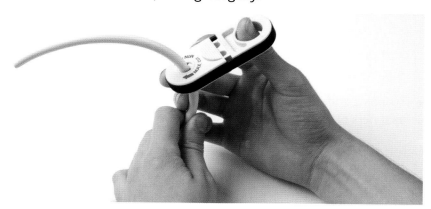

▲ Figure 4.151 Flexible cable sheath stripping tool for use on flex

The ringing tool is removed, the cable is bent to finish the cut and the sheath is then pulled off.

▲ Figure 4.152 Outer sheath being removed

Method 2 – There may be times when a ringing tool is not available, so it is important to know how to remove the outer sheath without this tool. Thermoplastic has a tendency to split when it is damaged and pressure is applied, and this can be used to help strip the sheath.

Bend the cable into a tight bend at the point where the sheath is to be removed. Using a sharp knife, gently score the top of the bend. The thermoplastic will split open like a little mouth (Figure 4.153).

▲ Figure 4.153 Bending cable to cause split in sheath

▲ Figure 4.154 Completed FP termination

Gently work the split until the inner cables are visible and then unbend, rotate and rebend the cable at about 90° from the first point. Repeat the scoring of the sheath until again the wires become visible. Repeat the bend and cut until the sheath can be removed, being careful, at all times, not to cut into the wires.

Once the sheath has been removed, the wires can be stripped and connected, as detailed in the next section, but remember that the conductors are fine strands and therefore need to be fully contained with a suitable terminal or they need to treated in some way, such as fitting ferrules, to make them stable.

Fire-resistant cable

Tools required

- stripping knife
- spanners or grips to suit gland size.

Safety considerations

- Cuts to hands from use of knife.

The use of gloves and eye protection is recommended.

Fire-resistant cables are stripped in a similar way to flexes. However, take care with the insulation on the inner cores as this is usually silicon rubber rather than PVC, and is easily damaged. Entry to enclosures is usually by means of a plastic gland.

Terminating single-core cables

Tools required

- Cable strippers.

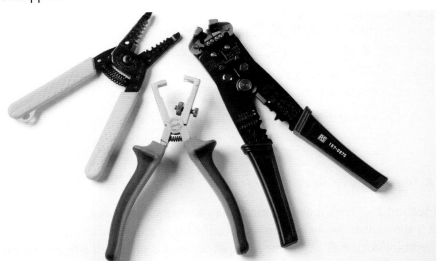

▲ Figure 4.155 Different types of cable strippers are available

Safety considerations

There is a risk of injury from slipping with cutters or pliers. The use of gloves and eye protection is recommended.

Method

As single-core cables do not have outer sheaths to remove because they are intended to be installed within trunking or conduit, the termination method is straightforward and requires the minimum of tools. This termination method will therefore also apply to the final connection of other cable types.

Use of wire-strippers is recommended for single-core conductors. Automatic wire-strippers tend to rip the insulation from the conductor and can damage it. Manual wire-strippers are preferred and these come in various forms, but all work on a similar principle.

Manual wire-strippers have two blades that cross one, another like scissors. Each blade has a notch so that together they cut around the conductor in the middle. In addition, wire-strippers also have some means of adjustment so that different sizes of conductor can be stripped.

Before wire-strippers can be used, they must be set to the correct size. This can be done using a scrap or off-cut of wire of the correct size. With the jaws of the wire-strippers together, turn the adjustment screw until the hole in the jaws is just bigger than the conductor to be stripped. Test the setting on the off-cut of wire, by placing the wire-strippers over the wire and squeezing the handles to close the jaws. Then slightly release the jaws and try to slide the insulation off, using the wire-strippers.

If the wire-strippers are correctly set, the insulation will come off easily and there will be no damage to the copper conductor. If the aperture is set too small, the insulation will slide off easily but the conductor may be damaged. If it is set too large, the insulation will not come off easily. A simple adjustment of the adjusting screw will correct these problems. Now, with the correct setting, the wires can be stripped safely.

Other types of wire-stripper will have pre-set stripping holes so that selecting the correct hole will ensure that the depth of cut is correct every time.

Once the conductors are stripped they are ready to be connected to the electrical equipment.

▲ Figure 4.156 Cable stripper jaws

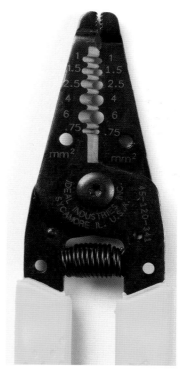

▲ Figure 4.157 Cable-strippers with pre-set holes

Terminating PVC/PVC flat profile cable

Tools required

Dependent on method used:

- electrician's knife
- side-cutters
- pliers.

▲ Figure 4.158 Pliers, side-cutters, electrician's knife (from left to right)

Safety considerations

- Cuts to hands from use of knife
- Injury from slipping with cutters or pliers

The use of gloves and eye protection is recommended.

Method

Before the conductors can be connected, the outer sheath of the cable must be removed.

First, identify how much of the sheath should be removed from the cable. The purpose of the sheath is to provide some mechanical protection for the insulation on the conductors. Too much sheath, however, takes up space within the accessory and will put excess strain on the conductors. The sheath should, therefore, be stripped back almost to where the cable enters the accessory, leaving only 10–15 mm, a thumb's width, of sheath within the accessory.

There is more than one way to remove the outer sheath. Each has its advantages and disadvantages but, whichever method is used, care must be taken to avoid damage occurring to either the conductors or the insulation around the conductors.

Method 1 – Having decided on the length of sheath required, and with the cable in place, score the sheath at the point to which it is to be removed. This can be performed with an electrician's knife. Care should be taken not to cut into the cable.

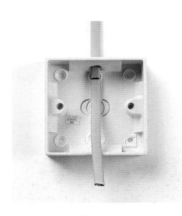

▲ Figure 4.159 Determining the point to which to strip the sheath

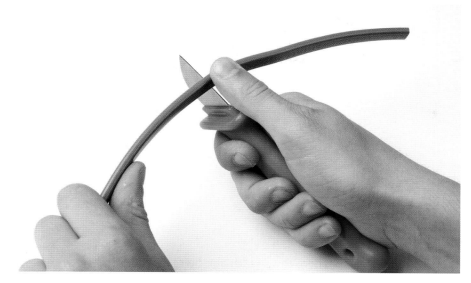

▲ Figure 4.160 Scoring the outer sheath

From the end of the cable, snip down the centre of the cable, using a pair of side-cutters. Split the line to one side and the neutral to the other.

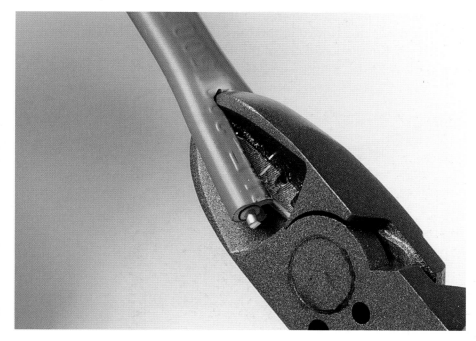

▲ Figure 4.161 Snipping the end of the cable

Grip each piece of split cable, conductor and sheath and pull them apart, tearing any uncut sheath as you go. Continue up to the score mark.

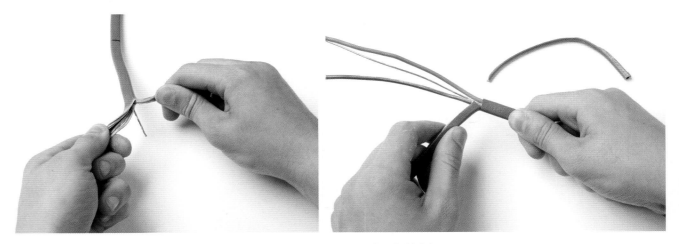

▲ Figure 4.162 Tearing down to the score mark (left) and removing the outer sheath (right)

Once the score mark is reached, the sheath can be removed from the wires. Then, holding the cable in one hand and with the thumb of the other hand placed on the sheath behind the score, jerk the torn sheath to make it break along the score mark, leaving a clean finish.

Method 2 – Snip down the end of the cable with the side-cutters, then use a sharp electrician's knife to slice down the cable. To do this, run the blade along the protective conductor. Note that this can damage the protective conductor or, even worse, the insulation of one of the live conductors if it is not done carefully.

▲ Figure 4.163 Running an electrician's knife along the cpc

Method 3 – Snip down the end of the cable and gently pull the cpc with pliers, almost like a cheese wire, to tear the outer sheath as far as the point to which it is to be stripped. Take care not to apply too much force or to damage the cpc, nor to strip too far.

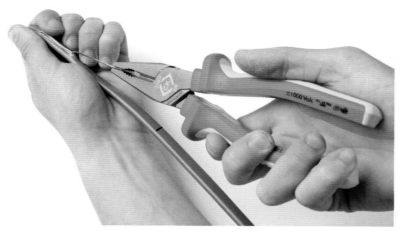

▲ Figure 4.164 Using the cpc to 'cut' the outer sheath

Once you have reached the desired strip position in the outer sheath, use side-cutters to cut away the outer sheath.

▲ Figure 4.165 Using side-cutters to cut away the outer sheath

Termination

Now the inner wires can be stripped in the same manner as for single-core wires, making sure that green and yellow sleeving is applied to the bare copper cpc. The sleeving must cover all of the bare cpc, apart from a small amount at the end, enough to go into the terminal.

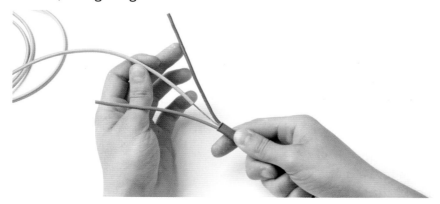

▲ Figure 4.166 Applying sleeving to bare cpc ready for final termination

The conductors can now be terminated into the electrical equipment, as described previously.

Terminating MICC cable

Tools required

- hacksaw or croppers
- MICC stripper
- pot wrench
- pot crimper
- combination pliers
- stripping knife
- spanners or grips to suit gland size
- insulation resistance tester
- low-resistance ohmmeter or a continuity buzzer

ACTIVITY

What size MICC could be used for a ring final circuit?

▲ Figure 4.167 Tools required to terminate MICC

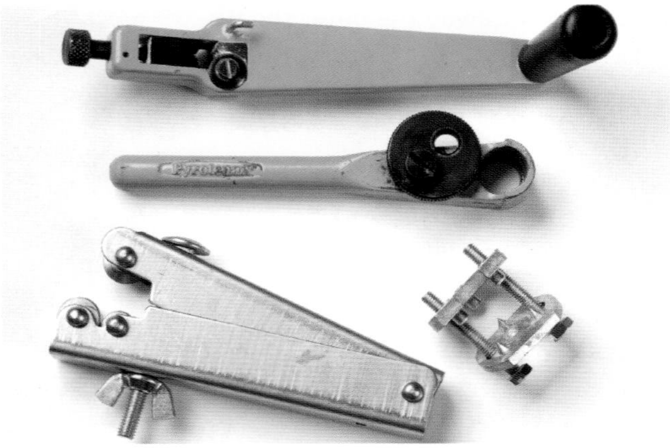

▲ Figure 4.168 Alternative MICC tools

Safety considerations

- Cuts to hands from use of knife.
- Cuts from the edges of the stripped copper sheath.
- Mineral insulation is in powder form, so could be an irritant if in contact with eyes.

The use of gloves and eye protection is recommended.

Method

The termination process for MICC cable is by far the most complicated of all the methods discussed here. These are the steps.

1 Prepare the cable.
2 Fit the shroud.
3 Remove the outer sheath, if applicable.
4 Strip the copper sheath.
5 Fit the gland to the cable.
6 Fit the pot to the cable.
7 Test the cable.

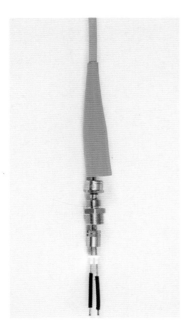

▲ Figure 4.169 MICC termination

8 Identify the cores.
9 Fit the gland to the electrical equipment or accessory.
10 Terminate the cores to the electrical equipment.

Pots, seals and glands are sized to fit a particular size of cable. A cable designated 2L1.5 has two cores, light-duty insulation (500V maximum) and a cross-sectional area of 1.5 mm^2. Termination components are chosen to match these specifications.

Step 1 Prepare the cable
To ensure easy stripping, the end of the cable must not be crushed in any way. The ideal way of cutting the cable is with a junior hacksaw; however a sharp pair of cable croppers can be used, with care.

Step 2 Fitting the shroud
Whilst the shroud can be fitted any time before step 5, it is best to cut the shroud top to the correct size before the outer sheath is removed. See the method of fitting the shroud, as described on page 304.

Step 3 Remove the outer sheath if applicable
Where the cable has an outer PVC sheath, it will need to be removed far enough to allow for termination but still within the length of the shroud. How much to strip can be determined by taking into account:

- the length of the termination tail
- the pot length
- the gland length
- an allowance to allow the crimper end to fit between the pot and the gland. This will vary with different crimp tools.

▲ Figure 4.170 Cutting the end of the cable

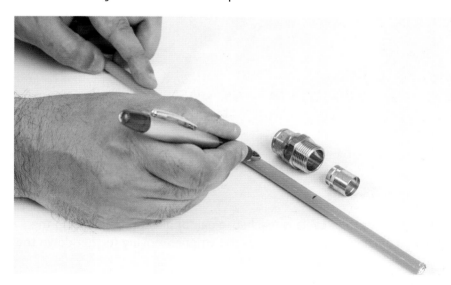

▲ Figure 4.171 Determining the length to strip

Step 4 Strip the copper sheath

The next step is to strip the copper sheath. Whilst there is more than one method of doing this, including using side-cutters, the easiest method is to use a specialist tool designed purely for this purpose.

▲ Figure 4.172 Different types of rotary strippers

Each of the tools works on the same principle. A small blade cuts into the metal sheath and peels off a small amount of copper sheath with each rotation.

▲ Figure 4.173 JOI stripper in use

When the correct length of cable is reached, a pair of pliers is used to grip the cable just ahead of the stripper. This stops the stripper moving further down the cable.

▲ Figure 4.174 Using pliers to end the stripping process

With the pliers in this position the stripper is turned again. As the stripper cannot move down the cable, it cuts a clean square edge to the cable. The sheath is pulled off and the cable is tapped gently to remove any powdered insulation that may be attached to the conductors.

Step 5 Fit the gland to the cable
Push the gland onto the cable but do not tighten it.

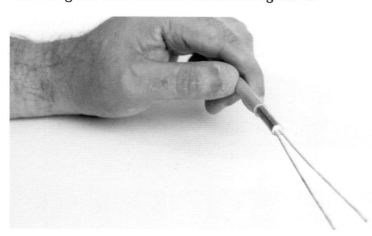

▲ Figure 4.175 Cable ready for fitting pot

Step 6 Fit the sealing pot to the cable
The sealing pot of the correct size is screwed onto the cable. A pot wrench is specifically designed for this job.

Using the type of pot wrench shown in Figure 4.176 requires that the wrench is tightened against the gland body, with the pot between the gland and the pot wrench.

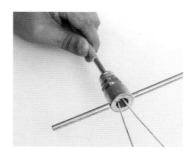

▲ Figure 4.176 Pot wrench that uses the gland

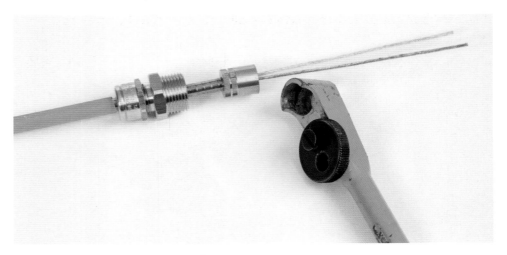

▲ Figure 4.177 One-handed pot wrench

This type of pot wrench does not require a gland to be fitted to the pot. The wheel on the pot wrench locks against the knurled edge of the pot, thus locking the pot to the wrench.

With either type, pressure needs to be applied to get the pot started. As the pot is turned, it cuts a thread on the cable sheath.

▲ Figure 4.178 Pot being fitted

Once the cable end gets to the base of the pot, stop turning.

▲ Figure 4.179 Inside the pot

If an earth tail pot is used, the earth tail must be aligned in the correct position, relative to the conductors.

▲ Figure 4.180 Alignment of earth-tailed pot

Once the pot is in the correct position, remove the pot wrench. Turn the cable so that the open end is facing downwards, then tap the pot to ensure that any loose material in the sealing pot falls out. Carefully inspect the inside of the sealing pot for any signs of debris.

The pot is now ready to be filled with compound. Place the sealing disc over the conductors and push it down to the bottom end of the pot to place the conductors into the correct position. Fill the pot from one side only, forcing the compound between the conductors and thus expelling any air in the pot and avoiding any air pockets.

The sealing pot must now be sealed with another plastic sealing disc. Place the disc over the conductors and press down to the mouth of the pot. Place the crimper over the pot and screw down until the crimper just touches the pot.

▲ Figure 4.181 Crimper in place ready to crimp

With a pair of pliers, pull gently on each conductor, in turn, to straighten it.

▲ Figure 4.182 Pulling the conductors

Screw the crimper down to force the sealing disc into the pot. Once the crimper is all the way down, unscrew the crimper but take care not to allow the pot to turn, as this will cause a short between the conductors.

Clean off any excess compound and fit stub sleeving to the conductors.

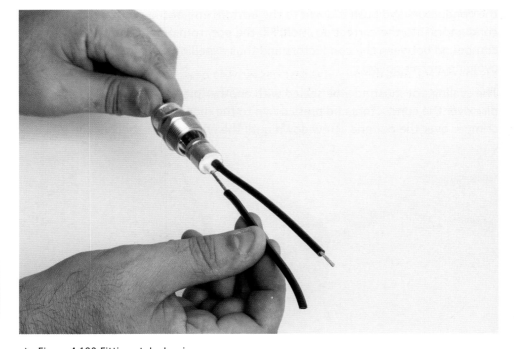

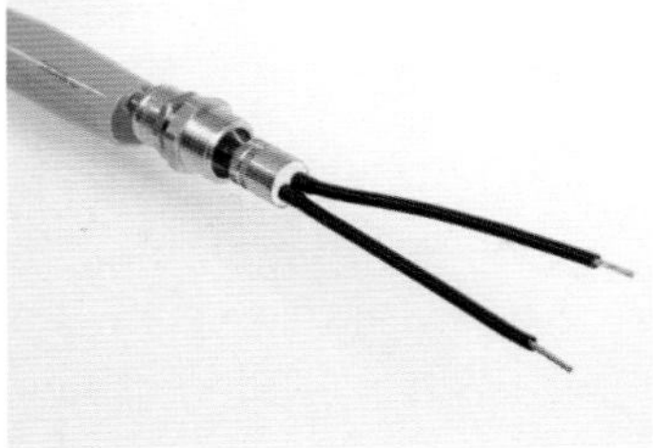

▲ Figure 4.184 Finished termination

▲ Figure 4.183 Fitting stub sleeving

Step 7 Test the cable

To ensure that there are no faults on the cable or cable termination, carry out an insulation resistance test between all conductor combinations and between each conductor and the sheath of the cable.

▲ Figure 4.185 Insulation resistance testing

Step 8 Identify the cores

The cores of an MICC cable will now need to be identified at both ends. This entails connecting one of the cores to the copper sheath at one end of the cable. Then, using a low-resistance ohmmeter or a continuity buzzer, at the other end of the cable test between the copper sheath of the cable and each of the cores until you identify which is connected to sheath. The cores are then marked up with tape or, if available in the termination kit, the appropriate coloured stub sleeving.

It is important that the previous test has already been completed to ensure that none of the cores is shorted to earth within the pots. This process will be repeated with each of the cores until all cores are identified.

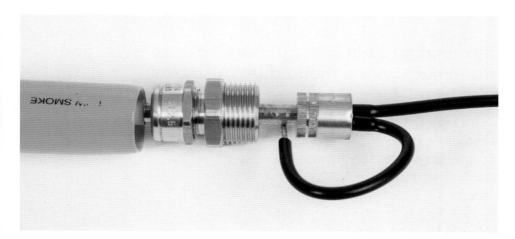

▲ Figure 4.186 Connecting one of the cores to the copper sheath at one end

Step 9 Fit the gland to the electrical equipment or accessory box

The gland is now fitted to the accessory or electrical equipment by means of a locknut or lock ring.

Using this method may mean that the pot protrudes into the accessory box, making the fitting of the accessory difficult. An alternative method is to fit a coupler to the accessory box by means of a male bush, then connect the gland to this, ensuring that the pot does not protrude into the accessory box.

Once the gland body is fitted tightly to the accessory, the back nut can be tightened, which causes the internal olive to compress and 'bite' into the copper sheath, forming a waterproof seal. After this stage, it is not possible to reposition the gland.

Step 10 Terminate the cores to the electrical equipment

The cable is now ready for final termination.

▲ Figure 4.187 MICC cable connected to accessory by means of a lock nut (left) and a coupler (right)

Terminating steel wire armoured cables (SWA) and aluminium wire armoured (AWA) cables

Tools required

- hacksaw (junior or senior, depending on size of cable) or SWA stripper
- stripping knife
- spanners or grips to suit gland size.

▲ Figure 4.188 Alternative tools for terminating SWA cable

Safety considerations

- Cuts to hands from use of knife or hacksaw.

The use of gloves and eye protection is recommended.

Method

Steel wire armoured cables are used wherever there is a need for a cable with added mechanical protection. They are found in many industrial installations and are used where cables need to be buried in the ground. Single-core armoured cable has aluminium armour rather than steel wire armour, to reduce induced circulating currents within the armour, an effect known as eddy currents. Larger cables may have aluminium conductors rather than copper conductors, to reduce costs.

Armoured cables are available with thermoplastic (PVC) and thermosetting (XLPE, cross-linked polyethylene) insulation. Whichever type of cable and whichever insulation type, the termination method is the same.

Whilst the armour of the cable is there for mechanical protection of the conductors, it may also be used as a cpc for the circuit. If the armour is not used as a cpc, it must still be connected to the main earth terminal (MET) of the installation, as the armour forms part of the electrical installation.

Armoured cables can be terminated with two different types of gland, BW for indoor use and CW for outdoor use. Each method is described below.

BW glands – these glands are for indoor use as they have no seal. The BW glands are available in two- and three-part configurations.

> **INDUSTRY TIP**
>
> When working with XLPE, you may find that it is generally stiffer than PVC.

▲ Figure 4.189 Two-part BW gland

The three-part gland has a cone ring as the extra component. The cone ring is tapered so that, as the gland is tightened, the cone ring is forced down onto the cone, tightening against the armour.

CW glands – these glands can be used outside. A CW gland has a seal, which seals the gland to the cable sheath. CW glands are available in three- and four-part configurations; again, the difference is the cone ring.

▲ Figure 4.190 Four-part CW gland

Whilst the gland may be terminated onto the cable before any connection to the equipment is made, there are occasions when armoured cable has to be installed between fixed points without any tolerance or slack. Larger cables are more difficult to handle, so careful marking out of the cable is important. The following method assumes that the cable is to be fitted to electrical equipment that has already been fixed to a wall.

These are the steps in the process of terminating an armoured cable.

1 Prepare the cable.
2 Fit the shroud.
3 Fit the lower part of the gland to the electrical equipment.

4   Ensure the gland is earthed as this will in turn earth the armour.
5   Cut the armour to the correct length.
6   Fit the upper part of the gland to the cable.
7   Mark up and remove the outer sheath.
8   Spread the armour strands.
9   Fit the gland parts together.
10 Remove the inner sleeve.
11 Terminate the cores to the electrical equipment.

Step 1 Prepare the cable

The cable is cut to length, allowing enough at the end to terminate into the electrical equipment. Armoured cables are cut with either a hacksaw or with ratchet cutters designed to cut armoured cable.

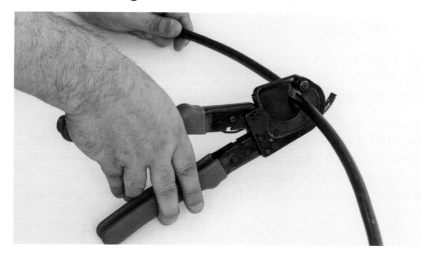

▲ Figure 4.191 Ratchet cutter in use

Step 2 Fit the shroud

Whilst the shroud can be fitted any time before step 8, it is best to cut the shroud top to the correct size before the outer sheath is removed. See the method of fitting the shroud, as described on page 304.

Steps 3 and 4 Fit the lower part of the gland to the electrical equipment and earth the armour

Steps 3 and 4 are integral in this method of termination. In other methods, where the gland is fitted to the cable before fitting to the enclosure, they are separate steps.

To earth the armour, the brass gland body (lower part) is passed through an earth tag and then into the accessory. It is secured in place with a lock ring or lock nut, using a pipe wrench. Before this can be done, a hole must be drilled into the accessory so that a bolt can be placed through the earth tag and the accessory. The earth 'fly-lead' can then be secured inside the accessory. This enables a connection to the installation earthing to be made.

INDUSTRY TIP

The armour of a SWA cable requires earthing, even if it is not used as the cpc for the circuit.

It is important to ensure that the paint is removed from the accessory at the point of connection between the earth tag and the casing. Also, make sure that the bolt is installed with the thread and nut on the inside of the accessory, as this will make the connection tamper-proof.

▲ Figure 4.192 Gland body correctly fitted to steel enclosure

The 'fly-lead' is made up of a suitably sized conductor, fitted with a ring-type crimp connector. This enables the securing bolt to act like a pillar terminal – the crimped end of the fly-lead is connected here, while the other end is terminated as a normal protective conductor. With the gland in place, the shroud can be brought up to cover the gland. There should be no strands of armour visible with the shroud in place.

▲ Figure 4.193 Armour gland with fly lead

An alternative to using the earth tag is to fit a special nut called a Piranha nut. This replaces the lock ring or lock nut and removes the need for drilling extra holes in the accessory. It also provides a means of connecting an earth lead to the gland.

▲ Figure 4.194 Piranha nuts

One side of the Piranha nut has small, sharp edges, like teeth, which bite into the enclosure as the nut is tightened, giving a good electrical connection. The nut also has threaded holes on each of the nut faces, to allow machine screws to be secured to the nut, thereby providing points for connecting the earthing fly-lead directly to the nut. This method is especially useful when terminating armoured cables into plastic boxes.

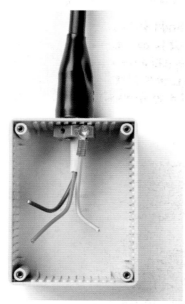

▲ Figure 4.195 SWA cable terminated into plastic box

▲ Figure 4.196 Purpose designed earth link plate

Step 5 Cut the armour to the correct length
Before step 5 can be completed, the cable needs to be marked at the correct position. For this, the cable is held against the gland body that has been fitted to the electrical enclosure.

A marking is made which is approximately the thickness of one of the armour wires above the base of the cone of the gland.

▲ Figure 4.197 Cable held against gland showing mark

Using a hacksaw, cut all the way around the cable, cutting through the outer sheath; continue cutting until each of the armour wires is cut to a depth of approximately half. Be careful not to cut too deep and thus damage the insulation of the conductors, but make sure that all of the armour wires are cut or nicked.

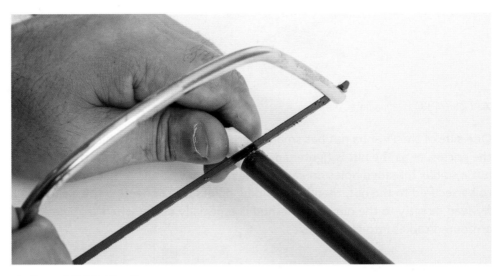

▲ Figure 4.198 Cutting the armour

This step may also be accomplished by using an armoured cable-stripper, which looks rather like a pipe-cutter but has a blade to cut the outer sheath and score the armour wires.

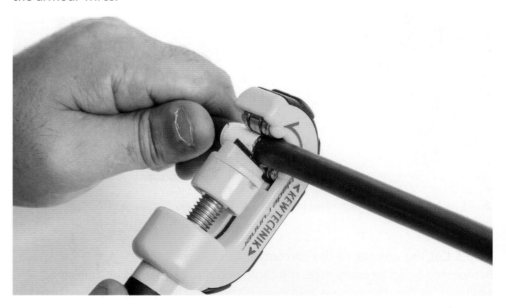

▲ Figure 4.199 Armoured cable-stripper in use

Remove the outer sheath, down to the cut point, by using a stripping knife. Hold the cable end in one hand, hold the knife across the cable at a slight angle and cut the outer sheath, pushing away from you.

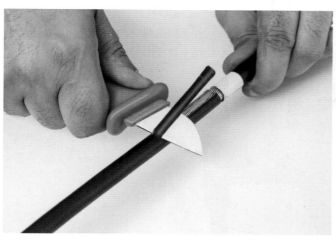
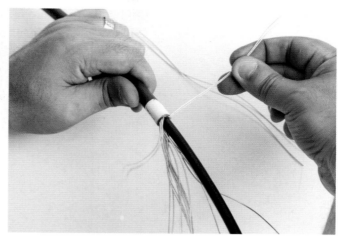

▲ Figure 4.200 Cutting away the outer sheath (left) and breaking the armours off (right)

Now remove the armour wires from the cable. This is achieved by bending the wires back and forth until they snap. Care must be taken to ensure that the armour ends under the sheath are not distorted, otherwise there may be problems when fitting the gland together.

Step 6 Fit upper part of the gland to the cable
At this stage the shroud should have been fitted to the cable. The upper part or parts of the gland should now be slid over the cable.

INDUSTRY TIP

To stop the gland sliding down the cable, the gland body can be pushed into the shroud or taped to the cable, with insulation tape.

▲ Figure 4.201 Shroud and gland body on cable

Step 7 Mark up and remove the outer sheath

Once again the cable is offered up against the gland body, now fitted to the electrical enclosure, and a mark is made just above the top of the cone.

▲ Figure 4.202 How to mark up the sheath

At the mark, the outer sheath is cut right round the cable, using a knife, a hacksaw or a SWA stripper. The small piece of sheath is then removed.

▲ Figure 4.203 Cable with sheath removed

Step 8 Spread the armour strands

The armour strands now need to be spread to fit over the cone of the gland. On smaller cables this can be accomplished holding the armour and rotating the cable against the direction of the wind of the armour. On larger cables, the strands may need to be spread apart, using a large, flat screwdriver. It is important to make sure that the ends of each armour strand are straight and that none of them crosses over another.

▲ Figure 4.204 Armour spread ready for fitting the gland together

Step 9 Fit the gland parts together

The cable is then passed through the gland body so that all the armour strands sit over the cone. The upper part of the gland is slid down the cable and tightened up to grip the armour. If a CW gland is fitted, the uppermost part of the gland is tightened onto the sheath of the cable to form a seal. Finally, the shroud can be fitted over the gland.

▲ Figure 4.205 BW gland fitted (left) and CW gland fitted (right)

Step 10 Remove the inner sleeve

The inner sleeve can now be stripped from the cores, in a similar way to stripping the outer sheath from a flexible cable. Care must be taken to avoid damaging the insulation of the inner cores.

> **INDUSTRY TIP**
>
> The inner sleeve is often referred to as 'bedding'.

In some cases, it is better to strip the inner core before fitting the gland together. Then, when passing the cable through the gland body, care must be taken to avoid damaging the insulation of the inner cores on the gland body.

▲ Figure 4.206 Cable fully terminated, CW glands

Step 11 Terminate the inner cores
The cable is now ready for final termination.

PILC cable

Paper insulated lead covered (PILC) cable is primarily used for power distribution by the distribution network operators (DNO). Therefore, the termination of this type of cable is specialised and will not be a subject of this section.

Termination of armoured or braided flex

Tools required

- stripping knife
- spanners or grips to suit gland size

Safety considerations

- Cuts to hands from use of knife.

The use of gloves and eye protection is recommended.

Method

The method for the termination of braided flexes is similar to that used for terminating armoured cables.

These are the steps in the termination process for braided flex.

Step 1 Strip the outer sheath

The cable outer sheath is removed at the correct position, ensuring that the outer sheath ends up inside the body of the gland. The outer sheath is removed, using a rotary stripper or an electrician's knife. Care must be taken not to cut too deep, as the braid of the flex is made of fine strands of wire.

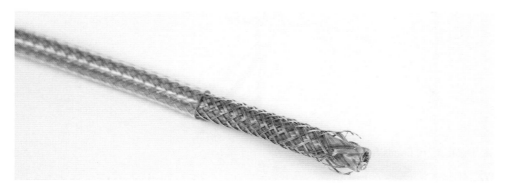

▲ Figure 4.207 Cable with outer sheath removed

Step 2 Fit the gland

Armoured or braided flex, such as SY flex, can be terminated using two different types of gland, CX and CXT. Both methods are described below.

▲ Figure 4.209 Terminating CX gland

▲ Figure 4.208 CX gland

- **Method 1** – the braid is fitted to a CX gland in much the same way as other armoured cables. This type of gland provides a higher degree of interference immunity than a CXT gland.
- **Method 2** – with a CXT gland, the braid is separated into two pigtails. The best way of doing this is by using a small screwdriver to separate the strands. The strands can then be made into two pigtails.

KEY TERM

CXT: the T in CXT means 'tail'.

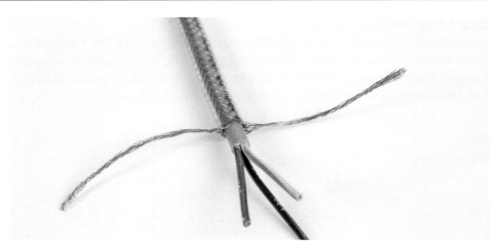

▲ Figure 4.210 Braids formed into pigtails

When the gland is fitted to the enclosure, the pigtails fit into slots in the gland and are sandwiched between washers and the locknut.

Whichever gland is used, the braid, even though it is not used as an earth, must be connected to the main earth terminal.

▲ Figure 4.212 CXT gland earthed

▲ Figure 4.211 Ready for final termination

Step 3 Remove the inner sleeve
The inner sleeve can now be stripped from the cores in a similar way to stripping the outer sheath from a flexible cable. Care must be taken to avoid damaging the insulation of the outer core.

Step 4 Terminate the inner cores
The cable is now ready for final termination.

Data cables

This section covers the termination of four types of data cables:

- Cat 5 & Cat 5e
- Cat 6
- fibre optic.

The termination of Cat 5, Cat 5e and Cat 6 data cable

The termination of Cat 5 and Cat 5e cable is the same; the termination of Cat 6 is very similar. These data cables are all twisted-pair type cables and usually contain four pairs (eight cores). Each pair is twisted to reduce **cross-talk** and each pair within the cable has a different number of twists per metre.

Within a data cable, the pairs are identified by:

- one cable taking a solid colour with a white stripe
- its pair being a white cable with the corresponding colour as its stripe.

When terminating these twisted pair cables, it is important that the pairs are not untwisted any more than is necessary to terminate the cores; otherwise cross-talk could be an issue.

For 'electrically noisy environments', shielded twisted pair cables (STP) are available. This section describes how to terminate unscreened twisted pair (UTP) cables that are commonly found within data installations. The table below shows the designations for the basic categories of twisted pair cables, with different configurations of screening.

▼ Types of twisted pair cable

Old name	New name	Cable screening	Pair screening
UTP	U/UTP	None	None
STP	U/FTP	None	Foil
FTP	F/UTP	Foil	None
S-STP	S/FTP	Braiding	Foil
S-FTP	SF/UTP	Foil, Braiding	None

Tools required

- rotary stripper (Cyclops)
- punch down tool

Removing the outer jacket

There is more than one method of removing the outer jacket of the data cable, but it is important to remove no more of the outer jacket than is absolutely necessary; about 50 mm is the ideal amount.

Method 1, using the integral ripcord – Some twisted-pair data cables come with an integral ripcord, which looks like a piece of thread. A pair of side-cutters is used to make a small cut in the end of the cable, to expose the end of the rip cord. With the ripcord held at 90° to the cable jacket, gently pull the ripcord so that it cuts the jacket.

KEY TERM

Cross-talk: a bleeding of signal from one conductor to another, through electro-magnetic induction. Twisting pairs of cable reduces this cross-talk.

INDUSTRY TIP

These cables are often installed in mini or maxi trunking to avoid clipping.

▲ Figure 4.213 Tools for terminating data cable

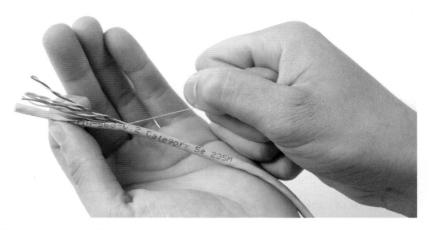

▲ Figure 4.214 Ripcord being pulled

Carry on pulling the ripcord until the intended length of the jacket to remove is reached. Pull the inner cores out from the jacket and use a pair of side-cutters to cut away the jacket.

Method 2, using a rotary stripper – The rotary stripper is passed over the end of the cable, to the point to which the outer jacket is to be stripped. The stripper is rotated to score the outer sheath, which is then removed from the cable.

▲ Figure 4.215 Rotating the stripper

The cable jacket is bent slightly, first one way and then the other, at the score point so that the jacket 'breaks'.

▲ Figure 4.216 Bending the cable sheath

The outer jacket can then be pulled off the cable. The cable is now ready for terminating.

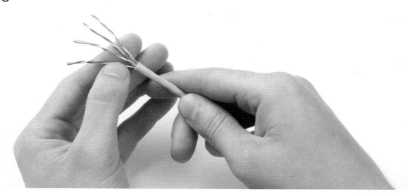

▲ Figure 4.217 Cable ready for terminating

Terminating the inner cores

As mentioned earlier, inside the cable there are eight colour-coded cores, which are twisted to form four pairs of wires. Each pair has a common colour theme, with one conductor of the pair having a solid colour and the other conductor of the pair having a stripe to match the theme. There are two wiring standards for the connection of an RJ45 connector for twisted-pair cables, T568A and T568B, the difference being the sequence of connection of the cores.

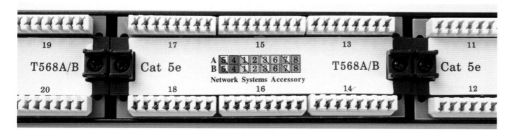

▲ Figure 4.218 Wiring connections shown on back of equipment

While it may not matter which standard is used, it is important to make sure that the same standard is used at both ends of the cable. It is also important to remember that:

- it is common practice for the UK and Europe to use T568B
- there may be a need to match data-cabling that has been installed on an earlier occasion.

Twisted-pair data cables are connected by means of insulation displacement connectors (IDC) which do not require the insulation of the cable to be stripped.

This type of termination is common in the telecoms part of the industry and is used where the conductor size is very small, so that it is impractical to use a screw-type terminal. The terminal has two blades, positioned to form a V-shape. As the wire is pushed into the terminal, the blades cut into the insulation until they make contact with the conductor. A special termination tool trims the wire to the correct length, as well as ensuring the wire is pushed far enough into the terminal. These terminals are not very good at withstanding vibration and so are only used for fixed, permanent connections.

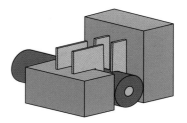

▲ Figure 4.219 Sectional view of insulation displacement terminal

This type of termination is used with solid-core data-cable.

Terminating to a data outlet plate – Place the data module on a solid surface. Untwist enough of the pair to terminate and lay the core across the appropriate IDC terminal. Using the punch-down tool, press down on each of the cores in turn, pushing the core into the terminal. Make sure the tool is the correct way round so that the tool cuts off any surplus conductor.

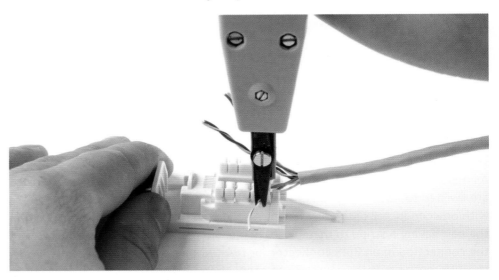

▲ Figure 4.220 Outlet being terminated

Start with the lower terminals, so that access is easier. Once all pairs are terminated, use a cable tie to secure the cable to the module.

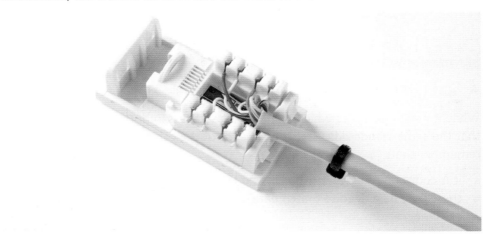

▲ Figure 4.221 Finalised module

Terminating to a patch panel – It is normal practice to leave enough slack on the cables to enable the patch panel to be pulled forward, out of the cabinet or rack, so that the terminations can be made on a solid surface. Trying to make the termination unsupported, in mid-air, is likely to result in faulty connections.

Terminate, using a similar method to that employed for terminating the data modules. If the cables approach the patch panel from left to right, then terminate the cables to the ports nearest the left-hand side first, to avoid terminated cables getting in the way of those yet to be terminated.

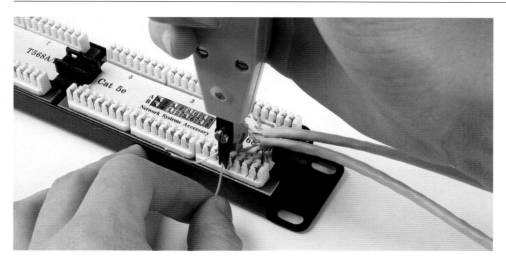

▲ Figure 4.222 Patch panel being terminated

Once a group of cables has been terminated, use a cable tie to secure them.

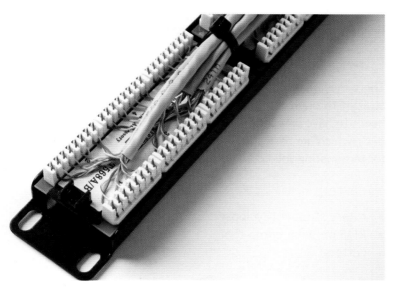

▲ Figure 4.223 Finished patch panel

Cat 6 cable

The method of termination of Cat 6 cable is very similar to that for Cat 5 and Cat 5e cable, but with some noticeable differences. The major difference between terminating Cat 6 cable and Cat 5 cables is the need to ensure that the pairs are separated from each other as much as possible. Within the cable, this is accomplished by means of a divider, which forms an X-shape, with each pair sitting in a separate section. When installing this cable, it is important to ensure that no 'kinks' or tight bends are introduced, to avoid the pairs coming out of their divider-section.

When terminating Cat 6 cables to a patch panel, the individual cables are kept separate from each other.

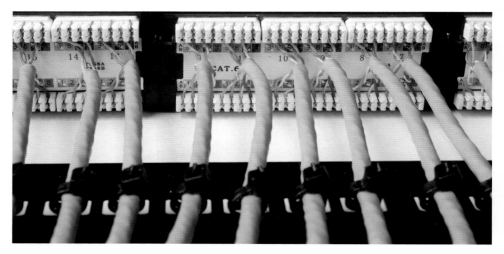

▲ Figure 4.224 Terminated Cat 6 patch panel

Once the cables have been installed they will require testing.

Data cables are always improving. For example, the Cat 7 cable has recently been introduced, which has a larger csa than Cat 5 and individual pair screening.

Fitting an RJ45 plug to a patch lead (Cat 5e)

There may be times when there is a requirement to make up a patch lead. A straight-through patch lead has the same connections at each end. The cable used needs to have cores that are flexible. The plug has insulation-piercing connectors (IPC), which are basically pins that pierce the insulation. An RJ45 plug is not suitable for terminating solid core cable; the termination will suffer premature failure.

Tools required

- rotary stripper (Cyclops)
- RJ45 modular crimp tool
- cutters

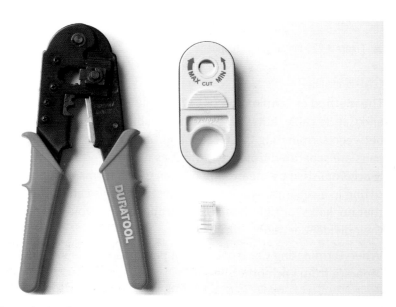

▲ Figure 4.225 Tools required to fit plugs to Cat 5 patch cable

Method

Use a rotary stripper to strip approximately 25 mm of the outer sheath of the cable. The method is the same as for solid-core data-cables.

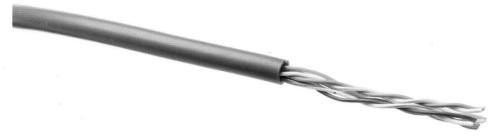

▲ Figure 4.226 Cable with the outer jacket stripped

Untwist each of the cores completely.

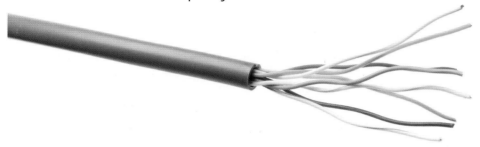

▲ Figure 4.227 Untwisted cores

Arrange the untwisted cores so that they are in the correct order for the wiring connection required. Once they are arranged, hold them between thumb and forefinger.

▲ Figure 4.228 Cores ready for trimming

Trim the cores to approximately 12–13 mm, using a pair of cutters. Make sure the cores are all the same length when held in the correct order.

Push the wires into the connector, making sure that they are still in the correct order and that they go right to the ends of the connector. The outer sheath should end up under the locking piece.

▲ Figure 4.229 Ready for crimping

Recheck that the cores are in the correct order. Carefully place the connector into the modular crimp tool and squeeze the handles tightly. The insulation piercing pins will pierce each of the eight wires and the locking tab that holds the outer plastic sheath will engage, securing the plug to the cable. Once both ends are terminated, the lead is ready to use.

▲ Figure 4.230 Plug being crimped

▲ Figure 4.231 Correctly fitted RJ45 plug

It is recommended that Cat 6 patch leads are purchased, due to the fact that hand-making a Cat 6 patch lead is difficult and the result is likely to be of a poor quality.

TIA wiring standards

RJ45 Pin	Wire colour (T568A)	Wire diagram (T568A)
1	White/Green	
2	Green	
3	White/Orange	
4	Blue	
5	White/Blue	
6	Orange	
7	White/Brown	
8	Brown	

▲ Figure 4.232 T568A standard

RJ45 Pin	Wire colour (T568B)	Wire diagram (T568B)
1	White/Orange	
2	Orange	
3	White/Green	
4	Blue	
5	White/Blue	
6	Green	
7	White/Brown	
8	Brown	

▲ Figure 4.233 T568B standard

Pins on an RJ45 plug are numbered from left to right, when the plug is held with the latch facing downwards and the pins on top of the plug.

▲ Figure 4.234 RJ45 plug

ACTIVITY

Look on the internet for the different types of UTP terminations available.

Fibre-optic cable

Fibre-optic cables are used in data and telecoms applications. The most common connector types are:

ST – straight tip

SC – subscriber connector

LC – lucent connector, a small form factor SC type connector with a locking mechanism developed by Lucent

FC – fibre connector, less common these days as prone to vibration loosening

▲ Figure 4.235 SC, LC and ST type connectors

Whatever type of connector is used, the termination process is similar. It is important that all dust is excluded from the termination. The environment when terminating fibre optics must be scrupulously clean. As the process of termination varies slightly from manufacturer to manufacturer, it is important to refer to manufacturer's instructions before terminating the fibre optic.

Tools required

- crimp tool
- die sets to suit type of connector
- stripper tools
- kevlar shears
- cable holder
- cleave tool

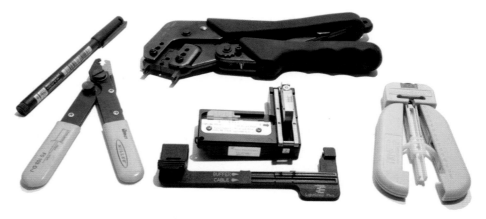

▲ Figure 4.236 Tools required for the termination of fibre optic cables

Safety considerations

Never look down the end of a connected fibre, as this could result in serious eye damage. The light transmitted is in the infra-red spectrum and therefore invisible to the eye, but can cause serious damage. Ensure both ends of the fibre are disconnected or that the transmitter is switched off.

When cutting the fibre, small glass shards are produced. If they get into the eye, they are very difficult to flush out. Safety glasses should be worn when terminating fibre-optic cables. Never eat, drink or smoke whilst terminating fibre optics, as this may lead to ingestion of glass particles.

Some termination kits may use adhesives in the termination process so attention will need to be paid to COSHH data sheets.

Method

The method described below uses Lightcrimp Plus SC connectors from Tyco Electronics. As the connectors are pre-polished, there is no need for polishing and adhesives are not used.

Each connector kit consists of:

1 connector assembly
2 crimp eyelet
3 inner eyelet
4 strain relief
5 connector housing
6 clear tubing.

To compensate for small-diameter fibres, the kit also comes with:

7 a bare buffer boot
8 small tubing.

Assembled on to the connector so as to protect the connector is:

9 a termination cover (front of connector)
10 a plunger dust cap (rear of connector).

INDUSTRY TIP
Beware of glass shards (splinters) when terminating fibre-optic cables.

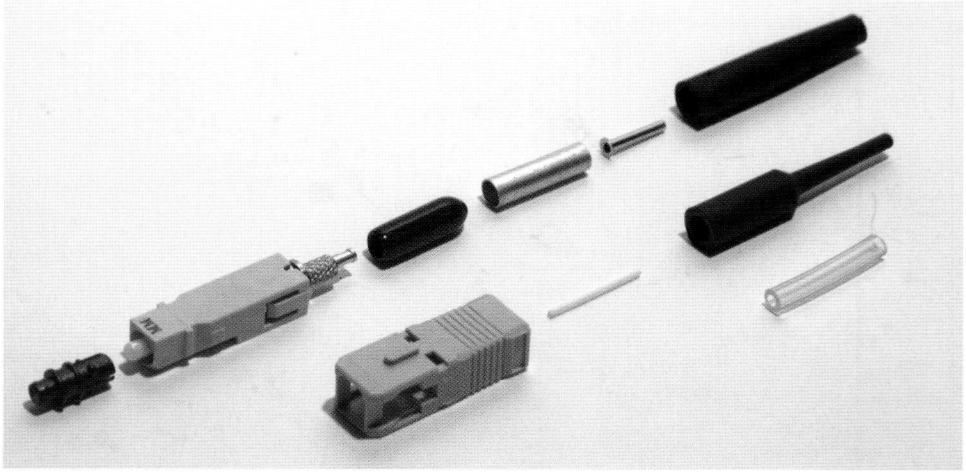

▲ Figure 4.237 Parts of Lightcrimp Plus SC connector

Step 1 – The termination of this fibre-optic cable requires only items 1, 5, 7 and 8 from the connection kit.

Step 2 – Before the final termination can be carried out, the cable must be prepared. The first step is to remove the outer cable jacket by scoring with a rotary stripper.

Step 3 – The cable is bent carefully so that the outer jacket splits. The outer jacket can now be removed by sliding it from the inner cores.

Step 4 – This cable has reinforcing strands to give added strength. These are now cut away with the shears.

Step 5 – The inner jacket is carefully scored with the rotary stripper in much the same way as was the outer jacket.

Step 6 – The inner jacket can now be removed. Notice the gel on the cable in the picture. This should be wiped away once the inner jacket has been removed.

Step 7 – The individual fibres are twisted together and must be separated.

Step 8 – The individual fibres of the cable are now separated, ready to be terminated.

▲ Figure 4.238

The next part of the process can be broken down into three sections:

- Preparation and stripping of the fibre
- Cleaving the fibre
- Fitting the termination.

Preparation and stripping of the fibre

 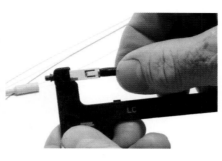 

Step 1 – Slide the bare buffer boot onto the fibre, small end first. It is best to do this now so that it does not get forgotten. Remove the plunger dust cap from the connector assembly and discard.

Step 2 – Push the connector assembly into the holder of the cable-holder assembly, with the termination cover facing out. Make sure that the connector butts up against the lip on the arm of the cable-holder assembly.

Step 3 – Insert the small (white) tubing into the plunger of the connector assembly until the tubing reaches the bottom.

Step 4 – Slide the fibre into the channel marked 'BUFFER'. Make sure that the tip of the fibre butts up against the end of the channel. Mark the fibre at each cross-slot of the channel.

Step 5 – Remove the fibre from the cable holder. The fibre is now ready for stripping.

Step 6 – Using the strip tool, strip the fibre to the first mark. It is recommended that you hold the strip tool at an angle to the fibre and strip the fibre in small sections.

Step 7 – Once the buffer has been stripped from the fibre, the fibre must be cleaned with an alcohol wipe to remove any fibre residue. The fibre is now ready for cleaving.

▲ Figure 4.239

Cleaving the fibre

Step 1 – Open the fibre clamp of the fibre-optic cleaver. Press the button, and slide the carriage back (toward the fibre clamp). Then, move the fibre slide back until it stops. Place the stripped fibre into the slot so that the end of the buffer is at the 8mm marking.

Step 2 – While applying pressure on the buffer, carefully slide the fibre slide forwards (towards the carriage) until it stops.

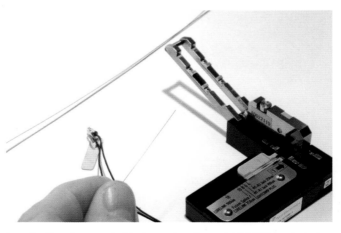

Step 3 – Gently close the fibre clamp, and slide the carriage forwards. DO NOT touch the button while sliding the carriage. Open the fibre clamp, and move the fibre slide back until it stops. Remove the cleaved fibre, and properly dispose of the scrap fibre.

▲ Figure 4.240

Fitting the termination

Step 1 – Open the cable clamp of the cable-holder assembly, and position the fibre (with the cleaved end facing the connector) inside the clamp. Move the fibre so that the end of the fibre is level with the front of the arm of the cable-holder assembly and, holding the fibre in place, close the clamp.

Step 2 – Carefully insert the fibre into the plunger of the connector assembly until the fibre reaches the internal fibre. Make sure that the remaining mark on the fibre enters the plunger.

Step 3 – The resultant bend in the fibre should hold the inserted fibre against the internal fibre. The fibre coating must enter the small tube that was installed in Step 4 of the 'Preparation and stripping' section. Make sure that the start of the fibre coating is not caught on the entry of the small tubing.

Step 4 – Squeeze the handles of the hand tool until the ratchet releases. Allow the handles to open fully.

With the connector assembly in the cable-holder assembly, position the ferrule or termination cover in the upper cavity of the front die and the plunger in the upper cavity of the rear die.

Gently push the fibre towards the connector assembly to make sure that the fibre is still touching the bottom and, then, slowly squeeze the tool handles together until the ratchet releases. Allow the handles to open fully and remove the connector from the dies.

Position the plunger of the connector assembly in the first (smallest) cavity of the front die, with the knurl against the edge of the groove in the die, and the ferrule or termination cover pointing in the direction of the arrow.

Slowly squeeze the tool handles together until the ratchet releases. Allow the handles to open fully and remove the connector assembly from the die.

▲ Figure 4.241

Step 5 – Slide the bare buffer boot over the plunger until the boot butts up against the connector assembly. Remove the connector assembly from the cable-holder assembly.

Step 7 – The fibre is now terminated, ready to be tested.

Step 6 – Align the key of the connector housing with the chamfered edges of the connector assembly. Slide the housing over the assembly until it snaps in place. DO NOT force the components.

INDUSTRY TIP

The arrows marked on the front die indicate the direction in which the ferrule or termination cover must be pointing when the connector is positioned in that cavity. For proper placement, and to avoid damage to the fibre, observe the direction of the arrows.

After the termination is complete, the ferrule end-face must be inspected for cleanliness, using a 200 × microscope. See the paragraph at the beginning of this section about the dangers of looking down a fibre-optic cable.

It can be seen from the foregoing that the process of terminating a fibre-optic cable is complex, but is a skill well worth obtaining. It is only by following the process and being scrupulously clean, that acceptable terminations can be made.

What needs bonding?

BS 7671:2018 The IET Wiring Regulations, 18th Edition requires that an extraneous conductive part be connected to the main earthing terminal (MET) by means of a suitably sized protective **bonding** conductor. An understanding of the term 'extraneous conductive part' helps when deciding whether an item requires bonding or not.

An extraneous conductive part is defined in BS 7671:2018 as:

- a conductive part
- liable to introduce a potential, generally Earth potential, and
- not forming part of the electrical installation.

By breaking down the definition in this way, it helps us come to a conclusion as to whether or not something requires bonding.
- A conductive part is generally metallic.
- The potential is usually taken as the mass of Earth, hence the capital E in 'Earth potential'.

Extraneous conductive parts would include such items as:

- metallic water installation pipes
- gas installation pipes
- other installation pipework and ducting
- exposed structural metalwork
- central heating and air-conditioning systems
- lightning protection systems.

When considering items on the list above, always use the criteria that define an extraneous conductive part to decide whether or not the item is an extraneous conductive part in a particular situation. For example, a structural steel beam supported on brick piers would not be regarded as an extraneous conductive part and would not require bonding, but a structural steel beam in contact with the ground is an extraneous conductive part that would require bonding.

Metallic parts of the electrical installation are *exposed* conductive parts rather than *extraneous* conductive parts and these require earthing.

▲ Figure 4.242 A clamp used for bonding extraneous conductive parts such as pipework

Installing protective bonding

When installing protective bonding to mains services, the following steps should be taken.

1 Select the correct size of protective bonding conductor.
2 Install the protective bonding conductor.
3 Make the correct choice of bonding clamp.
4 Install the bonding clamp.
5 Terminate the cable.

Cable sizes

The size of bonding conductor depends on the type of earthing system used.

TN-S and TT systems

For TN-S and TT systems the following criteria are used to determine the size of protective bonding conductors:

- half the size of the earthing conductor
- a minimum size of 6 mm^2
- a maximum size of 25 mm^2.

TN-C-S systems

For TN-C-S (PME) systems the size of protective bonding conductors is determined in accordance with the size of the supply neutral and Table 54.8 of **BS 7671:2018** The IET Wiring Regulations, 18th Edition (reproduced below).

▼ Table 54.8 of BS 7671:2018

Copper equivalent cross-sectional area of the supply neutral conductor	Copper-equivalent cross-sectional area of the PEN [main protective bonding] conductor
35 mm² or less	10 mm²
over 35 mm² up to 50mm²	16 mm²
over 50 mm² up to 95mm²	25 mm²
over 95 mm² up to 150 mm²	35 mm²
over 150 mm²	50 mm²

▼ Summary of protective bonding conductor sizes

Live conductors	Earthing conductors	Protective bonding conductor TN-S and TT systems	Protective bonding conductor TN-C-S systems
6 mm²	6 mm²	6 mm²	10 mm²
16 mm²	16 mm²*	10 mm²	10 mm²
25 mm²	16 mm²	10 mm²	10 mm²
35 mm²	16 mm²	10 mm²	10 mm²
95 mm²	50 mm²	25 mm²	25 mm²
150 mm²	95 mm²	25 mm²	50 mm²
240 mm²	120 mm²	25 mm²	50 mm²

* Where PME conditions apply, the earthing conductor will be 10 mm².

BS 7671:2018 does point out that local distributors' network conditions may require a larger conductor for PME systems.

Installation of main protective bonding conductors

A main protective bonding conductor of the correct size should be run from the MET to the point of connection to the service to be bonded. The conductor used must be of the correct colour: green and yellow.

Support of main protective bonding conductors

Section 522 of BS 7671:2018 requires that all cables are correctly supported throughout their length to avoid mechanical stresses on both the cable and the terminations. Gas, water and other service pipework should not be used as a method of support. Consideration needs to be given to this aspect, especially where the bonding cable is terminated to the service pipework to ensure that the protective bonding conductor does not become disconnected from the pipe clamp.

Point of connection

The connection of the bonding conductor to the service pipes should be:

▲ Figure 4.243 Clipped bonding conductor

- as near as practicable to the point of entry of the service into the building
- where practicable, within 600 mm of the service meter or at the point of entry if the service meter is external
- to the consumer's pipework – in the case of a water service, after the stopcock

- to hard metal pipework, not soft or flexible pipe
- before any branch pipework
- after any insulating sections.

The connection point is required to be accessible for future inspection and testing.

▲ Figure 4.244 Bonding to gas supply

Choice of bonding clamps

Bonding clamps must:

- meet the requirements of **BS 951:2009**
- be suitable for the environment in which they are to be installed
- be labelled in accordance with Regulation 514.13.1 of **BS 7671:2018** The IET Wiring Regulations, 18th Edition.

BS 951 clamps are designed only to fit circular pipes or rods. They are not designed to fit objects of irregular shape, to be attached to steel wire armoured cable or lead-sheathed cable. They are available in different lengths and different materials to suit different pipe sizes and different installation environments.

BS 951 clamps are designed for connection of bonding conductors of between 2.5 mm² and 70 mm². They also come in three standard band lengths, to suit pipes with diameters of:

- 12–32 mm
- 32–50 mm
- 50–75 mm.

BS 951 clamps are available to suit different environments and manufacturers have adopted a three-colour coding system to make selection easier:

- red for dry, non-corrosive atmospheres
- blue for corrosive or humid conditions
- green for corrosive or humid conditions, and for larger sizes of conductor.

HEALTH AND SAFETY

Always make sure all connections are electrically and mechanically sound. Bad joints are dangerous.

▲ Figure 4.245 Colour-coded BS 951 bonding clamps

Some extraneous parts, such as steelwork, are of an irregular shape but still require bonding. Various clamps are available from suppliers and these must be:

- electrically durable
- of adequate mechanical strength
- suitable for the environment where they are to be installed
- labelled in accordance with Regulation 514.13.1 of BS 7671:2018.

Labels for bonding clamps

BS 7671:2018 requires that the point of connection of every bonding conductor to an extraneous conductive part has a permanent and durable label fixed in a visible position stating 'safety electrical connection – do not remove' (Figure 4.246).

Installation of bonding clamps

The photograph in Figure 4.247 shows a bonding clamp as it appears when it first arrives. Note that the label is fitted this way for packaging; the slots in the label are only there for this purpose.

The six parts of the bonding clamp are:

1 strap
2 label
3 bonding clamp body
4 cable termination point and screw
5 locking nut
6 tightening screw.

Step-by-step instructions for fitting a BS 951 clamp

Step 1 – Preparing the bonding clamp for installation

- Remove the label from the strap.
- Remove the tightening screw and locking nut from the body of the clamp.
- Pass the tightening screw through the hole in the label and fit the locking nut.
- Refit the assembly to the clamp body, but leave the screw so that it is not protruding into the slot of the body.
- The clamp is now ready for installation.

▲ Figure 4.246 BS 7671:2018 permanent label

▲ Figure 4.247 A bonding clamp as supplied

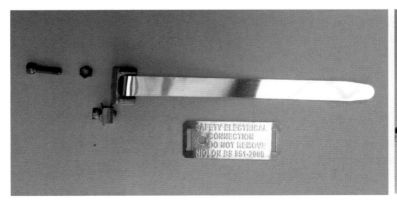

▲ Figure 4.248 Preparing the bonding clamp for installation

Step 2 – Preparing the pipe for installation of the bonding clamp

It is vitally important that the connection between the bonding clamp and the pipework is electrically sound. Tarnished pipework can be cleaned using wire wool; paint must be removed from painted pipework and the metal should be cleaned with wire wool.

▲ Figure 4.249

Step 3 – Installation of bonding clamp to pipe

Place the body of the clamp against the cleaned section of pipework.

▲ Figure 4.250

Pass the strap around the pipe and through the slot in the body of the bonding clamp.

▲ Figure 4.251

Pull the strap tight with one hand and, with the other hand, tighten the screw with a screwdriver.

Make sure that the locking nut is far enough up the tightening screw to prevent it locking against the body of the clamp.

The bonding clamp should now be tight against the pipe.

▲ Figure 4.252

The locking nut is now tightened against the body of the clamp.

▲ Figure 4.253

The bonding clamp is now installed ready for connection of the protective bonding.

▲ Figure 4.254

Termination of bonding conductors

Regulation 526.1 of **BS 7671:2018** The IET Wiring Regulations, 18th Edition requires that all electrical connections are:

- electrically durable
- of adequate mechanical strength.

The ideal method of terminating the bonding conductor is by means of a cable lug of the correct size for the cable.

▲ Figure 4.255 Cable to be terminated and correct size lug

Strip enough insulation from the cable so that the copper conductor meets the end of the cable lug, while the sheath of the cable fits tight to the base of the lug.

▲ Figure 4.256 Fitting the lug to the cable

The cable lug is then crimped to the cable using a proprietary cable crimper that suits the size of the lug.

▲ Figure 4.257 Crimped lug

Connection of bonding conductor to bonding clamp

Remove the cable termination screw.

Align the cable lug hole and reinsert the cable termination screw. Tighten with a screwdriver.

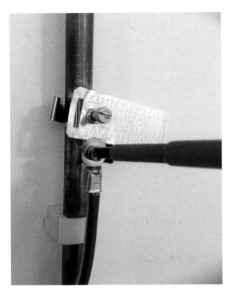

▲ Figure 4.258 Connecting the bonding conductor to the bonding clamp

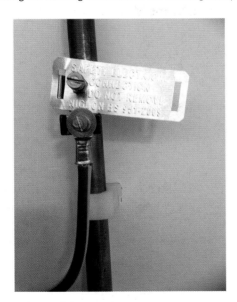

▲ Figure 4.259 The finished connection

Some of the larger bonding clamps are designed so that the cable can be terminated without the use of a cable lug. In this case, the cable is stripped back far enough to allow good metal-to-metal contact; the stripped end is inserted between the two connection screws and the screws are tightened equally to give a sound electrical connection.

▲ Figure 4.260 Bonding clamp for use without cable lug

Termination of more than one cable to a bonding clamp

Where two services are in close proximity to one another, it is acceptable to run one bonding cable to serve both services. If the bonding cable loops from one bonding clamp to another, the bonding conductor should be unbroken.

As shown in the photograph below, strip off enough insulation from the bonding cable to enable the cable to make metal-to-metal contact with the bonding clamp. A stripping knife would be the ideal tool to accomplish this task.

INDUSTRY TIP

There is a risk that if another trade removes a bonding conductor they may not bother to replace it. By using an unjointed bonding conductor for two services at least the end service is still bonded.

▲ Figure 4.261 Preparing cable to loop from one bonding clamp to another

▲ Figure 4.262 Parting the strands of the cable so that the connection screw of the bonding clamp can be fitted

▲ Figure 4.263 Passing the screw through the hole created

▲ Figure 4.264 Use a suitably sized screwdriver for tightening the connection screw for the finished connection

INDUSTRY TIP

Remember an extraneous conductive part is defined in BS 7671:2018 as a conductive part liable to introduce a potential, generally Earth potential, and not forming part of the electrical installation.

Testing of bonding conductors

Later in this chapter we look at how to carry out several tests that have to be performed on electrical installation work (starting on page 371). One of the tests is called 'Continuity of protective conductors'. BS 7671:2018 identifies that both main and supplementary bonding conductors are included in this test.

There are two methods of continuity testing; for protective conductors such as bonding conductors method 2 is preferred. The test method is covered on page 373. Here we will consider how to prepare the bonding prior to testing.

The purpose of the test is to ensure that the clamp and cable are securely and effectively connected to the extraneous conductive part.

Before the continuity test can be performed, the bonding conductor must be disconnected from the MET or, in the case of supplementary bonding, disconnected from the relevant circuit protective conductor (cpc). With one end of the bonding disconnected, clean a part of the extraneous conductive part close to the clamp. (If the extraneous conductive part has been painted then the paint must be removed from a small part to expose the metal.) Place a test probe in direct contact with the clean metal.

Now continue with the test method 2, using the long 'wander lead' (see page 375).

4 INSPECTION AND TESTING

It is a fundamental requirement of **BS 7671:2018** The IET Wiring Regulations, 18th Edition that 'during erection and on completion of an installation or an addition or alteration to an installation and before it is put into service, appropriate inspection and testing shall be carried out...'

Regulation 134.2.1 and Part 6 of BS 7671:2018 gives the requirements for inspection and testing.

Why carry out inspection and testing?

The reason for carrying out initial verification is to ensure that the installation is safe to be put into service and that it meets the requirements of BS 7671:2018.

The fundamental principles that must be followed when completing an electrical installation are that:

- good workmanship is employed
- proper materials are used
- equipment is installed in accordance with manufacturer's instructions.

It is these principles that are being checked when carrying out an initial verification.

Installation of electrical equipment

Compliance with standards

BS 7671:2018 requires that any installed equipment complies with relevant British or equivalent standards. Other standard numbers that you may see on equipment are those relating to European Norms (EN) or to the standards of the International Electrotechnical Commission (IEC). The same numbers are often used for equivalent standards, in combination with different prefixes. For example, the switch disconnector in a consumer unit may be marked as:

- IEC 947-3
- EN 60947-3
- BS EN 60947-3.

All three of these standards meet the requirements; in fact they are the same standard adopted in different parts of the world.

Damage to equipment

There should no damage to installed electrical equipment. Damage could result in the exposure of live parts and a consequent risk of shock.

Equipment suitable for its environment

Wherever a piece of electrical equipment is installed, it must be suitable for that environment. The specification for installed equipment is drawn up by the designer. When inspecting equipment for an initial verification, you are checking that the installed equipment conforms to the designer's instructions.

The designer will have taken into account:

- the environment where the equipment is to be installed
- the construction and materials of the building
- the intended use of the building
- who will use the building.

Equipment correctly installed

There are two aspects to consider when installing electrical equipment. Equipment should be installed:

- in accordance with the manufacturer's instructions
- in compliance with BS 7671:2018.

Where there is a conflict between the manufacturer's instructions and BS 7671:2018, the manufacturer should be consulted and will have final say. The reason for this is that the manufacturer will have designed their equipment to meet the requirements of all the applicable standards.

Remember the safe isolation procedure

Before any inspection and testing is carried out, the full isolation procedure should be carried out to ensure the installation is safe to be worked on, including inspection.

Figure 4.265 provides a reminder of the procedure.

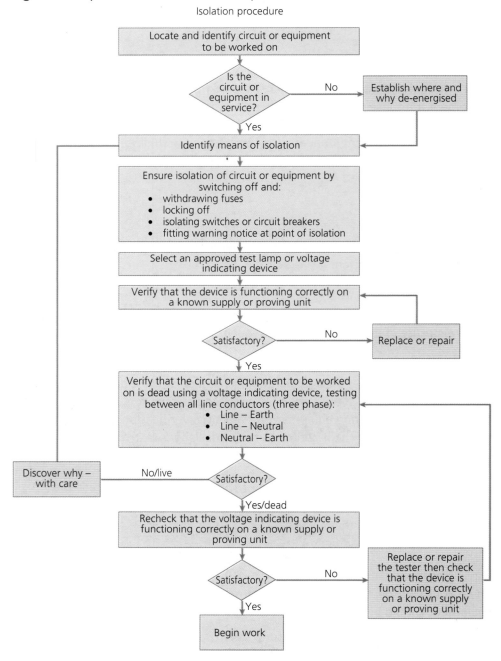

▲ Figure 4.265 Isolation procedure

The Schedule of Inspections for a new installation

At first sight, the Schedule of Inspections may appear very confusing. The document is discussed here, along with relevant Regulations from **BS 7671:2018** The IET Wiring Regulations, 18th Edition.

When completing a schedule of inspection for an Electrical Installation Certificate, boxes must be marked with either a tick for compliance, or as not applicable (n/a). No box should be marked as unsatisfactory and anything found as unsatisfactory should be corrected before any certification is issued.

The Item numbers are as detailed on the schedule of inspections. Areas not shown will be covered in greater depth in Chapter 6, *Inspection and testing,* of Book 2.

Item No	Description	Guidance (any numbers shown indicate regulation numbers or sections from BS 7671:2018)
1.0		**External condition of intake equipment** **(Note: any concerns or deviations should be reported to the DNO)**
1.1	Service cable	All of these items should be checked for their levels of basic protection against electric shock and how well they are fixed securely. That they are located for easy access for maintenance or proximity to other services such as gas. Inspectors should also look for signs of damage, tampering and that all tails remain insulated to all terminations.
1.2	Service head	
1.3	Earthing arrangement	
1.4	Meter tails	
1.5	Metering equipment	
1.6	Isolator (where present)	
3.0		**Automatic Disconnection of supply**
3.1		**Presence and adequacy of earthing and protective bonding arrangements**
	Earthing conductor	Suitably sized to be buried (TT as table 54.1) or sized for current capacity (TN Table 54.7). Continuous (if not visible throughout its length it may require testing to confirm this). Connections tight, suitably supported.
	Main protective bonding conductors	All extraneous parts connected by main protective bonding (MPB) to the main earthing terminal (MET) as (411.3.1.2). Sized correctly (544.1 or table 54.8), continuous (if not visible throughout its length it may require testing to confirm this). Connected at a suitable point at extraneous parts (within 600 mm of entry or on consumers' side of any valve, stop or meter and before any tee joints). Connections secure and the correct type of clamp used for the external influences (this may be identified by clamp colour coding).
	Provision of safety electrical earthing/ bonding labels at all appropriate locations	Suitable labelling 'safety electrical connection; do not remove' at any earth/bonding connection separate from any switchgear (514.13).
	RCDs provided for fault protection	Where an RCD is relied upon as fault protection (disconnection times are not met by a protective device as Z_s values are too high) the RCD is selected accordance with table 41.5 or 53.1. If RCDs are not relied on for fault protection and are provided for Additional Protection only, this would be marked as n/a.
4.0		**Basic protection**
4.1		**Presence and adequacy of measures to provide basic protection within the installation**
	Insulation of live parts	All conductors should be insulated with material suitable for the intended voltage. The insulation must cover conductors right up to the terminal or connection with no bare conductor showing (416.1)

→

	Barriers and enclosures	All barriers and enclosures provide adequate protection stopping contact with live parts. This would include checking the correct level of IP protection such as IP2X or IPXXB applied to all surfaces with exception of any accessible top surface which must be IP4X or IPXXD. Live parts must only be accessible by the use of a tool or key. There are some exceptions such as a ceiling rose, but these items must be located in a generally inaccessible location, such as a ceiling where a person requires steps to gain access (416.2).
5.0	**Additional protection**	
5.1	**Presence and effectiveness of additional protection methods**	
	RCDs	Where Additional Protection is provided, RCDs in accordance with 415.1 are used so it must be confirmed that the RCDs are rated with a residual current setting no more than 30 mA. The reasons for providing Additional Protection are covered later in this document.
7.0	**Consumer unit(s)/distribution board(s) (CU/DB)**	
7.1	Adequacy of access and working space for items including switchgear	The inspector must be satisfied that all switchgear and control equipment are fully accessible for use and maintainability. The particular requirements of section 729 may also need to be met. In dwellings, particular sections of the building regulations may also apply.
7.2	Presence of linked main switch(s)	Can the consumer unit/distribution board be isolated by one switching action? It must be clearly labelled (461.2).
7.3	Isolators	Suitable means of isolation provided for an installation and circuits. Devices should be capable of being locked in the off or open position (462.1).
7.4	Isolators, for every circuit or group of circuits and all items of equipment	462.2
7.5	Suitability of enclosures for IP and fire ratings	The housing must provide basic protection (see box 4.1) and must provide suitable fire ratings for the intended location (421.1). Any CU or DB installed within a dwelling must have a suitable fire rating such as a metal enclosure (422.1.201).
7.6	Protection against mechanical damage where cables enter equipment	Cables entering the enclosure must be provided with suitable mechanical protection such as glands, grommets or edging (522.8.5).
7.7	Confirmation that all conductor connections	Connections must be tight and checked to ensure that good connection to the conductor exists and terminations aren't clamped onto the insulation (526.1).
7.8	Avoidance of heating effects	Where conductors pass through metal enclosures, lines should be run with their corresponding neutrals to reduce heating effects caused by induced eddy currents (521.5.1).
7.9	Selection and correct ratings of circuit protective devices	Are the right type of fuses or circuit breakers used? Check for the correct rating (I_n see item 8.1) and short circuit/breaking capacity. Are they the correct type (gG or gM) of fuse or (B, C or D) circuit breaker for the intended application (411.4 and 432–433).
7.10	**Presence of suitable circuit charts, warning and other notices**	
	Provision of circuit charts or information	All information is displayed as required by 514.9
	Warning notices for isolation	All items intended for isolation are labelled as 514.11
	Periodic inspection and testing notice	Notice of the next periodic inspection date is displayed as 514.12.1
	RCD test notice	Where RCDs are located, an instruction notice for testing RCDs is present as 512.12.2
	Warning of non-standard colours present	Where older colours are present in existing parts of an installation 512.14.
7.11	Presence of labels to indicate the purpose of switchgear or protective devices	The purpose of all switchgear or controls should be clearly marked for its intended function. This includes all isolators and protective devices (514.1.1 and 514.8).

8.0		Circuits
8.1	Adequacy of conductors for current-carrying capacity	The inspector must be satisfied that all conductors are suitably sized for the intended load and voltage drop constraints. The inspector will be reliant on the designer's specification for this but a good level of experience is also required. The inspector must ensure correct coordination exists $I_b \leq I_n \leq I_z$ (523 and Chapter 43). This items links to item 7.8 of this document.
8.2	Cable installation methods	Are cables installed using suitable containment or management systems or where applicable, clipped direct. The method used must be compatible for the external influences present and the type of cable used (Chapter 52).
8.3	Segregation/separation of circuits	The inspector must be satisfied that any electrical circuits operating at ELV are adequately segregated from low voltage circuits unless the ELV circuits are insulated to the low voltage standards. An example may be a door bell circuit using bell wire must be segregated from lighting circuits. In addition, all electrical equipment, cables etc. is suitably distanced from any services that may affect the installation. An example may be hot water pipes near cables. If spacing is unavoidable, the electrical parts must be suited for the effect of the services. See section 528.
8.4	Cables correctly erected and supported	The inspector must be satisfied that all electrical equipment, cables and containment systems are suitably and securely installed. As an example; does a conduit have adequate saddles which are correctly spaced? (521–522).
8.5	Provision of fire barriers and sealing arrangements	The inspector must be satisfied that all electrical equipment provides suitable protection to stop the spread of fire/thermal effects in accordance with chapter 42. It must also be verified that elements of the building structure affected by the installation, such as trunking passing through floors, are adequately sealed (527.2).
8.6	Non-sheathed cables enclosed throughout	Does the containment system have the correct IP ratings where non-sheathed cables are used? (521.10.1 and 526.8)
8.7	Concealed cables adequately protected against damage	If cables are not in the zones of protection as detailed in 522.6.202, they must be protected by an earthed metallic covering. Also, where cables are concealed in a wall and are not provided with additional protection by an RCD as below (8.14), they must comply with this section. Cables in floors or above ceilings must comply with 522.6.201.
8.8	Conductors correctly identified	All conductors must be clearly identified by colour or marking. This includes sleeving of conductors. The inspector must be satisfied that all conductors are in accordance with section 514.
8.9	Presence, adequacy and correct termination of protective conductors	This is to verify that all protective conductors, other than those verified in 3.1 are selected (chapter 54) and correctly terminated to all equipment as required by 411.3.1.1.
8.10	Cables and conductors correctly connected and enclosed	All cable connections must be durable and of the correct type for the application (526).
8.11	No basic insulation of conductors visible outside enclosures	Sheathed cables must have the sheathing present up to the enclosure with no core insulation showing outside of that enclosure (526.8).
8.12	Single pole devices for switching and protection in line conductors only	The inspector must be satisfied that all single-pole devices control the line conductor of the circuit and not the neutral. This includes circuit breakers, fuses, switches etc. (132.14.1 and 530.3.2).
8.13	Accessories not damaged, securely fixed, correctly connected and suitable for external influences	The inspector must be satisfied that all electrical equipment is suitably selected and erected for all external influences it may be subjected to. A complete list of external influence can be seen in Appendix 5.
8.14		**Provision of additional protection by RCDs**
	Socket outlets at 32 A or less unless exempt	All socket-outlets rated up to 32 A must have Additional Protection by RCDs or if not installed, a socket has an appropriate label for a particular item of equipment or the designer has carried out a suitable risk assessment. The risk assessment should verify that the organization using the socket-outlets have trained staff, equipment registers and carry out regular checks and tests on equipment used on the socket outlets (411.3.3).

→

	Mobile equipment not exceeding 32 A for use outdoors	RCDs in accordance with 415.1 must be provided for all mobile equipment used outdoors not exceeding 32 A (411.3.3).
	Concealed cables in walls to a depth less than 50 mm	Cables concealed in a wall and within the zones of protection given in 522.6.202 must have RCD protection in accordance with 415.1 or have further impact protection as 522.6.204.
	Concealed cables in walls or partitions having metallic parts	Cables in partitions having metal parts other than screws, nails or similar, must have RCD protection in accordance with 415.1 or comply with 522.6.204 as stated in 522.6.203.
8.15	**Presence of appropriate devices for isolation and switching correctly located including**	
	Switching for mechanical maintenance	Any item of equipment that requires mechanical maintenance (non-electrical works such as cleaning filters, tensioning belts etc.) must have an appropriate switch located where it can be supervised during those such maintenance tasks. If they cannot be supervised, then the switch must be capable of being secured in the off position (464, 537.3).
	Emergency switching	Certain items of equipment may require a means of stopping the equipment in the event of an emergency such as a rotating machine. Stop buttons are an example and an inspection would determine that they are the correct type (latch able or key latched as an example) and located in a readily accessible position (465.1, 537.3).
	Functional switching or control	All current using equipment must have a switching device which controls the equipment or several items of equipment. An example would be a light switch (463.1, 537.3.1).
	Firefighter's switched	Certain equipment operating at high-voltages require control using a firefighter's switch located and in accordance with (537.4). Statutory Regulations may also require firefighter's switches in accordance with Chapter 53 such as petrol filling stations.
9.0	**Current using equipment (permanently connected)**	
9.1	Equipment not damaged, securely fixed and suitable for external influences	Equipment must be suitable for all the external influences it will be subjected to. This would include equipment being mechanically sound and having suitable IP protection such as IPX4 for equipment installed outdoors where splashes of water from all directions are likely (Appendix 5).
9.2	Presence of overload and undervoltage protection	Equipment, such as motors, must have an overload device set correctly for that equipment so if the machine jams, the overload device functions (Chapter 43). Undervoltage protection to motors may be provided by a contactor such as a DOL starter. This stops the equipment from suddenly re-starting should a power loss occur followed by sudden restoration of the power (445).
9.3	Installed to minimize the build-up of heat, restrict the spread of fire	Equipment must be suitable for the material it is mounted on and suitable when a risk of fire exists due to the stored or processed material as Chapter 42. It should be verified that the equipment is suitable by mark or specification. An example may be a luminaire mounted on wood should carry a mark such as a D within a triangle showing it has a limited surface temperature (see Table 55.3).
9.4	Adequacy of working space. Accessibility to equipment	All equipment, including connections, must be accessible for maintenance, inspection and have sufficient space for safe operation (132.12 and 513.1).
10.0	**Locations containing a bath or shower**	
10.1	30 mA protection to all LV circuits. Equipment suitable for zones. Supplementary bonding (where required)	Inspection of the equipment should be made to verify that it is suitable for the location and that all low-voltage circuits have suitable additional protection by an RCD to 415.1. Where required, all equipment should be linked by a cpc or supplementary equipotential bonding (701).
11.0	**Other special installations or enclosures**	
11.1	List all other special installations or locations present, if any	Any other special installation or location, as given in Part 7, should be listed on a separate sheet and the requirements for that specific section verified (Part 7).

Example check lists

Here are some examples of the type of checks that would be carried out for different scenarios.

Checks to make at a distribution board

Check that:

- the distribution board is correctly mounted – fitted in accordance with manufacturer's instructions, is firmly fixed to the wall and with the correct fixings
- all terminations are tight. Manufacturer's information will specify the correct tightness of connections at all terminals within the distribution board. Torque screwdrivers are available for just this task. The correct settings are available from the manufacturer's literature. Using a torque screwdriver ensures that connections are not over-tightened or too loose. Either situation could lead to problems
- all circuit breakers and switches are correctly installed
- cable sizes are as intended in the designer's specification
- the neutral and cpc conductors are connected in the same order as the line conductors or where this is not possible the conductors are labelled
- conductors are correctly identified: cpc are green/yellow, neutrals are blue, and so on
- conductors are correctly terminated so that the terminal screws make correct contact with the conductor cores, but not too much insulation is stripped off, exposing live conductors
- the protective device provides overload protection to the circuit conductors – for a cable rated at 17 A the appropriate circuit breaker would be 16 A
- the correct type of protective device, B,C or D, is fitted
- where RCDs are installed for additional protection, these do not have an $I_{\Delta n}$ rating of more than 30 mA
- the incoming supply conductors are connected to the correct terminals of the main switch
- internal barriers are in place
- the top surface of the distribution board meets the requirements of IP4X or IPXXD (the requirement of IP4X is that a 1 mm diameter wire shall not enter)
- the front, bottom and sides of the distribution board meet the requirements of IP2X or IPXXB (the requirement of IP2X is that a 12.5 mm diameter object shall not enter, or that a finger shall not be able to enter and touch live parts, so this requirement would extend to the correct fitting of blanks where circuit breaker ways are not used)
- the circuit charts are installed and contain all of the information required by Regulation 514.9.1 of **BS 7671:2018** The IET Wiring Regulations, 18th Edition
- these circuit charts are correct
- all necessary labels are installed, including:
  - periodic inspection and testing notice
  - RCD test notice
  - non-standard colours notice, in the case of an addition to an installation.

369

Checks to make on containment systems

Before cables are installed, check that:

- the containment system is suitable for the environment
- the containment system is complete
- the containment system is installed with enough saddles or fixings to support it
- all joints and accessories are tight
- there are sufficient draw-in or access points if required
- there are no rough edges that may damage the sheathing or insulation of cables
- in the case of metallic containment, it is correctly earthed, that the earthing is continuous and that any continuity links required are fitted
- that bends in containment are not so tight as to affect the minimum bending radius of cables.

Checks to make on circuit conductors

Check that circuit conductors:

- are correctly sized for current capacity and voltage drop
- are supported throughout their length to avoid strain on cables and terminations
- are suitable for the environment
- are not damaged
- are correctly identified (brown for line, etc.)
- are correctly terminated
- without an outer sheath (apart from bonding conductors) are enclosed within a containment system.

Checks to make on a protective bonding conductor connected to a service pipe

Check that the conductor is:

- the correct size
- coloured yellow and green
- correctly supported
- connected to the MET
- connected to a correctly installed BS 951 bonding clamp
- correctly terminated and that connections are tight.

Testing

Testing follows inspection and involves documenting test values for each circuit, which are recorded on a Schedule of Test Results. Many abbreviations are used on test record sheets and below is a guide to some of those abbreviations.

cpc   circuit protective conductor

csa   cross-sectional area

DNO distribution network operator

MET main earthing terminal

R_1 resistance of the line conductor in a final circuit

R_2 resistance of the cpc in a final circuit

r_1 resistance of the line conductor loop in a ring final circuit

r_n resistance of the neutral loop in a ring final circuit

r_2 resistance of the cpc loop in a ring final circuit

Testing an installation – the sequence of tests

When testing electrical installations, the prescribed sequence of tests is broken down into two distinct groups:

1 tests carried out before the installation is connected to the supply (dead tests)
2 tests carried out after the installation is connected to the supply (live tests).

This section deals with the tests to be carried out before the circuit or installation is connected to the supply.

BS 7671:2018 The IET Wiring Regulations, 18th Edition requires that these tests are carried out in a set sequence. The reasons for this are:

- to ensure safety
- that some tests rely on achieving a pass in a previous test to give an accurate result
- to minimise the amount of retesting if a particular test indicates a fault.

The prescribed sequence of tests *prior to connection to the supply* is as follows:

1 continuity of protective conductors, including main and supplementary bonding
2 continuity of ring final circuit conductors
3 insulation resistance
4 polarity.

Should any of the tests indicate a failure, the following procedure should be adopted:

1 the fault should be rectified
2 any previous tests that may have been influenced by the result should be repeated
3 the test that indicated the fault should be repeated.

For example, a failed insulation resistance test indicates that a fault exists between neutral and cpc. Once the fault has been corrected, the continuity of the circuit protective conductor test should be repeated as an open cpc may have been masked by the insulation resistance fault.

Instruments and instrument preparation

To carry out the 'dead' tests a low-resistance ohmmeter and an insulation resistance tester are required.

Low-resistance ohmmeter

A low-resistance ohmmeter is used to verify the continuity of conductors and the correct polarity of circuits.

Prior to carrying out any tests, the instrument must be checked to ensure:

- it is not damaged in any way
- it is still within its calibration date
- the batteries are OK
- the leads and crocodile clips are undamaged
- the instrument functions properly.

To ensure that accurate readings are obtained, the instrument test leads (and any long leads that are to be used for the tests) should be 'nulled'. Alternatively, a measurement can be made of the test leads' resistance so that this can be subtracted from the test measurements.

▲ Figure 4.266 Low-resistance ohmmeter

The low-resistance ohmmeter gives results in ohms (Ω). Normally, results are very low and to a resolution of two decimal places.

Insulation resistance tester

As its name suggests, an insulation resistance tester is used to measure insulation resistance.

Prior to carrying out insulation resistance tests, the insulation resistance tester must be checked to ensure:

- it is not damaged in any way
- it is still within its calibration date
- the batteries are okay
- the leads and crocodile clips are undamaged
- the instrument functions properly.

The insulation resistance tester gives results in megohms (MΩ). Insulation resistance testers have either a single voltage setting or multiple voltage settings. It is therefore important to ensure that the correct voltage range is selected.

Remember that the insulation resistance tester uses a voltage that can damage electronic equipment and may hurt the user.

In many cases, a low-resistance ohmmeter and an insulation resistance tester are incorporated into one test instrument, or are part of a multi-function tester. It is important, therefore, to ensure that the correct instrument setting is selected and that you are fully aware of the correct operation of the test instrument in each function.

▲ Figure 4.267 Insulation resistance tester

Testing continuity of protective conductors

The purpose of this test is to ensure that protective conductors are continuous (without breaks) and of a suitably low resistance.

The term 'protective conductor' encompasses the following conductors:

- the earthing conductor
- main protective bonding conductors
- supplementary bonding conductors
- circuit protective conductors.

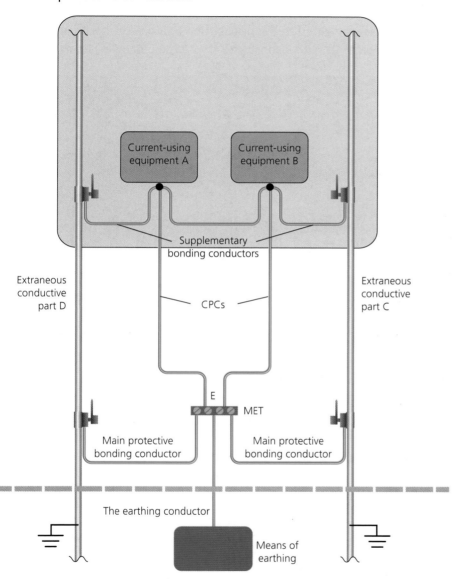

▲ Figure 4.268 Protective conductors

All of the protective conductors listed above must be verified for continuity – for which a low-resistance ohmmeter is used. It is important that the tests are carried out to meet the requirements of BS 7671:2018, rather than just to obtain test results.

Regulation 411.3.1.1 requires that where automatic disconnection is used as a means of protection against electric shock, every circuit must have a cpc which is run to and terminated at each and every point in the circuit. An exception is made for an insulated lampholder suspended from a point which has a cpc connected, for example an insulated pendant set.

Two test methods can be used to verify the continuity of protective conductors.

Test method 1

A temporary link is made between the line conductor and the protective conductor at the distribution board. For new circuits, it is best practice to do this before the circuit is connected to the distribution board. This can be achieved with a connector block or with a short lead that has a crocodile clip at each end.

▲ Figure 4.269 Connection at distribution board end

Using a low-resistance ohmmeter, a test is made between line and earth terminals at each point in the circuit. The highest measurement obtained is recorded on the schedule of test results as $R_1 + R_2$. The value of the reading depends on the length of the circuit and the cross-sectional area of the circuit conductors. Readings should increase as the test points become increasingly distant from the distribution board.

If a test at any point fails to provide a reading, check if any switch in the line conductor is switched to the 'on' position. If there is still no reading, this indicates that there is an open circuit in the line or protective conductor of the circuit; this will require further investigation and correction.

▲ Figure 4.270 Testing at point between line and cpc

EXAMPLES

1 A radial circuit is 35 m in length and is wired in 2.5 mm² live conductors with a 1.5 mm² cpc. What is the expected measurement of $R_1 + R_2$ for this circuit?

Circuit length   35 m

Resistance of 2.5 mm² with 1.5 mm²   19.51 mΩ/m

Expected resistance when testing $\dfrac{35 \times 19.51}{1000} = 0.68\,\Omega$

2 The measured $R_1 + R_2$ value for a circuit wired with a 1.5 mm² line and 1.5 mm² cpc is 1.95 Ω. What is the length of the circuit?

Measured $R_1 + R_2$ value   1.95 Ω

Resistance of 1.5 mm² with 1.5 mm² cpc   24.20 mΩ/m

Length of circuit therefore is $\dfrac{1.95 \times 1000}{24.20} = 81$ m

Values of resistance at 20°C can be found in the IET On-Site Guide Appendix I, Table I1. As measured values will be at a similar temperature, no adjustment has to be made for a change in resistance due to temperature.

Test method 2

Test method 2 makes use of a long lead or 'wander lead' to enable the inspector to reach between two points that are some distance away from one another. The resistance on the long test lead must be accounted for in the final test results. It can be nulled with the instrument test leads or its resistance measured and subtracted from the test results.

This method could be used to test circuit protective conductors which would result in a value for R_2 but it is more practicable to use method 1 when testing circuits.

Testing of main protective bonding conductors

Method 2 is the most appropriate method to test the continuity of main protective bonding conductors. The test should be carried out with a low-resistance ohmmeter and a long test lead. The instrument test leads and the long lead are nulled. The main protective bonding conductor is disconnected from the MET and one end of the long lead is connected to the disconnected protective bonding conductor. The long lead is then run out to the point where the other end of the protective bonding conductor is connected to the extraneous conductive part. A test is made using a low-resistance ohmmeter between the end of the long lead and the protective bonding conductor. A low resistance reading indicates continuity of the protective bonding conductor. On completion, the long lead is disconnected and the protective bonding conductor is connected to the MET.

ACTIVITY

The resistance of a long lead is measured as 0.89 Ω. The low-resistance meter gives a reading of 1.63 Ω. What will the value of R_2 be?

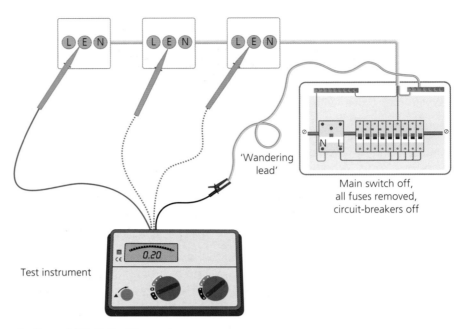

▲ Figure 4.271 Method 2

This method would also be used when testing the continuity of the earthing conductor.

When using this method, bear in mind that running a long lead out when testing could result in a trip hazard.

Which test method is most appropriate to use?

What is being tested?	Test method
Lighting circuit cpc	1
Radial socket outlet circuit cpc	1
Cooker circuit cpc	1
Continuity of earthing conductor	2
Continuity of protective bonding conductors	2
Continuity of supplementary bonding conductors	2

It can be seen from the table that when testing the continuity of circuit protective conductors, the most appropriate method is test method 1 and that when testing the continuity of a single protective conductor test method 2 is the most suitable method.

Parallel paths

So, what is meant by the term 'parallel path'? A parallel path refers to any conductive path that is not the intended earth path and which forms an alternative to the intended path. For example, a parallel path occurs where a separate cpc is run within metallic trunking or conduit which is not the intended cpc. When testing the continuity of the cpc, it would be necessary to eliminate this parallel path to ensure that only the intended cpc is tested. If the parallel path is not eliminated, the value of $R_1 + R_2$ will be artificially low. The parallel path can be eliminated by disconnecting one end of the conductor under test as shown in the diagram.

INDUSTRY TIP

Think of the parallel paths that can exist even in a domestic installation: all the bonding conductors, water and gas supplies to a boiler, cpc to immersion heaters and water pipes, washing machines, showers, and so on.

▲ Figure 4.272 Removal of parallel path during continuity testing

Continuity of ring final circuit conductors

The purpose of this test is to verify that the ring final circuit conductors form a complete ring without breaks or bridges. A ring final circuit with a break in the continuity of the conductors will continue to operate without any indication of an issue to the user, but could lead to overloading of the circuit.

For this test, a low-resistance ohmmeter is used. Before carrying out the test it is important to 'null' or zero the leads to get accurate test results. The ideal time to carry out this test is after the sockets have been connected but prior to connecting the ends of the ring final circuit to the distribution board.

This test consists of three separate steps. It is important that all three steps are carried out to ensure that all possible faults are identified.

Step 1

A test is carried out between the line conductors of each end of the ring final circuit at the distribution board.

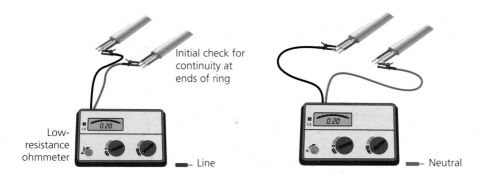

Initial check for continuity at ends of ring

Low-resistance ohmmeter

Line

Neutral

cpc

▲ Figure 4.273 Step 1 at the distribution board

This result is recorded as r_1 on the Schedule of Test Results. The test is repeated for the neutral conductor and the result is recorded as r_n. The results for the line loop and the neutral should not vary by more than 0.05 Ω as the line and neutral conductor will be of the same cross-sectional area.

The test is further repeated for the cpc and the result is recorded as r_2.

Where the cpc is of the same cross-sectional area as the live conductors, the readings will be the same. When using a cpc of a different size to the live conductors, such as in twin and earth cables, the expected reading for the cpc loop can be calculated from the following equation.

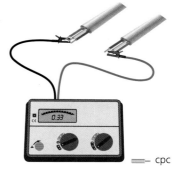

▲ Figure 4.274

$$\frac{r_1 \times \text{csa}}{\text{cpc csa}} = r_2$$

EXAMPLE

A ring final circuit is wired in 2.5 mm^2 live conductors with a 1.5 mm^2 cpc. The line loop (r_1) has been measured as 0.85 Ω. The expected reading for r_2 will be:

$$\frac{0.85 \times 2.5}{1.5} = 1.42\Omega$$

From the example, it can be seen that the ratio of the csa of the live and cpc conductors determines how much higher the reading is for the cpc loop. In this case, 2.5/1.5 means that r_2 will be 1.67 times higher than r_1.

A failure to get a reading on any of the loops indicates a break in the ring. A substantially different reading from the expected for any of the loops will need investigating.

Step 2

The open ends of the line and neutral are connected together so that the line of leg 1 is connected to the neutral of leg 2 and the line of leg 2 is connected to the neutral of leg 1. A test is carried out between line and neutral at each socket outlet.

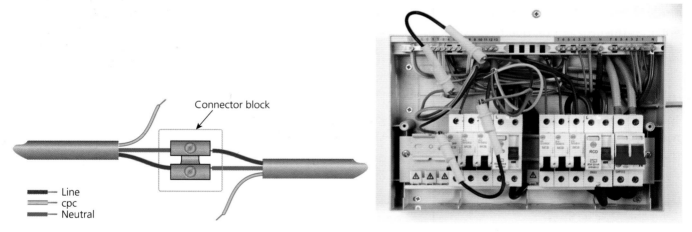

Connector block

Line
cpc
Neutral

▲ Figure 4.275 Step 2

The expected results for step 2 can be determined from the following equation.

$$\frac{r_1 + r_n}{4} = \text{L–N reading}$$

Following on from the example calculation in step 1,

$$= 0.43 \ \Omega$$

Step 3

The open ends of the line and cpc are connected together so that the line of leg 1 is connected to the cpc of leg 2 and the line of leg 2 is connected to the cpc of leg 1. A test is carried out between line and cpc at each socket outlet.

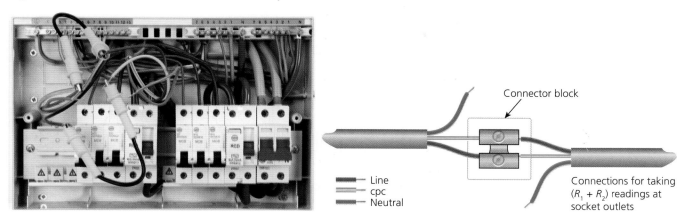

Line
cpc
Neutral

Connector block

Connections for taking $(R_1 + R_2)$ readings at socket outlets

▲ Figure 4.276 Step 3

The expected results for step 3 can be determined from the following equation.

$$\frac{r_1 + r_2}{4} = \text{L–CPC reading}$$

Following on from the example calculation in step 1,

$= 0.57\ \Omega$

The readings obtained at each socket outlet that is connected as part of the ring will be substantially the same. A socket outlet which is connected as a spur from the ring will give a different reading. Minor differences in readings may be obtained when testing at the fronts of the sockets due to connection resistances.

Test results

On completion of the test, the following results will have been recorded on the Schedule of Test Results:

- r_1, r_n, r_2
- $R_1 + R_2$
- polarity.

INDUSTRY TIP

Where there are a number of ring final circuits in a board, it is essential that each pair is correctly identified to prevent cross-connection between circuits.

EXAMPLE

The ring final circuit is wired in 2.5 mm² live conductors with a 1.5 mm² cpc. The design specification states that the circuit length is 85 m. What would the expected readings be for each step when testing the continuity of ring final circuit conductors?

Circuit length	85 m
Resistance of 2.5 mm² conductor	7.41 mΩ/m
Resistance of 1.5 mm² conductor	12.10 mΩ/m

Step 1

Calculate the resistance of r_1.

$$\frac{\text{m}\Omega/\text{m} \times \text{length}}{1000} = \frac{7.41 \times 85}{1000} = 0.63\,\Omega$$

As r_n is the same csa as r_1 the resistance will be the same.

Calculate the resistance of r_2.

$$\frac{\text{m}\Omega/\text{m} \times \text{length}}{1000} = \frac{12.10 \times 85}{1000} = 1.03\,\Omega$$

Step 2

Calculate the expected resistance (L–N) at each socket connected as part of the ring.

$$\frac{r_1 + r_n}{4} = \frac{0.63 + 0.63}{4} = 0.32\,\Omega$$

Step 3

Calculate the expected resistance (L–cpc) at each socket connected as part of the ring.

$$\frac{r_1 + r_2}{4} = \frac{0.63 + 1.03}{4} = 0.42\,\Omega$$

Note: values of resistance at 20 °C can be found in Guidance Note 3, Table B1 or in the IET On-Site Guide Appendix I, Table 11. As measured values will be at a similar temperature, no adjustment has to be made for the change in resistance due to temperature.

It should also be noted that the actual measured values of resistance may vary from the calculated values due to contact resistance at terminations.

Tested by:
Name (Capitals)..........
Signature.......... Date..........

Circuit details		1	Circuit number	2
		2	Circuit description	Ground floor sockets – Ring
Overcurrent device		3	BS (EN)	60898
		4	Type	B
		5	Rating (A)	32
		6	Breaking capacity (kA)	6
Conductor details		7	Reference method	C
		8	Live (mm²)	2.5
		9	cpc (mm²)	1.5
Ring final circuit continuity (Ω)		10	r_1(line)	0.85
		11	r_n(neutral)	0.85
		12	r_2(cpc)	1.42
Continuity (Ω) $(R_1 + R_2)$ or R_2		13	$(R_1 + R_2)^\star$	0.57
		14	R_2	N/A
Insulation resistance (MΩ)		15	Live-Live	>200
		16	Live-E	>200
Polarity		17		✓
Z_s (Ω)		18		
RCD (ms)		19	@ $I_{\Delta n}$	
		20	@ $5I_{\Delta n}$	
		21	Test button/ functionality	
Remarks (continue on a separate sheet if necessary)		22		

▶ Figure 4.277 Schedule of Test Results

Insulation resistance testing

The purpose of insulation resistance (IR) testing is to verify that the insulation of the conductors provides adequate insulation, is not damaged and that there are no shorts between live conductors or between live conductors and earth. This test is carried out before the circuit or installation is energised.

Testing is carried out with an insulation resistance tester set to the appropriate test voltage and the reading obtained will be in megohms (MΩ). The test voltages and minimum insulation resistances as given in **BS 7671:2018** The IET Wiring Regulations, 18th Edition are outlined in the table.

▼ Test voltages and minimum insulation resistances according to BS 7671:2018

Circuit nominal voltage	DC test voltage	Minimum insulation resistance in MΩ
SELV and PELV	250 V	0.5
Up to and including 500 V excluding SELV and PELV but including FELV	500 V	1
Above 500 V	1000 V	1

For circuits containing surge protection devices which cannot be disconnected, the test voltage can be reduced to 250 V but the minimum insulation resistance remains at 1 MΩ.

Pre-test checks

Prior to carrying out insulation resistance testing, check that:

- any previously inserted links have been removed
- the protective conductor of the circuit under test is connected to the MET; failure to do this will mean that faults to earth will not be identified as the test will only be carried out to the cpc of the circuit
- pilot or indicator lamps and capacitors are disconnected from the circuit; failure to do so will result in inaccurate readings
- voltage-sensitive equipment, such as dimmers, passive infrared detectors and touch switches, are disconnected and a link has been inserted to complete the circuit; failure to do so could result in damage to the equipment
- all loads, such as lamps, are disconnected from the circuit; failure to do so could lead to inaccurate results and, in the case of electronic loads, damage
- any switches that may create a break in the circuit have been switched to the 'on' position.

Insulation resistance tests are carried out between:

- live conductors
- live conductors and earth.

In the case of a single-phase circuit or installation this means testing between:

1 line and neutral
2 line and earth
3 neutral and earth.

KEY TERMS

SELV: separated extra-low voltage circuit. (Requirements for SELV and PELV can be found in Section 414 of BS 7671:2018).

PELV: protective extra-low voltage circuit.

FELV: functional extra-low voltage circuit (requirements can be found in Section 411.7 of BS 7671:2018).

ACTIVITY

IR tests on four circuits give L–N values of 20 MΩ, 100 MΩ, 3 MΩ and 5 MΩ respectively. What would be the overall reading?

INDUSTRY TIP

Many electricians carrying out IR tests and completing the Schedule of Inspections believe that neons, capacitors and starter coils will be damaged by the test. This is incorrect. They just give a false reading. It is electronic equipment such as dimmers and RCBOs that will be damaged.

In the case of a three-phase circuit or installation this means testing between:

1 line 1 and line 2
2 line 1 and line 3
3 line 1 and neutral
4 line 1 and earth
5 line 2 and line 3
6 line 2 and neutral
7 line 2 and earth
8 line 3 and neutral
9 line 3 and earth
10 neutral and earth.

The number of points of testing can be reduced by linking some of the conductors together. For example, if L1, L2, L3 and neutral were linked and a test carried out between these linked conductors and earth, that's four of the tests done in one go.

See examples of testing below.

Example 1: Testing a single lighting circuit

Carry out the pre-test checks. Using an insulation resistance tester with the test voltage set to 500 V, test between line and neutral. Operate any two-way switches and test again to ensure that all strappers are tested. Test between live conductors and earth. Operate any two-way switches to ensure there are no shorts between the strappers and earth. Make sure the readings are not below 1 MΩ. Record the readings on the Schedule of Test Results.

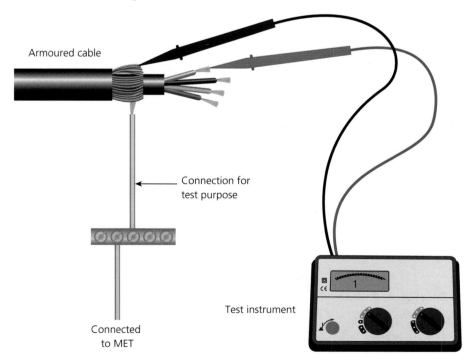

Armoured cable

Connection for test purpose

Connected to MET

Test instrument

▲ Figure 4.278 Insulation resistance testing of circuit with cpc connected to MET

Example 2: Testing a complete single-phase installation prior to connection of the supply tails to the DNO's metering equipment

Carry out the pre-test checks. Ensure that the main isolator of the distribution board and all circuit breakers are in the 'on' position. Ensure that all light switches are in the 'on' position, but with all lamps removed. Using an insulation resistance tester set to the 500 V test range, test between the line and neutral supply tails and record the reading. Link the line and neutral supply tails, test between these and the earthing conductor and record the reading. Remove any links.

1 Ready for insulation resistance testing
Main switch on, connection of earthing conductor to main earth terminal

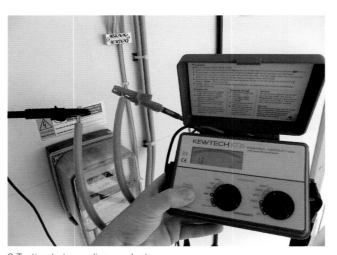

2 Testing between live conductors

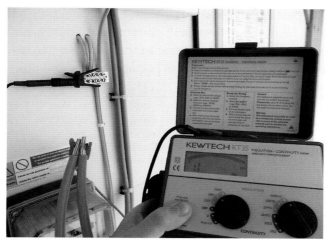

3 Testing between live and earth

▲ Figure 4.279 Insulation resistance testing on the tails of a distribution board

Polarity

Polarity must be checked to ensure that:

- all fuses, single-pole switches and circuit breakers are connected in the line conductor
- Edison screw (ES) lampholders are connected with the line conductor to the centre pin of the lampholder (not applicable to E14 and E27 types complying with **BS EN 60238** as per Regulation 559.6.1.8 of **BS 7671:2018** The IET Wiring Regulations, 18th Edition)
- socket outlets and similar accessories are correctly connected.

If a single-pole switch is connected to the neutral rather than the line conductor, a piece of electrical equipment would appear to be switched off but, in fact, would be live – an extremely dangerous situation.

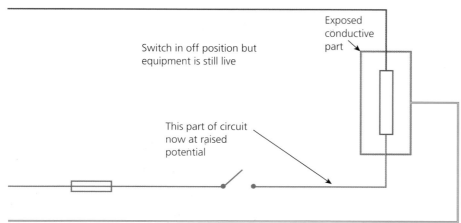

▲ Figure 4.280 Switch in neutral illustrating the circuit is not dead

If a fuse of a single-pole circuit breaker is installed in the neutral, in the case of a line–earth fault, the path of fault current flow would bypass the protective device. This would cause damage to the circuit conductors and, more importantly, make live the casings of Class I equipment.

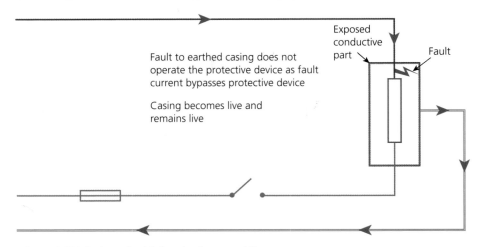

Fault to earthed casing does not operate the protective device as fault current bypasses protective device

Casing becomes live and remains live

Exposed conductive part

Fault

▲ Figure 4.281 Fault path with fuse in the neutral line

If a socket outlet were incorrectly wired, both of the unsafe situations above could exist.

If the outer screw thread of an ES lampholder is connected to the line conductor, anyone coming into contact with the screw thread of the lamp (while screwing in the lamp) could experience a shock. E14 and E27 lampholders are of an all-insulated construction, with the line and neutral not being connected to the lamp until the lamp is screwed fully home – thus, the person screwing in the lamp cannot come into contact with live parts.

Two methods of checking polarity can be used:

1 a test using a low-resistance ohmmeter
2 a visual check of core colours at all terminations throughout the circuit.

The test using a low-resistance ohmmeter is similar to the test used for verifying the continuity of circuit protective conductors. Here we will consider how this test would be carried out on a lighting circuit.

Incorrectly connected ES lampholder

Screw thread is live and accessible while lampholder is being screwed in

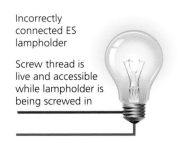

▲ Figure 4.282 Danger created by an incorrectly connected ES lampholder

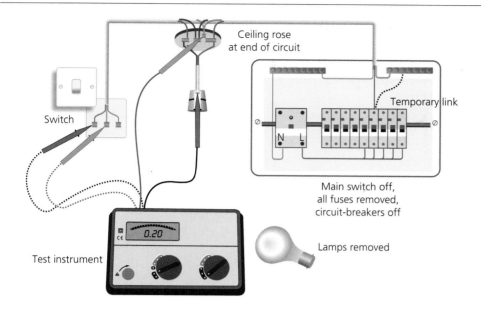

Note: The test may be carried out either at lighting points or switches

Remember to remove the temporary shorting link when testing is complete

▲ Figure 4.283 Polarity testing

A temporary link is made between the line conductor and the protective conductor at the distribution board. Using a low-resistance ohmmeter, and with the light switch in the closed position, a test is made between line and earth terminals at the light fitting. When the switch is opened, the low-resistance ohmmeter will indicate an open circuit and thus show that the switch has broken the line conductor. This test will be combined with a visual inspection to determine that the cpc is connected to the correct terminal.

In practice, this method would be incorporated into the testing of the continuity of the cpc.

It should also be noted that when testing the continuity of ring final circuit conductors by testing to the front of socket outlets, if satisfactory results are obtained in all three steps of the test, the polarity is also confirmed.

Following these tests, the installation or circuit is ready to be connected to the supply, after which a series of live tests are carried out before certification is completed and the installation handed over to the client.

Testing functionality of a circuit is different to functional testing of an electrical installation. At Level 2, you are required to be able to demonstrate simple 'dead tests' of circuits as described in this outcome. During continuity and polarity testing, switches and devices may be operated to see if they open and close circuits as intended. Any testing of circuits involving a supply must only be done under strict supervision.

Test your knowledge

1 Which of the following is a standard conduit size?

   A 15 mm

   B 20 mm

   C 22 mm

   D 28 mm

2 What tool is used to remove rough edges on the inside of a steel conduit which has just been cut?

   A Draw tape.

   B Holesaw.

   C Padsaw.

   D Reamer.

3 What tool must be used in PVC conduit when forming sets?

   A Bending spring.

   B Set square.

   C Draw tape.

   D Torque wrench.

4 What type of trunking bend is used to turn a trunking from a wall 90° onto a ceiling?

   A Flat bend.

   B Internal bend.

   C External bend.

   D Terminal bend.

5 What colour indicates the most suitable type of bonding clamp to be used in dry, non-corrosive areas?

   A Blue.

   B Green.

   C Brown.

   D Red.

6 What type of connection uses a compression tool?

   A Crimp.

   B Pillar.

   C Spliced.

   D Screw.

7 What is fitted to a PVC coated MICC cable first, when terminating into an accessory?

   A Pot.

   B Gland.

   C Shroud.

   D Seal.

8 What tool is used to score the armouring of an SWA cable when preparing to fit a gland?

   A Side cutters.

   B Electrician's knife.

   C Cross-cut saw.

   D Hacksaw.

9 What test instrument is used to check the polarity of MICC cable cores following termination at each end?

   A Insulation resistance tester.

   B Low-resistance ohmmeter.

   C Earth loop impedance tester.

   D Approved voltage indicator.

10 What test instrument ensures that cable cores are not short circuited?

   A Insulation resistance tester.

   B Low-resistance ohmmeter.

   C Earth loop impedance tester.

   D Approved voltage indicator.

11 Describe one situation where each of the flowing conduit fixings would be used.

   a Crampet.

   b Spacer bar saddle.

   c Distance saddle.

12 Determine the maximum number of 1.5 mm^2 cables that can be installed in a 20 mm conduit which is 3 m long having two bends.

13 Determine the smallest suitable trunking size capable of housing the following PVC stranded cables

   a 2×16 mm^2

   b 18×6 mm^2

   c 16×2.5 mm^2

14 Explain how to carry out an insulation resistance test on a new two-way lighting circuit which is securely isolated.

15 A ring-final test is undertaken and the following readings were obtained during step 1, end to end values:

$r_1 = 0.8 \ \Omega$

$r_n = 0.8 \ \Omega$

$r_2 = 1.34 \ \Omega$

What are the expected readings at each socket-outlet during:

   a Step 2 L-N

   b Step 3 L-E?

Practical task

In your training centre, ask your tutor for some part used reels of cable. Using an accurate low resistance ohmmeter, measure the cable resistance and determine the cable length using the resistivity formulae.

CHAPTER 5

COMMUNICATING WITH OTHERS

INTRODUCTION

Having a successful career in the electrotechnical industry involves more than technical skill and knowledge – you must also be a strong communicator. Employers understand that having a team able to communicate successfully can save huge amounts of time and money, improve safety and promote relationships. An understanding of whom to communicate with, including their roles, is also an essential skill when working on large projects.

Learning objectives

This table shows how the topics in this chapter meet the outcomes of the different qualifications. Please note that this chapter does not apply to the 8202–20, Technical Certificate in Electrical installation.

Topic	Electrotechnical Qualification (Installation) or (Maintenance) 5357	Level 2 Diploma in Electrical Installations (Buildings and Structures) 2365 Unit 308
1 Roles within the building services Industry	(105/005) 1.2 (114/014) 2.3	1.1; 1.2; 1.3
2 Information sources in the building services industry	(105/005) 1.3; 1.4	2.1; 2.2; 2.3; 2.4
3 Communication methods in work situations	(105/005) 1.1 (114/014) 2.3	3.1; 3.2; 3.3; 3.4

1 ROLES WITHIN THE BUILDING SERVICES INDUSTRY

Key roles of the site management team

It is important for everyone to understand the different *roles* and *responsibilities* in a site management team. Some of the roles and titles differ slightly depending on the individual organisation or the size of project. The following are examples of the main positions.

▲ Figure 5.1 The Gherkin, at 30 St. Mary Axe, London

Architect

The architect has traditionally been the individual associated with overall responsibility for a project. He or she is normally the one person **accredited** with the construction of a particular building. For example, the instantly recognisable tower known as The Gherkin in London was built by the renowned architect Sir Norman Foster (Figure 5.1).

The architect is usually employed by the client. In many traditional construction contracts, the architect takes the role of design team leader and is named as the person with authority to vary the works by use of an architect's instruction (AI). Modern, non-traditional contracts allow others (e.g. the client, project managers, etc.) to take the role of **contract administrator (CA)**, who is also allowed to vary a contract by instruction and therefore issue **variation orders (VOs)**.

The title 'architect' is a legally protected title, which is enforced by the Royal Institute of British Architects (RIBA). Other persons carrying out architectural design works are usually architectural technicians or architectural design consultants. These individuals have not normally completed the seven years of formal training required by the RIBA and are not officially entitled to use the title 'architect'.

The architect is usually responsible for all elements of architectural or building design including the specification of building form, structure, fire strategy, orientation, and interior and exterior design to meet client requirements, planning conditions and building regulation requirements. Generally, the architect does not carry out any special mechanical or electrical design works but is responsible for the final co-ordination of the building services fixtures in terms of reflected ceiling plans (indicating lights, grilles and architectural features).

In specialist projects or where the task requirements are challenging, the client may employ specialist designs from the other members of the design team, including structural engineers, landscape architects, fire engineers and specialist planning consultants. In these instances, the architect is normally the design team leader. Sometimes, where the architect is part of a large organisation, specialist engineers in these roles may be employed by that organisation.

Project manager

A project manager (PM) is usually a building professional in private practice, with specialist project management qualifications and experience. Many PMs take a course such as PRINCE2, which is a qualification recognised by Government departments for the project management of public projects. The PM's role is to ensure that the project is delivered on time, on budget and to all the correct standards.

The client normally appoints the PM to run projects and manage the works. The PM acts as the project leader and, in certain contracts, as the contract administrator responsible for the overall contract. The PM issues contract instructions and generally deals with the contractor and other professionals working for the client. The PM normally **chairs** construction project meetings between the client team and the contractor's construction team, reporting directly to the client or **funding body**.

On larger projects, there may also be a PM on the contractor's construction team. This person is normally site-based and has ultimate responsibility for the success or failure of the contractor's construction project. In order to ensure delivery, the contractor's PM is the leader of a **multi-disciplinary** site management team. In this role, the PM normally meets with the client on matters of contract, programme and cost.

The contractor's PM is usually focused on delivery of the project on time. In order to ensure a profitable outcome, the contractor's PM usually directs the contractor's team to ensure that the critical points on the construction programme are met. Often this involves putting pressure on contractors and sub-contractors to make up lost time, by requesting more labour or longer working times or, on the client, for information to ensure **work packages** are not delayed.

Clerk of works

The clerk of works (CoW) is another representative of the client. The CoW is responsible for ensuring that work is carried out on schedule and that the quality of the works meets both the specification and industry standards, including British Standards (BS) and European Normative standards (ENs).

INDUSTRY TIP

You can access these bodies via the following links.
- British Standards (BS): www.bsigroup.com/en-GB
- European Normative standards (ENs): www.cencenelec.eu/standards/DefEN/Pages/default.aspx

---

The CoW keeps a record of weather conditions and the labour resources on site to ensure that the work is carried out on schedule. Another responsibility is to ensure that adequate time and resources have been applied to particular tasks and that any shortcuts have not impacted on quality, commissioning or **warranties**. The CoW also issues site correction notifications to the contractor to indicate areas of work that do not meet the contract.

The CoW is not normally given the power by the client or by the terms of the contract to vary or extend the works. As a result, the CoW does not usually have any contractual connection to the trades on site though the CoW's role does include liaison and discussion with the different trade contractors with

KEY TERMS

Chair: the person who controls a meeting, and ensures that all the topics which need to be discussed (the agenda) are covered. The chair will also act to resolve disputes.

Funding body: an organisation or charity that provides money for a particular purpose. Sometimes construction projects may be funded, or partly funded, by a charity or organisation with an interest in providing a facility. For example, if new changing rooms were being constructed for a village football club, the funding body may be the Football Foundation.

KEY TERM

Multi-disciplinary: carrying out several different roles.

KEY TERMS

Work package: a 'collection' of work associated with one product on a project – for example, all processes associated with the design, manufacture and installation of the windows or of the sprinkler installation.

Warranty: a written promise to repair or replace any defects within a stated period of time. This could be one year after the completion of the contract. Sometimes, longer warranty periods are agreed.

respect to the quality of works and the number of workers. The CoW puts this information into reports to the client or the client's agent (architect or project manager) and the contractor.

Construction manager

The construction manager (CM) is part of the contractor's on-site team and reports directly to the contractor's PM.

The construction manager is responsible for ensuring that the construction element of the project runs to time and to the relevant quality standards, to ensure minimal reworking. The construction manager is responsible for all site activities, including inductions and site safety, and ensuring that contractor and sub-contractor **toolbox talks** take place. The sub-contractor has the most interaction with the construction manager as they spend a considerable amount of time out on the site.

The construction manager has normally been trained in general construction, either from a surveying, trade or construction background.

KEY TERM

Toolbox talk: a method of communicating safety issues (as a supplement to an induction) or of giving continuous training on particular techniques or methods.

▲ Figure 5.2 When people work together the work progresses better

Quantity surveyor

The role of quantity surveyor (QS) can exist in the client team and also in the construction team. In each team, the QS role involves similar responsibilities but from different perspectives.

The client's QS is normally a professional from a private practice of chartered quantity surveyors, which are generally regulated by the Royal Institute of Chartered Surveyors (RICS). The QS's role is to put together estimates by measuring and **quantifying** the designs from the client's design team prior to **tender**. The QS advises the client on budget allocation, selecting the method of

KEY TERMS

Quantify: estimate guideline costs.

Tender: an offer to carry out work, supply goods or buy land, shares or any form of asset for a fixed price, usually in competition with others.

procurement and **contract** documentation. In certain types of contract, they are also responsible for producing a **bill of quantity**. Once a project has been tendered, the QS reviews the priced tenders of the bidding contractors, checking the commercial terms and conditions and financial accuracy of each tender to ensure that the contractor has submitted an appropriate price.

The contractor's QS is normally employed within the contractor's business. Their role is to assess the tender documentation, bills of quantity, and so on, and check the costs that the estimating department has provided to ensure that the bid is competitive enough but still gives the company adequate profit.

The QS also checks the commercial terms of the contract. They review the contract, referring the documents to company lawyers if the terms are non-standard or appear too difficult to accept outright. This can result in negotiations over contract clauses between the contractor and the client, which are normally settled with a compromise that is acceptable to both parties.

Consulting engineers

Consulting engineers are highly qualified, specialist personnel required on complex projects. There are many different types of consulting engineers, who work on the structure or specific services.

Structural engineer

Structural engineers are professionally qualified (chartered) in the field of structural engineering and usually part of the client's design team.

The structural engineer is responsible for the structural stability of structures and ground works within the contract. Very early in the initial design stage, the structural engineer determines the **substructure** of the building. This involves the designer interpreting ground conditions to determine whether traditional foundations or piles are to be used to support the structure in the ground.

The structural engineer also designs the building support system, depending on the type and use of the building, as well as the requirements of the client. This work involves sizing and locating structural beams and determining the strength of walls and floors. On projects with no civil engineer, the structural engineer will also design features such as car parks, roads, drainage, water run-offs and environment protection measures.

Building services engineer

The building services engineer is also a professionally qualified member of the client's design team. Building services engineers are often **incorporated engineers**, although some are **chartered engineers**.

INDUSTRY TIP

Electrical apprentices who successfully complete the City and Guilds 5357, AM2 test and therefore complete their full apprenticeship are now eligible for EngTech recognition from the IET through a fast-track approval scheme specifically designed for the current apprenticeship.

KEY TERMS

Contract: a written agreement between two or more parties for doing something. A contract between a client and contractor is the agreement to carry out specific work for an agreed price.

Bill of quantity: a contract document comprising a list of materials required for the works and their estimated quantities.

Substructure: the type of foundations and support systems for the building, such as steel frames or pillars. In some modern buildings, the substructure is often hidden but some architects make features of the substructure.

KEY TERMS

Incorporated engineer: a specialist (also called an engineering technologist) who implements existing and emerging (new) technology within a particular field of engineering, entitled to use the title IEng. Incorporated Engineers are registered with the Engineering Council, which is the British regulatory body for engineers. It acts as proof that a particular level of training and experience has been met.

Chartered engineer: an engineer with professional competencies through training and experience, also registered with the Engineering Council. The title chartered engineer (CEng) is protected by civil law.

The building services engineer is responsible for the mechanical and electrical services in a project, often referred to as MEP.

- Mechanical services include heating, ventilation and air-conditioning systems.
- Electrical services include lighting, power and distribution, as well as telecommunications, fire alarms, alarm and communication systems, intercoms and disabled refuge communication systems.
- The 'P' in MEP stands for plumbing and public health engineering, including hot and cold pipework, above ground drainage and, in certain contracts, private underground systems prior to input into the public sewer.

The building services engineer designs any rainwater systems that run internally, whereas the architect designs the external rainwater systems (pipes on the exterior from gutter to ground level drain).

Building surveyor

Chartered building surveyors are building professionals regulated by the Royal Institute of Chartered Surveyors (RICS) and often employed on the client's team.

They are usually responsible for assessing existing buildings or renovation schemes, including valuation of property for lenders, scheduling of building defects, refurbishments and undertaking different types of building surveys.

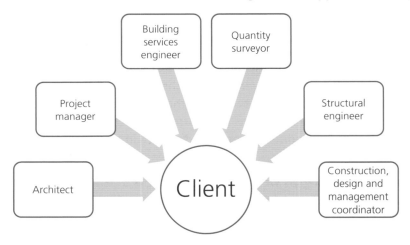

▲ Figure 5.3 Client–contractor relationship

Land surveyor

Chartered land surveyors are also regulated by the Royal Institute of Chartered Surveyors (RICS). Their work is generally for the client and they would normally be part of the client's design team.

They deal with the measurement of land, which may involve setting out structures on the land in accordance with the architect's drawings. They also often produce details of the areas surveyed, so that an architect can make an accurate design for a structure. In addition, they are often employed to determine the boundaries to properties, give valuations, and so on.

INDUSTRY TIP

More information regarding RICS can be found at www.rics.org/uk

ACTIVITY

Review the websites of different types of construction-related consulting engineers.

For example, you could start with the Association for Consultancy and Engineering (ACE) website, at: www.acenet.co.uk

Follow the links to visit the websites of various consulting engineers. Identify the different types of consultancy that operate on a local and national level. Look at how they differentiate themselves from the competition.

▲ Figure 5.4 Surveyors' equipment

Estimator

The estimator is part of the contractor's team and responsible for preparing the estimated price for the work within the **tender package** prepared by the client's quantity surveyor. This estimate has to be carefully prepared to ensure that the best prices are obtained for materials and labour from sub-contractors. The estimator is usually a former tradesperson, who has moved into the estimating team.

Buyer

As the name suggests, once a contract has been won, the buyer is responsible for purchasing all the materials from suppliers, within the costs stated in the tender by the estimator. Effective tendering involves offering the most economical prices the contractor can afford to offer. For this, the buyer must purchase or obtain the materials at the correct price, terms and conditions after the contract has been awarded. The buyer also has to ensure that the materials are delivered on time.

Contracts manager

The contracts manager is often a company director, or reports directly to a director, and works on the contractor team. Contracts managers are normally responsible for several construction projects so they are not normally part of the on-site team. As regular visitors to the site, their role is to liaise with the contractor's on-site management team, including the project manager and construction manager.

KEY TERM

Tender package: all the information a contractor needs to form a tender. A tender is a formal offer to carry out work for a cost. The client will send tender packages to many contractors and award the contract to the contractor who provides the best tender. The choice is not always based on price – it could be based on service provision or promised completion dates.

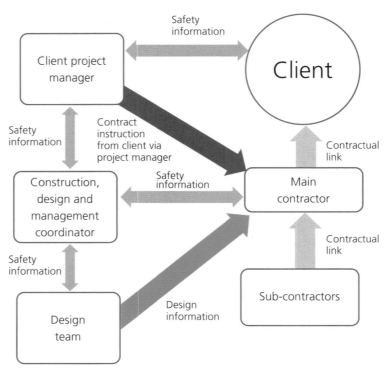

▲ Figure 5.5 Contractor and client team relationship

Key roles reporting to the site management team

Sub-contractors

Each sub-contractor is responsible for their package of work or construction discipline, such as substructure or ground work, joinery, brickwork, window installation, electrical, plumbing, plastering or decorating. They have to work safely, to the main contractor's construction programme, producing work that meets the specification, in a timely manner and accommodates the contractor and all the other sub-contractors working on site. In the majority of construction contracts, the sub-contractor is a 'domestic sub-contractor', who only works for a main contractor. Therefore, the sub-contractor can only be instructed by the contractor as they have no contractual link with the client or the client's team.

Site supervisor

The site supervisor can also be known as the general foreman. On smaller projects, they are main point of contact for all trades. On larger projects, this role is generally carried out by the construction manager, to whom the site supervisor reports. In some organisations, site supervisors are known as assistant construction managers.

Trade supervisor

The trade supervisor is normally the foreman or lead person for each sub-contractor. This person usually controls the individual trade operatives on site, within the general rules and requirements of the main contractor on site. They will usually be an experienced tradesperson with many years of site experience, able to interpret the requirements of their own off-site management team and integrate those requirements into the wider on-site requirements to ensure smooth running between the different trade contractors.

Trades

Each of the different trades and disciplines needs to liaise with the on-site management team and others. The following are examples of different construction trades:

- *Bricklayer* – builds brick and blockwork walls with different finishes.
- *Joiner* – carries out various tasks, including constructing stud walls (usually internal timber walls) and installing doors, frames, skirting boards, cupboards and other fitted fixtures.
- *Plasterer* – applies wet plaster and finishes or provides **dry-lined walls** with specialist taped joints or minimal skimmed finish, using a small amount of finishing plaster.
- *Tiler* – a finishing trade responsible for tiled finishes.
- *Electrician* – the trade contractor responsible for the installation of all electrical wiring including any wiring containment systems or specialist systems such as door entry security or fire alarms. On smaller projects, the electrician installs all specialist services or may employ a specialist installer for systems such as the intruder alarm, to meet any contractual or specification requirements. On larger projects, the electrical contractor is usually responsible for the low-voltage wiring and for providing wiring containment systems for the specialist installers of, for example, fire and security systems. These specialists are normally sub-contractors of the electrical contractor, who is responsible for managing them as well as their own staff.

KEY TERM

Dry-lined wall: an inner wall that is finished with plasterboard rather than a traditional wet plaster mix.

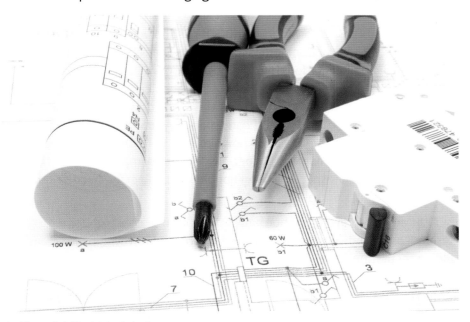

▲ Figure 5.6 Electrician's tools

- *Heating and ventilation (H&V) engineer* – carries out plumbing and heating installation works, normally in steel. (A plumber normally carries out installations in copper.)
- *Air-conditioning and refrigeration engineer* – closely linked to the electrical trades, although carrying out some pipework installation. Refrigeration engineers have to be aware of refrigeration legislation as certain gases are banned in the UK.
- *Gas fitter* – specialist qualified to work on gas and combustion equipment installations such as boilers. Gas fitters are registered with Gas Safe, which is a competent person scheme (CPS) that ensures registered tradesmen are qualified and competent to work on gas systems and boilers, within the legal requirements.
- *Decorator* – finishing trade responsible for decorated finishes to all surfaces, including woodwork.
- *Ground worker* – manual worker associated with levelling ground, excavations for substructure and foundation works, or providing trenches for pipes or incoming services. Ground workers are usually supported by a wide range of earth-moving equipment and excavation tools from a simple pick and shovel to complex hydraulic machines such as the standard backhoe (digger).

Key roles of site visitors

Building Control inspector

Every construction site is required to meet the requirements of the Building Act 1984 (as amended). The policing of this is the responsibility of the Building Control department of the relevant local authority. The planning consent may also specify additional requirements that have to be met, such as provisions for fire prevention or noise reduction.

The Building Control inspector or officer (BCO) has the power to give or deny approval of works, construction details and layouts that affect the building, in line with the Building Regulations. Normally, the BCO will refer to a particular approved Code of Practice (ACoP) relating to the technical area concerned. Each area of specialism is also referred to as the specific part in the Building Regulations that relates to it, with the corresponding approved documents. For example, Approved Document P (also known as Part P) relates to the electrical installation. An increasing number of specialist private companies now offer the services of an approved inspector. When appointed, this person has the same power as the Building Control officer.

The BCO or approved inspector normally comes from a construction background, as they deal mostly with building works. However, other areas of building regulation are classified as non-construction, as with Part P, which relates to the electrical installation. If the BCO does not have the relevant expertise, they must rely on the expertise of an approved inspector in that area.

INDUSTRY TIP

Access further information on Gas Safe at: www.gassaferegister.co.uk

INDUSTRY TIP

You can access the Building Act 1984 (as amended) at: www.legislation.gov.uk/ukpga/1984/55

You can access Approved Document P at: www.gov.uk/government/uploads/system/uploads/attachment_data/file/441872/BR_PDF_AD_P_2013.pdf

INDUSTRY TIP

The Building Control officer does not usually have electrical knowledge or experience and must therefore rely on the electrician to provide accurate and reliable information. Competent persons schemes (CPSs) address this shortfall by assessing each contractor's competence in advance of registering jobs with Building Control. More information relating to CPSs can be found at: www.gov.uk/building-regulations-competent-person-schemes

Water inspector

Every water supplier has a responsibility to supply **wholesome** water and to ensure that numerous checks are made on the water supply. No matter how many checks are made on the public supply system, incorrectly designed and installed private systems and connections put the public water supply at risk.

Since the introduction of the Water Supply (Water Fittings) Regulations 1999, water regulations inspectors employed by the water companies have statutory powers to impose improvement orders on the users of water supplies, who can face withdrawal of the public water supply if they do not comply. Water company operations are monitored by the Drinking Water Inspectorate (DWI) water inspectors, whose duties include ensuring that water continues to be wholesome.

HSE inspector

A HSE inspector is an officer of the Health and Safety Executive (HSE), a Government department. HSE inspectors have the right to enter any workplace without giving notice. They usually do make an appointment to visit, unless the element of surprise is thought necessary.

On a routine visit, an inspector would expect to look at the workplace, work activities and methods and procedures used in the management of health and safety, as well as checking compliance with health and safety law specific to that workplace. The inspector is **empowered** to talk to employees and their representatives and to take photographs and samples. An inspector may also call for a specific purpose, for example to follow up a complaint or incident.

If the inspector finds that safety regulations have been broken, depending on the level of severity, the inspector can carry out the following courses of action.

KEY TERM

Wholesome: term used in law to indicate water that is suitable for drinking.

INDUSTRY TIP

You can access the Water Supply (Water Fittings) Regulations 1999 at: www.legislation.gov.uk/uksi/1999/1148/contents/made

KEY TERM

Empowered: having the authority.

▲ Figure 5.7 Working safely is critical

- *Informal warning* – where the breach is minor, the inspector can explain best practice and legal requirements to the **duty holder** (usually the employer) on site, following up with written advice.
- *Improvement notice* – where the breach is more serious, this notice will say what has to be done, why and when the remedial action has to be completed by. The inspector can take further legal action if the notice is not complied with within the time period specified. The duty holder has the right of appeal to an industrial tribunal (an industry body with the power to settle disputes).
- *Prohibition notice* – where there is a risk of serious injury, the inspector is empowered to stop the activity immediately or after a determined period. Once this notice is applied, the activity cannot be resumed until remedial action has taken place. The duty holder has the right of appeal to an industrial tribunal.
- *Prosecution* – in certain circumstances, the inspector may consider prosecution necessary in order to punish offenders and deter other potential offenders. In some cases, unlimited fines and imprisonment may be imposed by the higher courts.

Electrical services inspector

The title 'electrical services inspector' is rarely used in the UK as the majority of installations are self-certified by the installing contractor. The inspector is normally Level 3 qualified, with specific qualifications in the inspection and testing of electrical installations.

Building services engineer

This is normally a professionally qualified individual who may have been a tradesperson prior to taking advanced qualifications and training. This individual will have qualifications to carry out, for example, estimating, contract management, design or the role of specialist **authorising engineer**. Most building services engineers have undergone a training programme that satisfies the requirements of the Engineering Council, the regulatory body for engineers and technicians in the UK.

2 INFORMATION SOURCES IN THE BUILDING SERVICES INDUSTRY

Legislation

Every industry has to comply with legislation that is specific to that industry. The construction industry is no exception: there are specific construction safety laws as well as the Construction (Design and Management) Regulations, which put duties on all stakeholders. However, much of the health and safety and employment legislation applies to all industries.

We looked at the Health and Safety legislation in great depth in Chapter 1, but there are many other regulations that influence everyday working in the electrical industry, which we must be aware of.

Employment Rights Act

The Employment Rights Act 1996 covers, but is not limited to, matters such as:

- employment particulars
- wages and payment
- disclosure rights
- time off
- reasons for dismissal
- redundancy
- employer insolvency.

▲ Figure 5.8 Employees' rights

INDUSTRY TIP

You can access The Employment Rights Act 1996 at: www.legislation.gov.uk/ukpga/1996/18/contents

Equality Act

The Equality Act 2010 came into force in October 2010 and replaces all previous legislation on equality. It sets out to ensure that everyone has fair and equal opportunity in the workplace in respect of the following protected characteristics (PCs):

- age
- pregnancy and maternity
- disability
- race
- religion or belief

INDUSTRY TIP

You can access The Equality Act 2010 at: www.legislation.gov.uk/ukpga/2010/15/contents

INDUSTRY TIP

You can access The Data Protection Act 1998 at: www.legislation.gov.uk/ukpga/1998/29/contents

ACTIVITY

Make a list of other documents or details of client's information that may be held and, for each, decide how it should be stored, used or disposed of.

KEY TERM

In breach of: breaking or failing to comply with.

ACTIVITY

Discuss of all the Non-statutory Regulations, Codes of Practice and Guides that an electrician would reference when designing, installing or testing an electrical installation.

- sex
- gender reassignment
- sexual orientation
- marriage and civil partnership.

Data Protection Act

The Data Protection Act came into force in 1998 and sets out how the data we hold for all clients, personal or otherwise, is stored, used, disposed of or shared. It means we have to be particularly careful on what we may leave around as we have lots of data for clients that should never be shared without that person's permission. Documents we use that contain client's information could include:

- job sheets
- test certificates
- correspondences such as emails and letters
- notes taken on notepads
- photographs.

Care must be taken at all times about keeping such information confidential.

Supporting documentation

There are many guidance and reference documents available from the HSE website to ensure that best practice is followed and that employers and all workers understand the basics of health and safety law. In addition, technical support and commentary is provided by manufacturers, professional institutions and experts in the relevant field. Figure 5.9 shows that statutory legislation (Acts and Regulations) are the most important type of documentation, as they represent legal requirements.

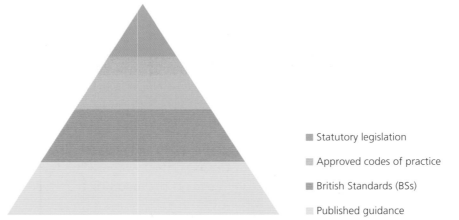

- Statutory legislation
- Approved codes of practice
- British Standards (BSs)
- Published guidance

▲ Figure 5.9 Hierarchy of work related legislation and guidance

However, the legislation is not always detailed or industry specific. The Approved Codes of Practice, British Standards and published Guidance give more detail and approved methods for achieving statutory compliance. These Non-statutory Regulations can often be stricter than the legislation. Being **in breach of** Non-statutory Regulations could result in the industry losing faith in your ability to perform, losing you customers and future tenders.

Information used in the workplace

The follow examples of information can be used in the workplace.

Job specifications

Job specifications are compiled in a variety of ways depending on the size, complexity and contractual arrangement of a job. Generally, the specification for a project includes the following sections.

Preliminary information

This first section includes details of the client, the contract administrator, the form of contract governing the project (e.g. JCT Minor Works 2011), and details such as the anticipated overall contract period.

In order to give some indication of all the different trades involved, a construction programme may also be referred to in the preliminaries. In large multi-trade projects the preliminaries also include the cost of site set-up and maintenance of items such as welfare and storage cabins, given as a cost per day or per week.

Scope of works

The preliminaries are normally followed by a scope of works. This section determines the extents of the specification (i.e. the extent of what the contractor has to quote a price for). If the contract is agreed, the scope of works will become the extent of works to be carried out.

For example, a simplified scope of works may be:

> 'The scope of works for this project is to install electrical systems within a new two-storey extension to an existing four-bedroom house. The works will include all wiring and containment systems for power, lighting and distribution. The works will not include the intruder alarm or fire detection systems.'

Although this scope of works immediately eliminates the fire and intruder alarm detection, there will be interfaces between systems that need to be defined. These are generally included in the particular specification and drawings that follow this section.

Specification of works

This section describes the particular works to be carried out. It gives as much detail as possible of what the designer and client require for that particular package of works. It may consist of simple descriptions or technical requirements laid down for the contractor (tenderer) to price and eventually work to.

Materials and workmanship

Also known as the standard specification, this section is usually the same for all projects. It comprises a list of requirements in terms of workmanship and selection of material grades and quality that applies to each job, along with any British Standards or industry codes of practice that may be applicable.

Schedules

This section includes drawing issue sheets indicating which drawings have been issued as part of the tender or contract. It also contains other schedules such as equipment schedules containing manufacturer selections, performance requirements, sizes and weights, and distribution board schedules.

Schedule of costs

This is usually the final document in the specification. The contractor (tenderer) has to complete it, specifying the cost of sections of work. Day work rates also have to be specified in case unmeasured or urgent works need to be carried out, which are not already contracted and cannot reasonably be priced or measured because of time pressures.

Plans or drawings

Most construction projects communicate contractual needs and design intent on drawings. These are provided in many formats, described in detail in Chapter 3, pages 156–164, and give different levels of detail depending on the purpose of the drawing and the technical requirements it needs to show.

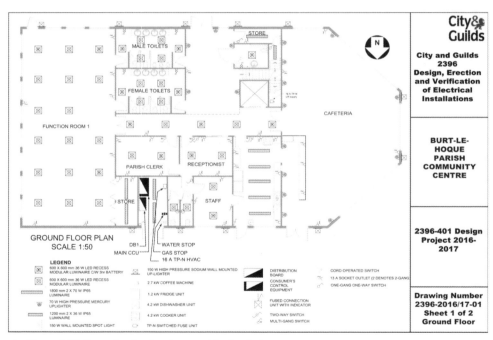

▲ Figure 5.10 Example of a drawing used for tendering

Work programmes

A work programme is initially drawn up by the client or the client's quantity surveyor to help determine the construction period and whether the scheme is feasible and affordable in the client's timeframe.

Once a contract has been awarded, the responsibility for the work programme lies with the contractor. On larger projects and in larger construction companies, planners are dedicated to producing large construction programmes.

Specialist software collates the detailed stages of hundreds of activities, individual start and completion dates, drying and curing times, materials procurement and so on, to find the most cost-effective timeline for achieving the construction, which usually involves the shortest period on site.

Time on site has a direct impact on the contractor's costs – quite literally, time is money. So it is important to determine when the number of personnel on site can be reduced and the welfare, cabins and storage reduced accordingly.

This lean and efficient programme includes key events and activities that are critical to successful completion. These events and activities make up the **critical path** and must be completed before the next set of activities can start. For example, a building needs to be watertight before wiring can be started, which makes being watertight a critical point on the critical path.

KEY TERM

Critical path: the sequence of key events and activities that determines the minimum time needed for a process such as building a construction project.

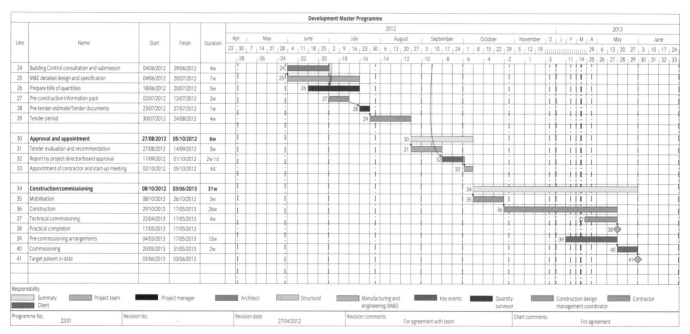

▲ Figure 5.11 Example of a work programme

Once a critical path analysis has been carried out, the project manager may decide there is too much risk in a particular critical point and ask the planner to re-plan activities to remove it from the critical path. This will involve building contingency time around that point, which will result in a slightly longer programme, but adds an element of realism so that if delays occur around the identified point they do not affect the critical path.

Delivery notes

Delivery notes are the documents that are attached to or delivered with any delivery. It is essential that delivery notes are checked against the delivered **consignments**, wherever possible at the time of delivery. This identifies if the wrong consignment has been delivered and if items are missing. The delivery note normally requires a signature to confirm the number of items delivered.

If a problem with the delivery is discovered later, responsibility for the problem can be difficult and costly to prove. However, missing components will also

KEY TERM

Consignment: a batch of goods.

affect performance on site. The delay in supply of even simple products such as bolt fixings to secure the structure can affect the timing on a project badly.

Delivery notes should be checked off and processed through the supervisor. They may then be sent to the administrators for processing or direct to the contractor's quantity surveyor for costing against the project.

Time sheets

Time sheets are a simple but very effective tool in assigning work costs to particular jobs or budgets. They indicate where labour costs have been expended, including where additional works have been carried out and need charging to the client, which is of vital importance to profit levels.

Time sheets that break work down into specific tasks can also be used by managers and estimators to indicate areas where the estimate was totally inaccurate. This information can then be used when pricing future works.

Once submitted, a time sheet is a legal document stating the hours worked on a task or project.

INDUSTRY TIP

It is important to fill in time sheets accurately as they are a way of determining how a client should be charged for labour and what payment is owed to you.

SMART ELECTRICAL

Industrial, Domestic and Commercial Contracts
Specialist in Domestic Rewires

Operative Name _George Jones_   Contract Name _Various_

Foreman Name _Alfie Harris_   Week ending _28/03/2013_

MONDAY

Job No	Address	Description of Works	Hrs
381	Smith Street	Assist JD rewire property	8

TUESDAY

Job No	Address	Description of Works	Hrs
381	Smith Street	Assist JD rewire property	8

WEDNESDAY

Job No	Address	Description of Works	Hrs
381	Smith Street	Assist JD rewire property	1
366	Water's Reach Development	Assist pulling in steel wire armoured cable	8

THURSDAY

Job No	Address	Description of Works	Hrs
366	Water's Reach Development	Fixing panels and terminating cables	9

FRIDAY

Job No	Address	Description of Works	Hrs
402	Local High School	Test & inspection of CDT rooms completed by J Hansen	6

Foreman Signature _A J Harris_   Week Total _40_

Figure 5.12 A sample timesheet ▶

Policy documents

Organisations, such as electrical contracting companies, will hold a range of policy documents setting out procedures that they and employees must follow. These could range from health and safety policies to quality control or employees' rights and disciplinary procedures.

The purpose of a policy is to provide a framework that can be followed so people working for the organisation can get on with the work they need to without having to constantly check or discuss the best way forward. Policies are written to ensure the organisation:

- meets legal requirements
- meets regulatory requirements
- improves quality
- promotes best practice.

In some situations, new employees are issued with company policies and have to sign them as a way of agreeing to work to those policies. If an organisation changes their policy, the employee is trained on those new policies and once again, signs them to agree to follow them.

Information given to customers

There is a range of information the client receives before, during and at the handover stage of a contract. Knowing the difference is essential in order to meet legal requirements as well as maintaining good relationships.

Quotations

Quotations are generally offers to provide goods and/or services for specific amounts of money. They may have general terms and conditions attached, which are normally printed on the reverse of a quotation. Typical terms and conditions are that the quotation is only valid for 30 days and that the title (ownership) of any fitted goods does not pass to the recipient until full payment has been made.

A quotation may also include specific terms, which define the detail of what has been quoted for and on what basis. An example would be to rewire a specific property as indicated on a specific drawing for a specific sum of money excluding VAT and that the property will be vacant with all fixings and furniture removed by the owner prior to work commencing.

If all the terms and conditions are met and the requirements do not change from those specified in the drawing, the quoted price will stand regardless of how long the contractor takes to carry out the work or how profitable the work is. If there is a change in the terms and conditions, the price should be adjusted and agreed accordingly.

Estimates

Estimates are generally provided when the job is not properly measurable due to time or physical restrictions and therefore a quotation cannot be given. The work is undertaken but may involve unforeseen difficulties that impact on the materials needed, time taken and costs incurred. The final invoice may therefore be different from the initial estimate. However, it is usual to give an explanation for why the price has changed.

▲ Figure 5.13 Invoices

Budget estimates

These are estimates that a client may request in order to plan a project or the viability of a project before any formal requests for quotations are made. It enables a client to either set aside a budget for work or decide if the work is necessary based on an estimated cost.

Invoices, applications for payment and statements

Invoices are formal requests for payment from one party to another. An invoice will normally contain the name and address of the customer, the trading name and office address of the contractor and any customer order or reference numbers. The registered office address and VAT registration number are given for limited companies. The invoice will also describe the works and give a breakdown of any non-fixed fee items.

Business invoices usually request payment within 30 days. In order to avoid delays in payment, it is important to ensure that the invoice is correctly detailed. Any areas of additional works should already have been agreed and be clearly identified on the invoice.

The contractor raising an invoice owes any tax or VAT to the Government from the date of the invoice regardless of the date of payment. In order to avoid paying out tax and VAT long before payment, especially where payment terms are 60 or 90 days, the contractor can make an *application for payment*. This is identical to an invoice, giving the expected tax and VAT payment date but not the VAT number. The client sends the application for payment through the payment processing system and then requires a tax invoice to be provided before or on the date of payment.

A *statement* is a notification of outstanding invoices or transactions. Where a contractor and client have a long-standing relationship, it is possible for several invoices to be outstanding over a payment period. A statement is therefore issued to act as a reminder and reduce the possibility of confusion about what has and has not been paid. Alternatively, where there is a credit account arrangement, the contractor will build up a number of transactions before issuing an invoice for payment. During the billing period, the unpaid transactions can be itemised on a statement.

Statutory cancellation rights

Cancellation rights give the consumer the rights, within a specified period of time, to cancel a contract without liability or having to take legal action and to get back any money paid. In the UK, the Cancellation of Contracts made in a Consumer's Home or Place of Work etc. Regulations 2008 detail what is and is not excepted. As a general guide, the following are typical areas where statutory cancellation rights exist:

- where the consumer cannot examine the purchase before they buy (i.e. distance selling by mail order, phone or internet)
- where the trader can unduly pressurise customers to buy
- where the consumer has little opportunity to compare different traders or read the 'small print'.

The Citizens Advice Bureau provides advice and up-to-date publications on consumers' cancellation rights, the different cancellation periods for different sectors and how cancellation must be communicated.

Handover information

Handover information needs to be delivered in order to achieve **practical completion**. As buildings have become more sophisticated, operation and maintenance documentation, for example, has become even more important. Handover information should include:

- a design philosophy and statement to say how the building has been designed and how it is to be operated, often now part of the building logbook
- Building Regulations tests and approvals, including the sign-off documents from the Building Control officer or approved inspector
- building leakage test certificates from construction
- the energy performance certification and registration information
- electrical test information
- lighting compliance certification (e.g. compliance with **CIBSE** Lighting Guide LG7)
- emergency lighting test certification
- fire alarm certification
- emergency voice communication (EVC) or disabled refuge system certificate
- construction plans at a scale of 1:50 or 1:100 as appropriate

KEY TERMS

Practical completion: the point at which a construction project is virtually finished, when the last percentage of monies are paid by the client, the responsibility for insuring the construction transfers to the client and the architect or contract administrator signs the certificate of practical completion. At this point there will still be a few minor snags to sort out (e.g. scratches in paintwork) or insignificant items to be completed.

CIBSE: the Chartered Institute of Building Services Engineers. Access more information at: www.cibse.org

- building envelope (fabric of the building) details
- structural plans with floor loadings, any specialist support information for plant, etc.
- mechanical and electrical equipment layouts at a scale of 1:50 or 1:100 as appropriate
- plant and specialist area layouts at a scale of 1:50, or even 1:20, depending on complexity of the systems
- additional information relating to the cleaning, maintenance and deconstruction of any of the systems or components and any disposal requirements at the time of writing the handover information.

It is most important that the documents are up to date and specific and relevant to the building or project being handed over.

Company working policies and procedures

Companies set out working policies and procedures so that all workers can expect to be treated fairly and with respect, know how they can put right a problem should one arise and understand what measures or sanctions will be imposed should a breach of policies and procedures occur.

Working policies and procedures set out a company's expectations on matters such as:

- *behaviour* – language and attitude should promote safety and client relationships, for example
- *timekeeping* – poor timekeeping by one person can affect many members of staff and ultimately affect costs and completion targets
- *dress code* – should promote a professional image.

These working policies become part of the terms and conditions of employment. They are often provided in a staff handbook, although such a handbook is not required by law. An employee accepts the terms and conditions of employment when they sign a contract of employment or attend work without challenging any of the terms of the contract.

Contract of employment

A contract of employment is a legally binding agreement between employer and employee. It can be a verbal or written contract, including various terms and conditions. Any changes to the terms of employment will probably not be notified in a new contract of employment, but in the staff handbook, in a letter or on the company notice board.

If the contract is verbal, the employer must give a statement of the main terms and conditions within two months of the start of employment, under the provisions of the Employment Rights Act 1996.

The statement would include:

- holidays entitlements
- scale of pay

- basic working hours
- pensions and pension schemes
- notice of rights.

Health and safety policy

All companies with five or more employees are required by law to set out their health and safety policy in writing. Related issues detailed in this policy, or separately, include:

- bribery
- bullying and harassment
- disciplinary and grievance procedures
- equality and diversity
- maternity, paternity and adoption
- pay
- redundancy.

It is a legal requirement for employers to set out disciplinary rules and grievance procedures in writing and make sure that every employee knows about them and the available courses of action if they are unhappy with a disciplinary decision. Although related, the bullying and harassment policy is often a separate document because of the many issues and sensitivities that have to be considered.

Limits to personnel responsibility

Within company policy there is likely to be a statement of the **limit of authority** for different individuals and groups within the organisation. Limits of authority will vary depending on the responsibility and experience of the employee and the duties expected of them by the company.

Ultimately the employer is responsible for the acts or omissions of their employees, including any financial or legal repercussions. On a project, the employer has a duty to ensure that all personnel are competent and capable of completing the tasks they are required to carry out in their role. As a result, it is in the employer's interests to set limits of authority appropriately.

Apprentices or those undertaking work experience, under the age of 18, are classed as 'young workers' who will not be permitted to make decisions that are beyond their capabilities. Their authority is limited to common sense, basic health and safely matters, and those areas where they have been adequately trained.

A supervisor, with Level 3 qualifications and more training and experience than trainee staff, would have considerably more authority. Supervisors are responsible for ensuring that projects run smoothly, for developing risk assessments and method statements, and for ensuring that agreed assessments and statements are in place and adhered to. Supervisors also have financial responsibilities in ordering materials and requesting labour to respond to work programmes.

▲ Figure 5.14 Signing a contract

KEY TERM

Limit of authority: many decisions and actions need to be made on a daily basis, but they must be made by the right person. Dangerous or costly situations could arise if somebody made a key decision without the right knowledge or experience to take that decision. If somebody does this, they are exceeding their limit of authority.

INDUSTRY TIP

More information can be found on the HSE website relating to young workers at: www.hse.gov.uk/youngpeople/index.htm

3 COMMUNICATION METHODS IN WORK SITUATIONS

Communication in the workplace is essential for the safety and wellbeing of staff and the **profitability** of a job. Various methods of oral and written communication are used in different circumstances.

Oral communication

Oral communication (also known as verbal communication) is essentially speech. Oral communication can take many forms on a job or on site, such as a spoken instruction from a colleague, pointing out a potential improvement, or a verbal request on site to stop work on the grounds of health and safety. Oral communication with clients is also crucial make sure the job is carried out as expected. This will involve questions and descriptions. Oral communication may not always be face to face – it could be made via telephone, a site radio, or a public address system (e.g. the client asking for a particular piece of work to be started immediately).

Positive customer relationships must be maintained at all times as these are an essential part of business development and **customer retention**. Effective communication with customers is a key part of this – the customer should feel confident in your capabilities, that you understand their wishes, and are well-informed about plans and the progress with or any delays to the job.

Oral communication is less formal than written communication. Although oral communication can be **legally binding**, it is difficult to record and prove, and it can easily become confused and lead to mistakes. Therefore, most important oral communications are followed up in writing, usually referring to the date and time of the original communication.

ACTIVITY

While you are at your training centre, imagine you are a client who wants some items of equipment installed at various heights within the building or room you are in. This could be things like socket outlets at 500 mm above the floor, a particular type of light located in a particular location, and so on. Make a list containing around ten items.

Now orally communicate these requirements to a friend or colleague at the centre – but they must not write them down. Several minutes later, ask the friend or colleague to repeat the requirements back to you. How much did they remember?

Written communication

Written communication is any form of non-verbal communication in a written form. Written communication can come in many forms and using many different mediums (channels of communication), including letters, emails and SMS text messages.

Most companies now communicate by email, whether on a **formal** or **informal** basis. This is a highly effective method of communicating clear information, with the option of attaching pictures and drawings. Emails are useful to document agreements and plans between multiple **recipients** – this means lots of the key stakeholders, regardless of where they are in the world, can be simultaneously kept up-to-date with general progress reports or any problems and delays to work. Large attachments, such as plans or images can also be sent.

Letters are considered the most formal method of communication, due to the effort involved in typing, printing and posting them. They are used in the most formal situations, such as offering an appointment, terminating a contract or following up a less formal email.

Written communication is a reflection on your business and once sent, cannot be erased. Time and care should be taken to produce clear, quality and accurate written communication – a badly worded or misleading email or letter can damage years of business building.

Methods of communication for people with different needs

When choosing the best methods of communication, people's specific needs should be considered carefully, to ensure the work is completed as agreed.

Some clients or businesses you communicate with may not have English as their first language. Friendliness, patience and being receptive to body language are excellent tools for breaking down language barriers. Be considerate and think about how you would feel trying to communicate in a non-English speaking country.

INDUSTRY TIP

Different dialects or accents can sometimes cause communication difficulties. You can politely request the speaker to repeat what has been said.

KEY FACT

When dealing with a person using English as a second language, it is important to be polite and patient, and to check understanding where necessary. Use of dialect and colloquial terminology – like very informal, regional, or slang terms – should be avoided.

Communicating orally with most people with physical disabilities should be no different to communicating with non-disabled people. Special provision might need to be made for those with a visual impairment, such as communicating verbally, rather than via email. Written communication may be more appropriate for those with a hearing impairment, or using face-to-face communication methods to ensure the speaker's lips are clearly visible for lip readers to follow. The key is to be aware and make adjustments without fuss.

KEY TERMS

Formal: official, important.

Informal: unofficial, relaxed.

Recipient: a person who receives.

INDUSTRY TIP

Do be aware that email chains with multiple recipients can quickly become complex: with many replies, this can lead to information overload, crucial information being overlooked by key stakeholders, and can eventually have a detrimental effect on relationships.

ACTIVITY

Imagine you have been awarded a contract to rewire a house, which will take six days. For three of those days, there will be no power to any of the circuits. Write an email to your tutor or a friend, as if they were the client, detailing the implications of the loss of services and what they need to consider during this time.

ACTIVITY

Write a list of equipment or a set of instructions on a sheet of paper – something that contains several points or items. Pass on the information to each person in turn but using a range of different communication methods such as text, email, oral, handwritten, and so on.

Once everyone has passed on the information, compare the end product with the original. Is it the same?

INDUSTRY TIP

If you are unfortunate and encounter a dissatisfied and angry client, remember, they are probably angry with the company and you should try not to take it personally. Look for ways you and your company can improve processes and systems, to avoid similar situations in the future.

When communicating with those with learning difficulties, it is important to remember that everyone has different communication needs, and language must be kept at a level that will be understood by the individual. If there is any doubt, it might be useful to communicate in conjunction with or through a relative or advocate, who can give feedback on whether the communication was appropriate.

Actions to take to deal with conflicts

Conflicts between clients and site operatives should be rare if clear communication is maintained. However, if problems occur and communication breaks down, the company needs to ensure that the client is not offended and everything is done to resolve the situation.

It is important that the correct person, such as the site supervisor or manager, deals with a client who is angry with the organisation or company rather than the operative personally. In employment law, this could be deemed as harassment due to the client's unreasonable response to a breakdown in communication from the company.

Where a dispute is clearly between individuals, each party must be heard in order to find common ground or grounds for reparation (ways to repair the situation). The dispute could be something simple such as the result of poor communication or misunderstanding of something said.

Where the dispute cannot be resolved, it may be necessary to remove the operative from the project and away from any further conflict, without blame (assuming no blame exists). In these situations, it is important to confirm politely to the client that the operative has been moved with no blame attached, in order for both parties to move on.

Where conflict exists between co-workers, there is usually a defined route to follow within the company's disciplinary and grievance policy. This normally involves an opportunity to resolve matters informally, where the two parties air their views and problems, and hopefully reach a resolution. Where this is unsuccessful, the company must get involved formally. It needs to avoid the accusation of constructive dismissal (when an employee resigns as a result of the employer making things difficult for the employee) if one of the parties resigns (quits) because of the dispute.

The company will try to resolve the issue. However, it might be appropriate to put both parties on a first-stage warning to stop the activity that is causing the problem. Disciplinary action will need to be taken if offensive behaviour continues, in line with the bullying and harassment policy.

ACTIVITY

Try this activity with a small group. One of the group accuses the other of stealing a personal tool but the person being accused believes the owner has simply lost that tool.

Consider the best forms of communication to resolve this conflict and consider the best approach by others so the conflict doesn't get worse. Discuss this as a group.

The effects of poor communication

Poor communication between operatives is likely to create a difficult working environment and lead to losses in productivity and a low team morale. This also risks stifling staff ideas for improvement in working methods and practices.

Management has a responsibility to communicate certain matters such as safety information. Effective communication is a two-way process – both or all parties need to be clear. Where there is misunderstanding, there is a risk of mistrust and where there is mistrust, there is a risk of conflict. Therefore, poor communication from management can lead to mistrust in the wider team infrastructure, or in the organisation as a whole, impacting on productivity and safety.

Customer relationships are an essential part of business development and customer retention – it is essential to have effective, meaningful communication with customers. Good communication between the company and customer ensures that invoices are paid on time and that both parties are prepared to do business together in future.

HEALTH AND SAFETY

Communication can also take the form of noises or alerts, such as sounding the horn on a forklift truck to warn people it is coming. Lack of such communication can risk the health and safety of those on site.

ACTIVITY

What types of questions should be asked on a customer satisfaction form used to gauge how an organisation and their employees performed when, for example, they rewired a house?

Test your knowledge

1   Who is responsible for fixing skirting boards on a building site?

   A   Bricklayer.

   B   Painter and decorator.

   C   Carpenter and joiner.

   D   Plasterer.

2   Which of the following would describe a company that is employed by a main contractor to work on a building site?

   A   Sub-contractor.

   B   Half-contractor.

   C   Supplement-contractor.

   D   Partial-contractor.

3   Who is the expert responsible for ensuring that a building core is strong enough to support a building when in full use?

   A   Architect.

   B   Structural Engineer.

   C   Quantity Surveyor.

   D   Bricklayer.

4   Which Act places responsibility on a company to keep records of customers safe?

   A   Official Secrets Act.

   B   Environment Protection Act.

   C   Data Protection Act.

   D   Equalities Act.

5   What would be given to a potential customer to provide a cost for work where it is not possible to know exactly how long the job will take?

   A   Quotation.

   B   Estimate.

   C   Tender.

   D   Invoice.

6   Which of the following would be given to a client during handover?

   A   Tenders.

   B   Quotation.

   C   Certification.

   D   Bill of quantities.

7   Which of the following would be included on a contract of employment?

   A   Required overtime hours.

   B   Holiday destinations.

   C   Redundancy notice.

   D   Basic working hours.

8　What is effective verbal communication based on?

　A　Seeing and writing.

　B　Speaking and listening.

　C　Touching and listening.

　D　Writing and talking.

9　What is the term used to describe a situation where an employee resigns because an employer is bullying them?

　A　Constructive dismissal.

　B　Deconstructive resignation.

　C　Unlawful leaving.

　D　Destructive redundancy.

10　What form of communication best ensures an instruction has been fully understood?

　A　Oral.

　B　Email.

　C　Text.

　D　Letter.

Appendix: About your qualification

INTRODUCTION TO THE ELECTROTECHNICAL QUALIFICATIONS

You are completing one of the following qualifications:

- Level 2 Diploma in Electrical Installation (2365-02)
- Level 2 Technical Certificate in Electrical Installation (8202-20)
- Level 3 Electrotechnical Qualification (installation or maintenance) (5357-03).

These qualifications aim to equip you with the practical skills and technical knowledge necessary to help you gain employment within the industry, or progress your training within the current apprenticeship route.

Alternatively, you may be trying to progress to the Level 3 Technical Certificate (8202-03) as a means of progressing to further education at Level 4 and above.

Whatever route you wish to take, each of these qualifications fully integrate, allowing seamless progression.

HOW TO BECOME AN ELECTRICIAN

To become a fully recognised electrician, you must complete the following:

- 5357-03 knowledge units
- 5357-03 on-site performance units
- Achievement Measurement Test 2 (AM2S) synoptic end test.

In addition, on completion of the above, you may also gain EngTech recognition from the Engineering Council through a fast-track process, as the on-site performance units measure the behaviours the Engineering Council requires for this recognition. That means you can use the EngTech abbreviation after your name, showing professional recognition.

If you are not currently enrolled on the Electrotechnical Qualification (5357-03), do not worry. You can transfer your skills gained through the Diploma or Technical Certificate towards the 5357 knowledge units, meaning you will not need to complete all of the assessments.

The easiest way to explain your progression route is to show each qualification side by side.

INDUSTRY TIP

For more information on EngTech, visit the IET's pages at: www.theiet.org/membership/profreg/engtech/index.cfm

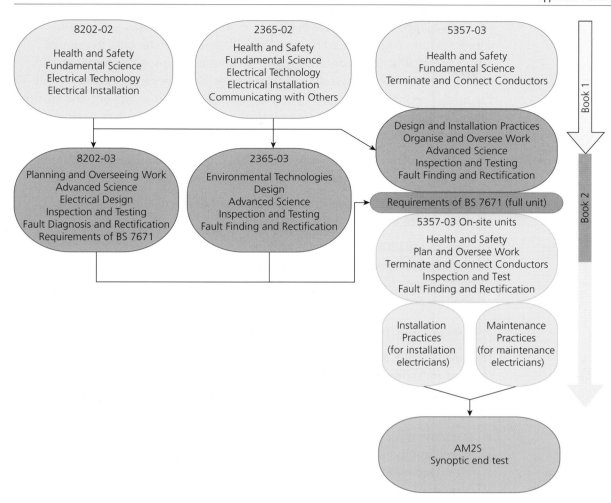

▲ Figure 6.1 Qualification progression routes

Why are there different routes?

The route you take into a career as an electrician will depend on a number of factors, such as your age, employment status and where you live within the UK.

- Learners who are 16–19 and not employed full time within the electrical industry would enrol on 8202-20.
- Adult learners wishing to become electricians would enrol on 2365-02.
- Apprentices of any age working within the electrical industry would enrol on 5357-03.

How to achieve your qualification

The requirements for successfully obtaining your qualification depend on which programme you are enrolled on.

8202-20

Level 2 is assessed using one on-screen multiple choice examination and one practical synoptic assignment.

For the synoptic assignment, a typical brief might be to install a cable and wiring system. You will need to draw on skills and understanding developed across the qualification content in order to consider the specific requirements of the particular wiring system and related electrical principles, and carry out the brief. This includes the ability to plan tasks, such as marking out and cutting cables, and apply the appropriate practical and hand skills to carry them out using appropriate tools and equipment.

You will also demonstrate that you are following health and safety regulations at all times by drawing upon your knowledge of legislation and regulations.

The exam draws from across the entire content of the qualification, using multiple choice questions to:

- confirm breadth of knowledge and understanding
- test applied knowledge and understanding – giving the opportunity to demonstrate higher-level integrated understanding through application, analysis and evaluation.

2365-02 and 5357-03

The Level 2 2365 and the corresponding units of 5357-03 are assessed using a range of methods including practical assessments, written examinations and online multiple choice examinations. Each examination, whether written or multiple choice, has a Test Specification. These show the following information:

- assessment method e.g. multiple choice or written
- examination duration
- permitted materials e.g. closed or open book
- number of questions
- approximate grade boundaries.

The grade boundaries may be subject to slight changes to ensure fairness, should any variations in the difficulty of the test be identified.

The Evolve platform

All City & Guilds multiple choice online tests are taken using the Evolve platform. The following points explain how you can navigate the Evolve tests.

1   Signing in.

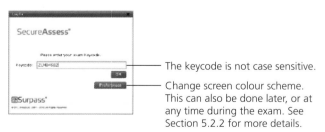

The keycode is not case sensitive.

Change screen colour scheme. This can also be done later, or at any time during the exam. See Section 5.2.2 for more details.

2   Once you have signed in using your keycode, you can change preferences. Some people prefer to read text using different colours or different backgrounds. For example, some people with dyslexia find reading black text on a yellow background much easier. If you do not set preferences,

> **INDUSTRY TIP**
>
> The way in which knowledge is assessed by each test is laid out in the Test Specifications. These can be found on the qualification pages at www.cityandguilds.com, by searching the qualification number at the top of the screen.

the default will be black text on a white background. You can change your preferences at any point during the examination.

3   You will then be shown a screen with your details. Ensure they are correct as you wouldn't want to pass an exam for someone else!

4   The welcome screen will give you options to either take a practice familiarisation session, or to carry on and start your examination. The clock does not start until you start the actual exam so if you are not familiar with the features of an Evolve test, use the practice session.

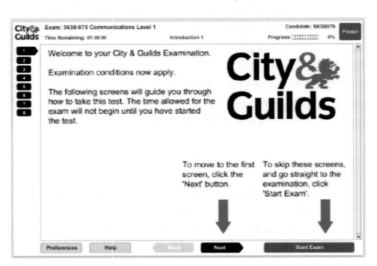

5   When you are taking the examination, the screen has several features.

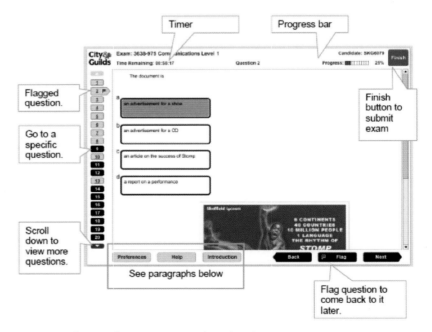

- *Timer* – shows the time remaining for the examination.
- *Progress* – gives you an indication of how much of the examination you have completed.

- *Finish* – only click this button when you are sure you have completed the examination.
- *Question numbers* – the numbers at the side show the individual question numbers. These indicate the questions that have been answered and the questions that have not been answered. They also indicate any questions that have been flagged (see below). You can navigate through the questions by clicking on each number or by using the 'back' and 'next' buttons at the bottom of the screen.
- *Flag* – if there are any questions you had doubts with, click the 'flag' button and a flag will appear by the question number. This means you can easily navigate back to that question should you have time.
- *Preferences* – throughout the examination you can change the screen preferences.
- *Help* – this offers help with the features of the system (not with the examination questions, unfortunately!)
- *Introduction* – this takes you back to the welcome screen.

Glossary

Accident an unplanned event that results in injury or ill health, damage, or loss of business.

Algebra the branch of mathematics that uses letters and symbols to represent numbers, to express rules and formulae in general terms.

Analogue instruments that use magnetism to measure electrical quantities.

Arc an electrical breakdown of a gas that produces an ongoing plasma discharge, resulting from a current through a normally non-conductive medium such as air.

Asbestos a rock-like mineral that can be used to strengthen materials or provide good fire resistance.

Atomic theory the study of atoms and electrons (also referred to as electron theory).

Audit to conduct a systematic review to make sure standards and management systems are being followed.

Authorising engineer an engineer, usually chartered, appointed to check the knowledge and experience of those operating or working on specialist systems, to make sure that all the employer's legal duties are met.

Average voltage the peak value multiplied by 0.637.

Base load basic amount of electricity.

Battering or slope the angle of the trench walls of an excavation that prevents the walls collapsing, measured in relation to the horizontal surface.

Bend a British Standard 90° bend. One double set is equivalent to one bend.

Bill of quantity a contract document comprising a list of materials required for the works and their estimated quantities.

Bit the part of the drill that does the cutting.

Business opportunity the opportunity to make profit from the work or contract.

Cable basket a lightweight containment system that normally clips together.

Carbon emission the polluting gas given off from the burning of fossil fuels such as gas, oil or coal.

Carbon footprint the amount of carbon dioxide emitted from burning fuels during the production and transportation process. This may be offset by the amount of carbon released naturally by decay. For example, biomass fuel is made from plant material that would release carbon dioxide as it rots naturally if left, so burning it efficiently and gaining use from it as a fuel means this fuel has a reduced carbon footprint, and is more carbon efficient.

Chair the person who controls a meeting, ensures that all the topics that need to be discussed (listed on the agenda) are covered. The chair also acts to resolve disputes.

Chartered engineer a registered engineer with professional competencies through training and experience.

Chuck the part of the drill used for holding the drill bit.

CIBSE the Chartered Institute of Building Services Engineers.

Civil law law that deals with disputes between individuals and/ or organisations, in which liability is decided and compensation is awarded to the victim.

Class II Equipment equipment having basic insulation and a further reinforced layer of insulation around live parts, meaning the risk of faults is minimal.

Closed circuit a complete circuit connected to a source of energy. If the circuit contains a switch and the switch is off (open), it becomes an open circuit.

Common denominator a denominator that can be divided exactly by all of the denominators in the question.

Competent person recognised term for someone with the necessary skills, knowledge and experience to manage health and safety in the workplace.

CompEx Scheme the global solution for validating core competency of employees and contract staff of major users in the gas, oil and chemical sectors. This covers both offshore and onshore activities.

Compliance the act of carrying out a command or requirement.

Composite cables multi-cored cables, where the cores are surrounded by a sheath providing mechanical protection.

Computer-aided design (CAD) software specialist software for producing drawings, including a library of symbols and features, such as walls and doors, that can be dragged and dropped into the drawing. In the days before CAD, a draughtsperson would have to painstakingly draw every detail on every drawing by hand.

Conductors materials that allow the movement of electrons and therefore current.

Consignment a batch of goods.

Containment system conduit or trunking providing a level of mechanical protection to cables.

Contamination the introduction of a substance that should not be there, such as water, oil or dust.

Contract a written agreement between two or more parties for doing something. A contract between a client and contractor is the agreement to carry out specific work for an agreed price.

Contract administrator (CA) the person named in the contract with the contractual power to change matters or items of work that will cause a contract variation.

Corrosion the breaking down or destruction of a material, especially a metal, through chemical reactions.

Criminal law decides if someone is guilty of a criminal act.

Critical path the sequence of key events and activities that determines the minimum time needed for a process, such as a construction project.

Cross-talk a bleeding of signal from one conductor to another, through electromagnetic induction. Twisting pairs of cable reduces this cross-talk.

CRS this means 'centres', indicating that a given dimension on a drawing is measured between the centres of components.

Datum line a reference point or line from which multiple measurements are made.

Digital instruments that use electronic components to measure electricity.

Discrimination arrangement of protective devices that ensures the device on the supply side of the fault operates before any other device.

Diversity allows the designer to use skill and experience to calculate what the full load on an installation may be given that some circuits operate at certain times (e.g. lighting) while others operate for very short times (e.g. hand driers in toilets). The result of applying diversity to the maximum demand is usually used to determine the supply characteristics.

Dry-lined wall an inner wall that is finished with plasterboard rather than a traditional wet plaster mix.

Duty holder any person or organisation holding a legal duty, in particular with regard to health and safety regulations (i.e. Health and Safety at Work Act etc. 1974). The person in control of a danger is the duty holder. This person must be competent by formal training and experience and with sufficient knowledge to avoid danger. The level of competence will differ for different items of work.

Earth Earth with a capital E represents the potential of the ground we stand on.

Efficiency the amount of useful work done by a machine compared to the energy supplied, expressed as a percentage.

Effluvia emissions of gas, or odorous fumes given off by decaying waste.

Electricity the flow of electrons as they move from atom to atom.

Electrolyte a chemical solution that contains many ions. Examples include salty water and lemon juice. In major battery production, the electrolyte may be an alkaline or acid solution, or a gel.

Electrons negatively charged parts of an atom, which orbit the nucleus. ('Electron' is the Greek word for amber, which is a material that is easily electrified by static.)

Enabling Act allows the Secretary of State to make further laws (regulations) without the need to pass another Act of Parliament.

Environment the land, water and air around us.

Equal potential when the voltage between any two parts of a system is within safe touch voltage levels, usually 50 V A.C. depending on the location.

Fatality death.

FELV functional extra-low voltage circuit (requirements can be found in Section 411.7 of BS 7671:2018).

Fossil fuels fuels found deep under ground that are mined or drilled for. They take millions of years to form, so once used they cannot be replaced.

Funding body an organisation that provides money for a particular purpose, such as a construction project.

Granular soils gravel, sand or silt (coarse-grained soil) with little or no clay content. Although some moist granular soils exhibit apparent cohesion (grains sticking together forming a solid), they have no cohesive strength. Granular soil cannot be moulded when moist and crumbles easily when dry.

Hazard anything with the potential to cause harm (e.g. chemicals, working at height, a fault on electrical equipment).

Hazardous substance something that can cause ill health to people.

Hot work work that involves actual or potential sources of ignition and carried out in an area where there is a risk of fire or explosion (e.g. welding, flame cutting or grinding).

HSS high-speed steel.

Hypotenuse the longest side of a right-angled triangle, which is opposite the right angle.

I_2R losses power losses caused by the current heating the cable and the heat energy being lost.

Incorporated engineer a specialist (also called an engineering technologist) who implements existing and emerging (new) technology within a particular field of engineering, entitled to use the title IEng (which acts as proof that a particular level of training and experience has been met).

Indices the plural of index.

Induction a current in a conductor induces (produces) a magnetic field around it, and a magnetic field induces (produces) a current in a conductor.

Insulators   materials that resist the flow of electrons and therefore current.

IP code   international protection code used to identify accessories and equipment according to their resistance to penetration and water ingress.

Isolation   disconnecting from and separating all sources of electrical energy.

Joule   the unit of measurement for energy, defined as the capacity to do work over a period of time.

Landfill   burying waste in large holes in the ground.

Liability   a debt or other legal obligation in order to compensate for harm.

Limit of authority   if somebody makes a key decision, without the right knowledge or experience to take that decision, they have exceeded their limit of authority. Dangerous or costly situations could result.

Live conductor   a conductor intended to be energised in normal service.

Load profiles   when and where electricity will be needed.

Manual handling   the movement of items by lifting, lowering, carrying, pushing or pulling by human effort alone.

Marking out   when measurements are identified on the walls and ceilings to ensure that equipment and accessories are installed in the correct place.

Materials   the accessories, components and cables acquired specifically for use during the installation process.

Materials take-off sheet   a list of all the materials required for an installation, and their quantities, to enable accurate stock control and costing of materials. Sometimes also referred to as a 'materials list', this can be used to record materials used to date and yet to be used on a job.

Maximum demand   the full load current of an installation assuming all circuits are drawing their design current at one time. As most installations do not actually do this, designers apply diversity.

Mean   the average of a set of numbers. To calculate the mean, add up all the numbers of the set, then divide by how many numbers there are in the set.

Mechanical advantage   a measure of the increase in force gained by using a tool.

Method statements   instructions on performing tasks safely in accordance with the risk assessment.

Nanotechnology   where a material or substance is created by changing matter at an atomic or molecular level.

National Standard   based on International Standards produced by the International Electrotechnical Commission (IEC), member nations create their own versions specific to their needs. Other CENELEC countries use the term 'rules' rather than 'standards' or 'regulations'. For example, the national wiring standard in the Republic of Ireland is the National Rules for Electrical Installations (ET101).

Near miss   any incident that could, but does not, result in an accident.

Negligible   small or unimportant and therefore not worth considering.

Nosings   the front edge of a step. If they are not square, they can lead to people slipping down the step. Many steps are fitted with additional metal strips to avoid the step from wearing.

Nuclear fission   creating energy by splitting atoms.

Oscillating   moving back and forth in a regular pattern.

Passive infrared (PIR)   a sensor that detects movement.

Pattress   the recessed container behind an electrical fitting, such as a socket, sometimes referred to as a back box.

Peak to peak   the value of voltage or current between the positive peak and negative peak of the waveform.

PELV   protective extra-low voltage circuit.

Periodic time   the amount of time to complete one full cycle.

Permit-to-work (PTW)   a written safe system of work produced to support the safe completion of potentially dangerous work and support communications between different persons.

Pi   the ratio of a circle's circumference to its diameter. It is a constant value used to determine the properties of a circle.

Plant   tools and machinery.

Pollution   contamination of the natural environment causing change to that environment.

Power tool   any tool that uses electrical energy, from batteries or from a mains supply, in order to work.

Practical completion   the point at which a construction project is virtually finished, when the last percentage of monies are paid by the client, the responsibility for insuring the construction transfers to the client and the architect or contract administrator signs the certificate of practical completion.

Protection methods   methods given in BS 7671:2018 for protection against faults.

Pythagoras' Theorem   for a right-angled triangle with sides of lengths a, b, and h, where h is the length of the hypotenuse, $a^2 + b^2 = h^2$.

Quantify   estimate guideline costs.

SWA   steel wire armoured

T   teslas

t   time (measured in seconds, s).

TN-C-S   Earth neutral-combined-separate

TN-S   Earth neutral-separate

TP-N   triple pole and neutral

TPI   teeth per inch

TT   system in which the consumer provides the earth connection

v   velocity (measured in metres per second, m/s)

V   voltage (measured in volts, V)

VA   volt amperes

VAT   value added tax

W   work (measured in joules, J, or Nm)

Wb   weber

WEEE   Waste Electrical and Electronic Equipment Regulations 2006

XLPE   cross-linked polyethylene

Test your knowledge: answers

CHAPTER 1

1   **C** The Health and Safety at Work etc. Act (HSWA) is an Enabling Act passed through Parliament.
2   **B** The Waste Electrical Electronic Equipment Regulations (WEEE) sets requirements for the control of waste electrical products.
3   **D** Employers must provide suitable personal protective equipment (PPE) for all employees.
4   **B** Accident books must be kept for a period of 3 years after the last entry date made in them.
5   **C** RAMS stands for Risk Assessments and Method Statements which are carried out for all tasks on site.
6   **A** PPE is considered the last resort when considering risk reduction methods. All other forms of removing or reducing the risk must be considered first.
7   **B** The Control of Substances Hazardous to Health (COSHH) is intended to control substances such as adhesives, to minimise the risk to health.
8   **A** The three key parts of the fire triangle are sources of ignition, sources of fuel, and sources of oxygen.
9   **B** Site transformers are used to transform the voltage to 110 V AC for the supply to power tools on a site.
10 **D** To prove that a circuit has been isolated correctly, the presence of voltage is checked on the load side of the circuit protective device to ensure the circuit is 'dead', i.e. no voltage present.
11 The person in control of the danger is the duty holder. This person must be competent by formal training and experience and with sufficient knowledge to avoid danger. The level of competence will differ for different items of work. In this example, this would normally be the supervising electrician.
12 Any three examples which may include: COPA, EPA, EA, HWR, PPCA, WEEE
13 **1** identify the hazards, **2** decide who may be harmed and how, **3** evaluate the risks and decide on precautions, **4** record the findings and implement them, **5** review the assessment and update if necessary.
14 **1** Elimination, **2** Reduction, **3** Enclosure, **4** Remove persons, **5** Reduce contact
15 Supply leads should be manufactured to relevant standards (BS or BS EN).
They should be suitable for the environment.
They should be free from cuts or fraying.
There should be no visible exposed conductor insulation (damaged sheath) or exposed live conductors.
There should be no signs of damage to the cord grip.
There should be no joints evident.

CHAPTER 2

1   **C** 2.3×10^{-3} (milli) would be the same as moving the decimal point three times to the left.
2   **A** In order to determine the sine of an angle, the opposite side is divided by the hypotenuse. Remember SOH CAH TOA.
3   **B** The unit of measurement for charge is the coulomb (C).
4   **C** As gravity, g, is 9.81 m/s, and force is determined by $F = m \times g$, so $F = 9.81 \times 10 = 98.1$ N.
5   **D** A copper atom has 29 electrons. Remember, different elements have a different number of electrons.
6   **D** If the resistors are in series, the total resistance is 36 Ω so, applying Ohm's law $I = \dfrac{V}{R}$ then $\dfrac{200}{36} = 5.56$ (to 2 d.p.)
7   **D** When two conductors are placed together carrying equal current but in opposite directions, the equal magnetic fields rotate in opposite directions and cancel each other out leaving no magnetic field.
8   **A** If a wire is wound around a core creating a electromagnet and the fingers of the right hand follow the direction of current flow, the thumb will point to the magnet's North pole.
9   **C** Transformer cores are made from thin metal sheets known as laminates which reduce eddy current flow.
10 **D** A diode will only allow current to flow in one direction.

11 As

$$R = \frac{\rho L}{A}$$

When transposed

$$L = \frac{RA}{\rho}$$

So

$$\frac{0.8 \times 1.5 \times 10^{-6}}{0.0172 \times 10^{-6}} = 69.8$$

12 The final overall cost for each supplier is:
- Lights and Shades: £116.00
- Square Wholesaler: £112.80
- Dark No More: £112.20

so the cheapest is Dark No More.

13 a As the image shows a parallel circuit, the total resistance is found by

$$\frac{1}{R_t} = \frac{1}{R_1} + \frac{1}{R_2} + \frac{1}{R_3}$$

So

$$\frac{1}{R_t} = \frac{1}{3} + \frac{1}{3} + \frac{1}{3} = 1\,\Omega$$

b Then applying Ohm's law using the total resistance:

$$I = \frac{V}{R} \text{ so } \frac{5}{1} = 5\,A$$

14 A Peak-to-peak value.
B Half cycle/period.
C Peak value.
D 1 period or cycle.

15 a If the resistors are parallel and the supply voltage is 100 V, the value of voltage across each resistor is the same as the supply, so the value V1 is 100 V.

b First the total resistance is needed, so

$$R = \frac{V}{I} \text{ so } \frac{100}{25}\,4\,\Omega$$

So if

$$\frac{1}{R_t} = \frac{1}{R_1} + \frac{1}{R_2} + \frac{1}{R_3}$$

then

$$\frac{1}{4} = \frac{1}{10} + \frac{1}{10} + \frac{1}{R_3}$$

So

$$\frac{1}{R_3} = \frac{1}{4} - \left(\frac{1}{10} + \frac{1}{10}\right) = 20\,\Omega$$

CHAPTER 3

1 **D** The IET On-Site Guide is non-statutory and provides technical detail on how to carry out inspection and testing.

2 **C** A circuit diagram shows detailed connections between components.

3 **B** The actual scale is 1:50, so if 1 mm = 50 mm then 200 mm (20 cm) is 50 times bigger so 200 × 50 = 10 000 mm or 10 m.

4 **D** 400 000 V or 400 kV is a transmission voltage. A and B are distribution voltages, and C is the voltage at which electricity is generated.

5 **C** Solar is a renewable energy source utilising the Sun's energy. The others listed are fossil fuels, producing CO_2 when used.

6 **D** Extraneous parts are parts that do not form part of the electrical installation but may provide a path to earth.

7 **A** In any TN system, a circuit supplying socket-outlets less than 63 A, or fixed equipment less than 32 A, must disconnect within 0.4 seconds.

8 **A** Where additional protection is required, the maximum rating for an RCD is 30 mA.

9 **D** The total earth fault loop impedance uses the symbol Z_s.

10 **C** The main switch for an installation is the means used to isolate the entire installation by an electrician, as it controls the whole installation and can be secured in the off position.

11 The wording on the label must read 'Safety Electrical Connection. Do Not Remove.'

12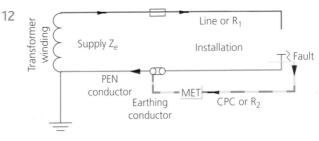

13 a 6   b Unlimited   c 2.5 mm²

14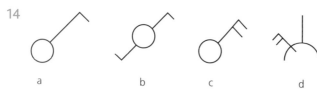

a   b   c   d

15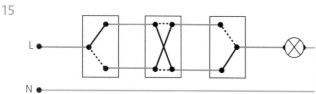

CHAPTER 4

1 **B** 20 mm is a standard conduit size.
2 **D** A reamer is used to remove any rough edges from the inside of a conduit.
3 **A** A bending spring is inserted into a PVC conduit when forming any bends to ensure the conduit keeps shape.
4 **B** An internal bend is used to ensure the lid is on the underside of the ceiling.
5 **D** Red indicates that the clamp is suitable for these areas.
6 **A** A crimp requires a compression tool to secure the crimp to the cable conductor.
7 **C** The shroud must be fitted first otherwise it will not fit over the gland.
8 **D** A hacksaw is used to score the armouring of an SWA cable.
9 **B** A low-resistance ohmmeter is used to check the polarity of the cores.
10 **A** Insulation resistance testers use a high voltage to ensure that there are no insulation faults leading to short circuits.
11 A crampet is used to fix conduit to wood or into concrete and usually used to secure conduit which will be in concrete such as floor screeds.
   A spacer bar saddle is used in most common situations where the wall is even and dry.
   A distance saddle is used where a wall is not even or is damp.
12 Using the IET On-Site Guide Tables E3 and E4, 20 mm conduit factor = 233, 1.5 mm² cable factor = 22. So $\frac{233}{22} = 10.59$
   The maximum is 10 cables.

13 Using the IET On-Site Guide tables E5 and E6, factors for the cables are:
   $47.8 \times 2 = 95.6$
   $21.2 \times 18 = 381.6$
   $12.6 \times 16 = 201.6$
   Total = 678.8
   So a suitable trunking is 75 × 25, having a factor of 738.
14 Ensure all lamps are removed. Using insulation resistance tester set to 500 V, at DB on load side of protective device, test between L-N, L-E, N-E. Ensure test is carried out in all possible switch positions to test all strappers. All results should be above 1 MΩ. Ensure circuit is reinstated ready to be energised.
15 Step 2 L-N: $\frac{0.8 + 0.8}{4} = 0.4\,\Omega$
   Step 3 L-E: $\frac{0.8 + 1.34}{4} = 0.54\,\Omega$

CHAPTER 5

1 **C** A carpenter/joiner
2 **A** A sub-contractor is employed by the main contractor.
3 **B** A structural engineer would design the main building core to ensure it suitable supports the building.
4 **C** The Data Protection Act places a duty on companies to keep records of customer details safe and secure.
5 **B** Where the work involved has many unknown factors, an estimate of the costs would be given. As an example, if the work was in an occupied office, hold ups may exist where drilling etc. has to be done at times where disruption is minimised. This could end up taking much more time than anticipated.
6 **C** Certification would be handed to a client during the handover process.
7 **D** The required basic working hours must be included on a contract of employment.
8 **B** Verbal communication is only effective if both parties are speaking and listening alternately.
9 **A** Constructive dismissal.
10 **A** When communicating orally, body language and/or responses give a good indication that instructions are understood.

Index